CANADA'S FLYING HERITAGE

CANADA'S
FLYING
HERITAGE

by Frank H. Ellis

UNIVERSITY OF TORONTO PRESS

Copyright Canada, 1954
by University of Toronto Press
Printed in Canada

Revised and reprinted, 1961
Reprinted, 1962, 1968, 1973
First paperback edition, 1980
ISBN 0-8020-6417-5

Library of Congress Catalog Number: 55-1357

FRANK H. ELLIS is a pioneer Canadian pilot, who built and flew his own aircraft at Calgary prior to August 1914, thus qualifying as a member of the *Early Birds* of America. He subsequently spent eight years in military and commercial flying in Canada, ranging from Ontario to Manitoba and farther west. He became the first Canadian to parachute from a plane in Canada, when he jumped at Crystal Beach, Lake Erie, in July 1919.

After his retirement from active flying, Mr. Ellis continued the building of airplane models as a hobby, and in 1936 published a book for boys on this subject, entitled *Duration Flying Models* (Hamilton). This book was followed by many articles on flying topics—more than a hundred—which have appeared in *Canadian Aviation, The Beaver, Aero Digest, Canadian Geographic Journal*, and elsewhere. These articles created widespread interest, and Mr. Ellis has in his possession many letters from pioneers of flying thanking him for his efforts in rescuing from obscurity this important chapter of Canadian history.

So much material came into Mr. Ellis's possession—photographs, records of almost-forgotten flights and early experiments—that he resolved to embark on a major project covering the whole of Canada's early flying history. He began his research in 1941, writing over 6,000 letters to persons in Canada, the United States, and other parts of the world. Much of the information he gathered in this way is unique and would have been impossible to locate a few years later.

Mr. Ellis first submitted his manuscript to a commercial publisher, who was keenly interested in the project, but after a lengthy exploration of ways and means came to the conclusion that he could not finance it, in view of its length, and the extremely large number of illustrations. When Mr. Ellis wrote to the University of Toronto Press about the manuscript, the publisher willingly turned it over to us, feeling that, as a non-profit organization, the Press was in a better position to seek the support that would be needed.

At the request of the Editor of the Press, Professor T. R. Loudon, Head of the Departments of Civil and Aeronautical Engineering of the University of Toronto, examined the manuscript. Professor Loudon was enthusiastic in his comments. Dr. J. H. Parkin, Director of the Division of Mechanical Engineering, National Research Council, was also consulted, and agreed that this was "a historical document of great value." Mr. T. M. "Pat" Reid, Aviation Sales Manager of Imperial Oil Limited, who knew of the project at a very early stage, encouraged the author by his personal interest and by supplying material from his own experiences. Mr. C. H. "Punch" Dickins, Sales Director of De Havilland Aircraft, confirmed the Press's own feeling when he said, "It will be a contribution to the industry of real value."

The cost of the project was estimated at many thousands of dollars. The Press was prepared to subsidize publication as generously as its resources would permit, but the greater part of the cost remained to be covered.

In February of 1952, Imperial Oil Limited offered to subsidize a major portion of the production expenses. This decision was based largely on the recommendation of Mr. T. M. Reid, whose interest in the project had been continuous. The University of Toronto Press joins the author in making public acknowledgment of this generous and imaginative contribution of Imperial Oil Limited.

During the preparation of the manuscript for publication, the University of Toronto Press drew generously on the help of such readers as Mr. W. B. Burchall, long closely connected with aviation and for many years Secretary of the Air Industries and Transport Association of Canada; Professor T. R. Loudon, who re-read the entire manuscript and was consulted on many points; and Dr. J. H. Parkin, who also read the manuscript at an early stage and checked many details. To these readers the publication of this book, which they did so much to encourage, will, we know, be adequate recompense for their expenditure of time and knowledge.

To the great sorrow of all, "Pat" Reid will not see this book in published form. The air tragedy which brought his death occurred only a few weeks after a final editorial conference with the author.

Pat Reid was a flying companion of early pilots in every part of Canada. *Canada's Flying Heritage* is as much a memorial to him as to Canadian aviation itself.

UNIVERSITY OF TORONTO PRESS

SECOND EDITION

In preparing the second edition of this book, the author has taken the opportunity of adding new information, and of rounding off the story with a brief account of modern developments. Being the story of the past, the text as a whole remains untouched.

Readers will be interested to know that following the death of J. A. D. McCurdy, Mr. Ellis is Canada's senior living pilot.

Acknowledgments

ALTHOUGH photograph albums, scrap books, and similar records are prized possessions of the owner, as the years go by they usually fall into the possession of others. Relegated to basement or attic storage, they are too often forgotten or destroyed.

A historian is fortunate who is able to locate some of these hidden treasures of the past, and I deeply value those that have been made accessible to me. It has only been through the co-operation and help of many persons that I have been able to tell so full a story and include so many photographs. For all information and photographs supplied, and for permission to reproduce the latter in this book, I am properly grateful.

A number of those to whom I addressed inquiries extended help far beyond anything I could possibly have expected. I wish to express my special thanks to the following:

Lieutenant Harry M. Benner, Hammondsport, N.Y.; Helen Black, New York; Sir Arthur Whitten Brown, Swansea, Wales; Mr. Arthur C. Carty, London, Ont.; Mr. William Wallace Gibson, San Francisco; Mr. Paul F. Groome, Regina; Mr. George Hewson, Toronto; Colonel Ernest Jones, Clifden, Va.; Colonel Douglas G. Joy, Winnipeg; Doctor Theo. A. Link, Toronto; Mr. W. D. MacBride, White Horse; Mrs. J. C. Mackenzie-Grieve, Victoria, B.C.; Mr. Ernest Maunder, St. John's, Nfld.; Mr. J. R. O'Malley, Ottawa; Dr. J. H. Parkin, Ottawa; Mr. Richard Pearce, Toronto; Mr. Arthur G. Renstrom, Washington; Mr. Antoine Roy, Quebec City; Mr. Douglas G. Shenstone, Ottawa; Mr. William M. Stark, Vancouver; Mr. G. A. Thompson, Vancouver; Mr. John B. Underwood, Grass Valley, Calif.; Mr. Irenée Vachon, Montreal; Mr. Wolfgang Von Gronau, Stuttgart, Germany.

Many others assisted me most generously by lending photographs and data in their possession. To this group my sincere thanks are due:

Mr. Carlton C. Agar, Vancouver; Mr. A. H. Allardyce, Winnipeg; Flying Officer H. C. Anderson, R.C.A.F.; Mrs. W. M. Archibald, Creston, B.C.; Mr. Jack Austin, Toronto; Mr. J. R. Ayling, Bulawayo, South Africa; Mr. A. "Matt" Berry, Edmonton; Mr. Leslie Bishop, Winnipeg; Mr. J. H. Blackburn, Edmonton; Mr. W. E. Boeing, Seattle; Mrs. B. J. Bourchier, Montreal; Mr. A. Bowman, Saskatoon; Mr. Walter Brookins, Los Angeles; Mr. Norman Bruce, Calgary; Mr. H. A. Bruno, New York; Mrs. Alys McKey Bryant, Washington; Mr. W. B. Burchall, Ottawa; Mr. Maurice Burbidge, Lethbridge; The Bureau of Aeronautics, Washington; Mr. G. A. Cameron, Victoria; Mr. F. Campbell, Edison Institute, Dearborn, Mich.; Mr. J. Innes Carling, London, Ont.; Miss Nellie Carson, Saskatoon; Mrs. Rene Cera, Woodbridge, Ont.; Mr. M. Carbonneau, Montreal; Clarke Steamships Company Limited, Montreal; Mr. E. M. Coles, Hamilton, Ont.; Mr. E. J. Cooper, Westmount, Que.; Mrs. Esmée Cruickshank, Vancouver; De Havilland Aircraft of Canada, Limited, Toronto; Mr. S. Pete Derbyshire, Edmonton; Mr. E. Dickey, Revelstoke, B.C.; Mr. W. R. Drew, Seattle, Wash.; Mr. Jack V. Elliott, Houston, Tex.; Mr. H. W. Firth, Dawson City, Y.T.; Mr. Donald G. Fisher, Kingston, Ont.; Mrs. Jeanne Fothergill, Montreal; Mr. William Fray, Kelowna, B.C.; Mr. Fred Furguson, Winnipeg; Mr. Y. Galitzine, Luton, Bedfordshire, England; Mr. Philip C. Garratt, Toronto; Mr. R. A. Gibson, Ottawa; Mr. Walter Gilbert, Chilliwack Lake, B.C.; Mr. W. H. Golding, Saint John, N.B.; Mrs. Marion Gorman, Hillsboro, Ore.; Mr. E. R. Grange, Toronto; Mr. Edmund Greenwood, Westmount, Que.; Mrs. J. J. Guest, Parrsboro, N.S.; Wing Commander F. H. Hitchins, R.C.A.F.; Mr. Basil D. Hobbs, Montreal; Mr. H. Hollick-Kenyon, Edmonton; Mr. Joe Holliday, Toronto; Mr. Clair G. Horton, San Antonio, Tex.; Mr. E. C. Hoy, Chicago;

ACKNOWLEDGMENTS

Mr. Donald R. Jacques, Vancouver; Mr. C. E. A. Jeffery, St. John's, Nfld.; Mrs. Eloise Jenkins, Ottawa; Mr. Emil Kading, Toronto; Mr. William Kahre, Rimouski, Que.; Mr. Aubrey Keif, New York; Mr. Eric Knowles, Saskatoon; Mr. Jean Marie Landry, Quebec City; Group Captain Z. L. Leigh, R.C.A.F.; Dr. Richard U. Light, Kalamazoo, Mich.; Mr. Robert A. Logan, Lake Charlotte, N.S.; Mr. "Ace" Pepper, Lucky Lake, Sask.; Mr. Preston Mark, Miscou Harbour, N.B.; Mr. Glenn L. Martin, Baltimore, Md.; Mr. Didier Masson, Ciudad Chetumal, Mexico; Major J. S. Matthews, Vancouver; Mr. W. H. "Wop" May, Edmonton; Mr. Grant W. G. McConachie, Vancouver; Mr. Edwin Mona, Vancouver; Mr. A. P. Nadeau, Chandler Station, Que.; the United States Navy; Mr. H. A. "Doc" Oaks, Port Arthur, Ont.; Mr. Milo E. Oliphant, Chicago; Mr. Brian A. Peck, Métis Beach, Que.; Mrs. G. H. R. Phillips, Toronto; Mr. Horace A. Porter, Saint John, N.B.; Mrs. E. A. Pritchett, Fort Worth, Tex.; Mr. B. A. Rawson, Winnipeg; R.C.A.F. Records Office, Ottawa; the late T. M. "Pat" Reid, Toronto; Mr. Hugh A. Robinson, Washington; Mr. C. G. Rogers, Sydney, N.S.; Mr. Roland Rohlfs, New York; Mr. Norman Ruse, Regina; Mr. F. Maclure Sclanders, Saint John, N.B.; Mr. William E. Scripps, Detroit; Mr. Murton A. Seymour, St. Catharines, Ont.; Mr. C. W. Shaffer, Sidney, Neb.; Mrs. A. J. Shelfoon, Trenton, Ont.; Mrs. Evelyn Smith, Calgary; Mr. Charles Smutney, Chicago; Mrs. D. E. Spotwood, Ste Agathe des Monts, Que.; Mr. W. A. Straith, Winnipeg; Major General St. Clair Streett, Dayton, Ohio; Miss Margaret E. Sullivan, Sturgeon Falls, Ont.; Mrs. C. R. Tanner, Calgary; Mr. William Templeton, Honolulu; Mr. Robert W. Thom, Collingwood, Ont.; Mr. J. H. Tudhope, Montreal; Mr. W. R. Turnbull, Rothesay, N.B.; Mr. A. K. Tylee, Toronto; United States Air Force; Mr. Romeo Vachon, Ottawa; Mr. F. C. Weber, North Battleford, Sask.; Mr. Happy Wells, Chilliwack, B.C.; Flying Officer R. B. West, Brandon, Man.; Mr. Charles Willard, Los Angeles; Mr. H. A. Wills, Cochrane, Ont.; Mr. H. A. Wilson, Westmount, Que.; Mr. Fred Whitt, Cleveland, Ohio; Miss Madge Wolfenden, Victoria, B.C.; Mr. Dennis K. Yorath, Edmonton.

To another group who assisted greatly by the loan of photographic material, I extend sincere thanks:

Mr. E. T. Applewhaite, Prince Rupert, B.C.; Miss B. Bannerman, Edmonton; Mr. Chester Bloom, Arlington, Va.; Squadron Leader R. S. Booth, Beaminster, Dorset, England; Mr. Aden Bowan, Saskatoon, Sask.; Mrs. Alan Butler, Dunstable, Bedfordshire, England; Mr. Lyall Canning, Parrsboro, N.S.; Mr. S. L. de Carteret, Ottawa; De Havilland Aircraft Limited, Hatfield, Hertfordshire, England; Mr. W. Austin Denehie, Chicago; Commander C. P. Edwards, Ottawa; Miss Florence Francis, Parrsboro, N.S.; Group Captain Elmer G. Fullerton, Calgary; Mr. R. H. Gale, Montreal; Mr. Lester D. Gardner, New York; Chief of Police and Mrs. J. H. A. Gilbert, Sherridon, Man.; Mr. G. H. Gillis, Halifax; Air Vice-Marshall A. E. Godfrey, Ottawa; Mr. A. Donald Goodwin, Montreal; Mr. Stuart Graham, Montreal; Mr. W. Graham, Vancouver; Mrs. Beatrice Hamilton, Revelstoke, B.C.; Mr. B. R. J. "Fish" Hassell, Rockford, Ill.; Mr. G. Herring, Ottawa; Mrs. M. A. Leveson-Gower, Vancouver; Mr. George Martin, Vancouver; Mrs. Sinclair MacKay, West Vancouver; Mr. George J. Mickleborough, Toronto; Mr. A. B. Mitchell, Selkirk, Man.; Mr. J. H. Moir, Revelstoke, B.C.; the National Geographic Society, Washington; the New Brunswick Publishing Company, Saint John, N.B.; Mrs. Ruth Law Oliver, San Francisco; Mr. Lew Parmenter, Montreal; Mrs. Mabel Ely Pierce, Santa Rosa, Calif.; Public Archives of Canada, Ottawa; Mr. E. L. Richardson, Lynnmour, B.C.; Mr. Robert Rideout, Vancouver; Lieutenant Colonel Scott-Griffin, Toronto; Short Brothers, Limited, Dublin, Ireland; Mr. T. W. (Tommy) Siers, Montreal; First Lieutenant R. E. Sliker, Washington; Mr. Orme Stuart, Prince Rupert, B.C.; Mr. G. Swanson, Winnipeg; Mr. Walter Thompson, Montreal; Mr. J. C. Turpin, Barnstable, Mass.; Miss Noreene Underwood, Stettler, Alta.; United Kingdom Information Office, Ottawa; U.S. Signal Corps, Washington; Mr. Vanoni, Montreal; Mr. Victor Vernon, Huntsville, Ala.; Mr. Lawrence S. Vibert, Miscou Plains, N.B.; Vickers-Armstrongs, Limited, London, England; Dr. J. C. Webster, Shediac, N.B.; Mr. Clifford P. Wilson, Winnipeg; Mr. J. A. Wilson, Ottawa; Mr. J. Charles Yule, Calgary; Mr. John Zmurchyk, Calgary.

Many persons with whom I was in touch did not have photographs but were able to supply helpful data. Deep appreciation is extended to the following for their help and consideration:

Mr. C. Adrin, Montreal; Mrs. Katherine Beardmore. Kirkfield, Ont., Mr. Fred Biggs, Stettler, Alta.; Mr. Emery Boucher, Quebec City; Mr. T. Ralph Bulman, Vernon, B.C.; Miss Thelma Burns, Portage la Prairie, Man.; Captain Fred Clarke, Vancouver; Wing Commander K. B. Conn, Ottawa; Miss B. Conway, Edmonton; Miss Betty Cooper, Edmonton; Mr. T. S. Corless, Quesnel, B.C.; Mr. Morley W. Coxworth, Q.C., Davidson, Sask.; Mr. R. J. Daverell, Regina; Mr. C. H. "Punch" Dickins, Toronto; Mrs. Betsy Flaherty, Vancouver; Air Commodore A. Fletcher, London, England; Mr. Arthur Flury, Berne, Switzerland; Dr. Charles G. Fraser, Toronto; Mr. F. C. Galbraith, Calgary; Mr. F. P. Galbraith, Red Deer, Alta.; Mr. Howard Gallagher, Vancouver; Mrs. Frank Gostick, Edmonton; Group Captain Roy S. Grandy, St. John's, Nfld.; Mr. John Grierson, Hamburg, Germany; Mr. Carter Guest, Ottawa; Lieutenant Colonel H. E. Hartney, Washington; Mr. D. C. Harvey, Halifax; Mr. F. P. Healey, Hamilton, Ont.; Mr. George E. Herman, Truro, N.S.; Mr. G. E. Hess, Milwaukee, Wis.; Mr. W. J. "Bill" Hill, Sault Ste Marie, Ont.; Mr. W. A. Hunter, Telegraph Creek, B.C.; Mr. Jack Hyde, Sault Ste Marie, Ont.; Irvin Air Chute, Limited, Fort Erie, Ont.; Mr. F. C. Jennings, Ottawa; Mr. K. Johannesson, Winnipeg; Mr. Sidney W. Johns, Saskatoon; Mr. Walter Johnson, Waterways, Alta.; Mr. J. L. Johnston, Winnipeg; Mr. Frank Jones, Vancouver; Mr. Thomas King, Golden,

B.C.; Mr. H. J. Lassaldine, Windsor, Ont.; Mr. R. E. Lawson, Montreal; Air Vice-Marshal Robert Leckie, Ottawa; Mr. Jerome Lederer, New York; Mr. J. J. Lefebvre, Montreal; Mr. G. E. Marquis, Quebec City; Mr. E. E. Massicotte, Montreal; Miss K. Mills, Ottawa; Colonel T. C. Macaulay, Severna Park, Md.; Mr. J. G. Macphail, Ottawa; Squadron Leader H. F. McClellan, St. Johns, Que.; Mr. H. O. McCurry, Ottawa; Mr. G. R. McGregor, Montreal; Mr. A. D. Newton, San Francisco; Mr. Norman C. Pearce, Toronto; Dr. E. Guthrie Perry, Winnipeg; Mr. W. V. B. Riddell, Ottawa; Mr. Tom P. Sigsworth, Calgary; Mr. Charles Skinner, Willow Bunch, Sask.; Miss Alice Stevenson, New York; Mr. J. R. Strickland, Calgary; Miss Marion E. Thompson, Toronto; Mr. George T. Underwood, Chehalis, Wash.; Mr. R. S. Walker, Ottawa; Mr. W. S. Wallace, Toronto; Mr. and Mrs. G. Watson, Winnipeg; Mr. R. L. Whitman, London, Ont.; Squadron Leader J. Scott Williams, Ottawa; Mrs. Elizabeth Young, Kelowna, B.C.

Many other persons were kind enough to reply to thousands of inquiries which I sent out. Some were unable to lend direct assistance, but offered suggestions and gave addresses which, when followed up, frequently opened up new avenues of search. To all who replied to the multitude of letters which were mailed I also wish to express sincere thanks, and especially to the following, all of whom had their part in helping to piece the story together:

Mr. V. S. Bennett, St. John's, Nfld.; Mr. Geoffrey Bowring, St. John's, Nfld.; Mr. W. C. Butler, Toronto; Mr. David A. Clarke, West Vancouver; Mr. E. E. Clawson, Charlottetown, P.E.I.; Squadron Leader P. A. Cumyn, Ottawa; Miss Annie F. Donohoe, Halifax; Mr. B. Douglas, De Havilland Aircraft Company, Johannesburg, South Africa; Mr. David Drinnan, Stettler, Alta.; Mr. Richard Finnie, Carp, Ont.; Miss Florence M. Gifford, Cleveland, Ohio; Miss Jean C. Gill, Charlottetown, P.E.I.; Miss Marjorie Gordon, Winnipeg; Miss Doreen Harper, Fredericton, N.B.; Mr. N. E. Harve, Regina; Mr. J. P. Hodgson, New Westminster, B.C.; Mr. Willard E. Ireland, Victoria; Hon. R. B. Job, St. John's, Nfld.; Air Vice-Marshal G. O. Johnson, Ottawa; Mr. D. B. Johnston, Kamloops, B.C.; Dr. W. Kaye Lamb, Ottawa; Mr. K. H. Ketcheson, Davidson, Sask.; Mr. Roy Knabenshue, Arlington, Va.; Mr. W. G. C. Lanskail, Nelson, B.C.; Air Vice-Marshall T. A. Lawrence, Barrie, Ont.; Mr. Stanley B. Lee, Armstrong, Ont.; Mr. Charles R. Lorway, Sydney, N.S.; Mrs. Phyllis Medhurst, Lethbridge, Alta.; Mr. Charles M. Moffitt, Troy, Ohio; Major Ian C. Morgan, Montreal; Mr. A. G. Muir, Hamilton, Ont.; Miss Helen MacDonald, Winnipeg; Mr. Reuben MacDonald, Charlottetown, P.E.I.; Mr. A. F. "Sandy" MacDonald, Toronto; Mr. Robert Mackay, Calgary; Mr. J. M. McConnell, Montreal; Mr. R. McIntyre, Sydney, N.S.; Miss B. McKay, Pennfield Ridge, N.B.; Mr. S. A. McKay, Pennfield Ridge, N.B.; Mr. Harold W. Nelson, Sudbury,

Ont.; Professor F. W. Pawlowski, Pau, Basses-Pyrénées, France; Rear-Admiral Albert C. Read, Washington; Mr. George C. Rooke, Regina; Mr. J. R. Ross, New Westminster, B.C.; Mr. Wilfred Rutledge, Port Arthur, Ont.; Mr. E. A. Saunders, Halifax; Mayor A. W. Shackelford, Lethbridge, Alta.; Mr. K. Shaw, Montreal; Mr. John F. Stephen, Oyama, B.C.; Mr. L. J. Tripp, Toronto; Mr. J. A. Vopini, Davidson, Sask.; Miss Margaret E. Warren, Toronto; Mr. M. Wood, Cochrane, Ont.; Mr. W. J. Workman, Amos, Que.

A debt of thanks is owed to a few personal friends, several of whom, while on service during World War II, devoted precious leave in various cities to digging up a number of items which until then had completely eluded me. For help so willingly given, it is a pleasant duty for me to extend thanks to Mr. Met Chapman (R.C.N.), West Vancouver; Mr. Davin Bill (R.C.A.F.), West Vancouver; Mr. Johnny Hailstone (R.C.A.F.), West Vancouver; Mrs. Lillian Harrison, Vancouver; Mr. Harry Parker (R.C.A.F.), West Vancouver; and Mr. James Wilson, Dartmouth, N.S.

A large number of newspapers, libraries, archives, and private firms are represented by the names of individuals already mentioned, but there are, in addition, several professional photographers whose pictures require acknowledgment. They are: Berger Studio, Hillsboro, Ore.; Mr. L. B. Foote, Winnipeg; Mr. Yousuf Karsh, Ottawa; Keystone View Company, New York; London News Agency Photo Limited, London, England; McDermid Studios, Edmonton; Mr. Mitchell, Selkirk, Man.; Mr. W. V. Ring, Calgary; Mr. Harry Steele, Winnipeg.

My debt to the postal authorities must not be overlooked. In the lengthy period, extending well over a decade, during which the contents of this work were gathered, not one piece of mail went astray through postal error. To the Post Office and to the faithful mailmen who plodded to our door through all weathers, summer and winter, a word of praise is surely due.

It will be noted that assistance has been received from a multitude of sources. Every effort has been put forward to secure reproduction rights and to check acknowledgments carefully. Should any omission or error have occurred, it would be appreciated if such were brought to my attention.

In spite of the efforts I have made myself to check all details of the story, and the valuable assistance I have received from others, it is

ACKNOWLEDGMENTS

probably inevitable that some errors have occurred. Besides apologizing in advance for them, I wish to earnestly request anyone who finds what appears to be a mistake to communicate with me. I hope I have cleared up a considerable number of misconceptions that had crept into a history so little of which is recorded, and I trust that the publication of this volume may bring to light others. Since so much of this story is still part of the recollection of living men, this is certainly the time to set the record straight.

Once the story was gathered and put to-gether, much remained to be done before it finally took shape as a published book. To my publishers, University of Toronto Press, go my sincere thanks for the efforts and enthusiasm so fully extended in editing and publishing this book.

To Imperial Oil Limited, whose financial sponsorship enabled this work to be published, I would say this. I hope that through their generosity, Canada's past in the air will emerge from obscurity, and many Canadians will have the opportunity to learn of their country's great flying heritage. No country can tell a finer story.

F. H. E.

Contents

PART 1

PIONEERS OF THE AIR, 1907-1914

1. Wings rise in the East

ONE of these days the owner of a lot in suburban Calgary digging in his back yard on a sunny spring day is going to turn up on his spade a rusty length of wire. When he reaches down to yank it out of the ground he will marvel at its toughness. Perhaps he will find attached to the end that he finally unearths an even rustier turnbuckle or strut fitting. There will be no label to inform him that he has stumbled on a tiny fragment of the history of Canadian aviation.

Perhaps, as a boy forty years ago, he had travelled the old trolley line across the prairie from Calgary to Bowness Park, thrusting his head out of the window to get a better look at the flying machine sitting in the grass two hundred yards away, and at the two young fellows who were tinkering away at the engine. Maybe, a year or two later, he made the trip again only to see the crumpled remains of what had been the *West Wind,* pride and joy—and despair—of Tom Blakely and myself: the flying machine that for two brief years was the centre of our dreams of glory—the cause of many sleepless nights, and of many exciting days.

The part that Tom Blakely and I played in the history of Canadian aviation is practically nil. Long since we have gone our separate ways. And if we had never met, the progress of flight would not have been retarded by so much as a day. Yet the interest that drew me to answer his ad in a Calgary paper in 1913 has survived, and out of that interest has come this book.

I suppose that the memory of that wrecked machine weathering away to nothing in the prairie weeds has something to do with it all.

In 1927 an organization was formed in the United States called the *Early Birds.* To qualify for membership a non-American must have piloted an airship, balloon, glider, or airplane prior to August 4, 1914. In the early thirties I was accepted as an Early Bird. Shortly afterwards I began searching for other Canadians who might be eligible for membership. I was driving a bus in West Vancouver at that time, so most of my search was conducted by correspondence. My contacts and letters grew over the years. I wrote an article or two, including one on the history of flying in British Columbia for the *British Columbia Historical Quarterly* in 1938. Correspondence and articles multiplied, my search for information to fill in the gaps grew wider, one thing leading to another, until somewhere along the way it occurred to me that I had the makings of a history of Canadian aviation. But from the first day I thought of such a book, I was determined to record not only the significant events of Canadian aviation, but also the more modest part played by those forgotten flyers who flew by guess and by God or with calculating caution—for the sheer love of flying —in the early days.

It takes your breath away a bit to think that there are aircraft today that can leave their own sound chasing through the air behind them. If anyone had dared to predict that kind of performance in those times we'd have known what to call him. Even in 1914 it was still an accomplishment just to get in the air, circle back, and get safely down again.

What people thought about flying at the turn

F. W. (CASEY) BALDWIN

of the century is difficult for me to know, as I was only seven at that time, living in Nottingham, England. But kite clubs were common then, and kite-flying was almost a national sport —I don't know how many spools of milliner's thread I filched out of my mother's shop to fly our kites. We used to let them out until they disappeared in the clouds, and then let go. When we weren't flying kites we were reading—there was plenty of science fiction for boys. We read about all kinds of fantastic aerial machines; I recall one that was equally at home on land, under water, or in the air. In my later teens— about 1911—I remember collecting a set of cigarette cards put out by Players featuring *The World's Aeroplanes*, illustrated in colour. I was model-crazy at the time, and constructed small models of most of them. Among others, I copied four machines designed and built by the Aerial Experiment Association, including the famous *Red Wing*.

In the brief description on the back of these cards there was nothing to tell the reader that this association had originated in Canada, or that two of its designer-pilots were Canadian-born and Canadian-trained engineers. To attempt, however, to define too closely which achievements in the early history of flying were specifically Canadian, and which American, would run contrary to the whole spirit of that time. In fact it would be difficult to find a better example of international friendship and co-operation than the story of the Aerial Experiment Association, or a more truly international figure than the senior partner of the group, Alexander Graham Bell. The inventor of the telephone was a Scot by birth and education, who lived for a time in Canada, and later became an American by naturalization. He could even be considered a part-time Canadian, for he owned a summer home in Canada. At any rate, he was unquestionably the father of Canadian aviation.

Dr. Bell had long been interested in the possibilities of flight, and for many years had quietly conducted experiments with large kites of ingenious and original design in an effort to learn what types of lifting surfaces were most effective. He was on very friendly terms with Professor S. P. Langley, whose pioneer experiments in model-flying are a part of American history, and, as far back as 1894, Professor Langley had visited Dr. Bell at Beinn Bhreagh, the latter's summer home at Baddeck, situated on the shores of beautiful Bras d'Or Lake, Cape Breton Island, N.S.

In May 1896, Dr. Bell visited Professor Langley, and was present at Quantico, Va., on the 6th, when Langley's miniature, steam-driven "aerodrome" made a most successful flight. So impressed was Dr. Bell with what he saw that he continued his own experiments with renewed vigour. These were concerned entirely with

MRS. ALEXANDER GRAHAM BELL

kites, and were conducted in the vicinity of Baddeck. From the nature of their design—their many triangular lifting surfaces of waterproofed paper or silk were attached row on row to light frames—they were called "tetrahedral" kites, and when flown proved to be light, strong, and very stable in the air.

One of Bell's first papers on the subject, written in 1899, was published in the National Academy of Science magazine, under the title of "Kites with Radial Wings."

In that same year the Wright brothers had written Langley, then Secretary of the Smithsonian Institution, asking for literature and information on flying. In 1902 they published a paper describing their experiments with a wind tunnel. Before the year was out, a Canadian engineer, Wallace Rupert Turnbull, had built the second wind tunnel in a large barn adjacent to his home at Rothesay, a short distance from Saint John, N.B. In it he made scientific tests on aerofoils and wing angles, the first on the continent. Trained in mechanical and electrical engineering at Cornell University and the University of Berlin, Turnbull was the first Canadian to tackle the purely theoretical aspects of aeronautics. In a treatise entitled "Research on the Forms and Stability of Aeroplanes," published in the *Physical Review* of March 1907, he emphasized the lack of true lateral stability in airplanes of that day because of their straight-built wings. He advocated the upward inclination of each wing from its centre section outward to its tip, creating what was termed a dihedral angle, by which lateral stability could be substantially increased. Turnbull's paper had a strong influence on subsequent design.

Meanwhile, two events had occurred which had a major effect on the progress of flight. On December 8, 1903, Langley's steam-powered "aerodrome" stood poised for its final test on its unique platform—the roof of a houseboat floating on the Potomac. In the presence of an expectant, awe-struck crowd of celebrities and sightseers, the huge machine gathered speed on its short tracks, reached their end, and plunged ponderously into the river. On the 17th of the same month, at Kitty Hawk, N.C., in the presence of his brother Wilbur and only five spectators, Orville Wright made the world's first successful controlled flight in a heavier-than-air machine.

What were the problems that faced a would-be flyer in the first decade of this century?

CYGNET I
On its first flight, December 6, 1907

There were three major challenges. First he needed the time and the money to design and build a machine. Langley, largely financed by the United States government, had been provided with ample supplies of both. His failure dried up the pool of public funds for many years to follow and convinced the public at large that flying machines were dangerous, impractical, expensive toys, with which no sober successful business man would dream of associating his name, let alone his pocket-book. The second problem was to find or design an engine that was light, yet powerful enough to drive the machine up to the critical speed required for it to become airborne. Finally, having built his machine and installed an adequate motor, the would-be flyer had to learn how to keep it aloft once it was in the air, and to control its flight. In the next three years the Wrights added substantially to their flying experience on this continent, but it was 1906 before Santos Dumont, a Brazilian, made the first flight in Europe.

During this period two young Canadians, Frederick W. (Casey) Baldwin and John A. D. McCurdy, were busy completing their engineer-

ing course at the University of Toronto. Mc-
Curdy's father had been Dr. Bell's secretary,
and the McCurdy home was at Baddeck, N.S.
In the spring of 1907, their graduation year, the
friends decided to spend the summer at Baddeck. McCurdy, having filled Baldwin's ears
with accounts of Bell's kite-flying experiments,
would have no trouble persuading his friend to
come.

The aging inventor was not slow in sizing up
the possibilities of the two young engineers.
One day that autumn in Dr. Bell's home, the
three were discussing the problems of flight—
mistakes of the past, possibilities of the future
—when a suggestion was made by Mrs. Bell.
Why not organize a company to put their collective ideas into effect? To prove that she
meant to be taken seriously she offered to meet
the entire cost of the undertaking. Thus, in a
quiet corner of Canada, the Aerial Experiment
Association came into being—and in one short
year became known the world over.

Dr. Bell invited an American, Glenn H.
Curtiss, to become a member, because of the
latter's knowledge of gasoline-driven engines.
Even at that time, Curtiss, as a designer and
manufacturer of motorcycles, was a recognized
authority in such matters. With an eye to the
future, the United States government asked that

an official observer be allowed to join and assist
the group. The request was granted, and Lieutenant Thomas Selfridge of the United States
Army was included. Thus the A.E.A. became
well balanced, and international in membership.

The first actual "flight" of the Association was
made by Selfridge, on whom the lot fell to be
the first to try his luck aloft on one of Dr. Bell's
huge tetrahedral kites. The young man lay full
length aboard the kite, *Cygnet I*, which first
rode the waters of Bras d'Or Lake on two flimsy
pontoons, while a sturdy launch, the *Blue Hill*,
towed the apparatus at sufficient speed to enable it to lift into the air. The kite rose quickly
to a height of 168 feet, holding very steady, and
for 7 minutes on December 6, 1907, Selfridge
sailed serenely into a chilling wind.

The kite was not equipped with controls of
any kind. A crew member on the launch had
been stationed at the end of the tow-line, with
orders to sever the rope with an axe as soon as
he saw the pontoons settle on the lake. In the
excitement, he forgot to do so. Before the line
was finally released, the kite and Selfridge were
dragged below the surface. The airman was
quickly rescued, but the kite was demolished,
and scattered wreckage floated all around.

As Curtiss's machine-shop was situated at
Hammondsport, N.Y., work and further experi-

THE *RED WING*
Just before its first flight on March 12, 1908

THE AERIAL EXPERIMENT ASSOCIATION

Left to right: Casey Baldwin, Thomas Selfridge, Glenn Curtiss, Dr. Bell, John McCurdy. *Extreme right*: Augustus Post of the Aero Club of America

ments were transferred to that point. All four of the engine-equipped aircraft which the Association subsequently built were made there, and tested either on the frozen surface of near-by Lake Keuka, or at the race track on the outskirts of Hammondsport.

By March 1908, their first machine, the *Red Wing*, so called from the colour of its surface covering, was ready for testing. After a number of experiments had been conducted with differently designed tail assemblies—mainly at taxiing speeds—a suitable design was decided upon. All members of the A.E.A. worked diligently on each craft made, and the different members in turn were credited with the main design of a machine. The *Red Wing* was credited to Selfridge, although it made only two flights and he had no opportunity to fly it.

Baldwin had the first whack at it. On March 12, 1908, the craft was pushed on its runners (it had no wheels) on to the frozen surface of Lake

Keuka, where it was checked with loving care for its first battle against the force of gravity. Casey Baldwin took his place in the pilot's seat, an ordinary kitchen chair with the legs removed. He was partially protected from the wind by a canvas-covered framework, and, with the motor at full throttle, the *Red Wing* quickly moved ahead. After a brief run it lifted easily on the light breeze, and sailed through the air for a flight of 319 feet before Baldwin brought it down to a good landing on the ice. Great was the jubilation among the members of the A.E.A. This was the first publicly announced flight in America, and when Baldwin soared into the frosty air that March day, he rose to fame as the first Canadian subject ever to fly a heavier-than-air machine.

A second flight by Baldwin of 120 feet ended in a crack-up, and although the airman suffered no injury, the machine was quite badly damaged. It was therefore dismantled, and construc-

THE *JUNE BUG*

During its flight on July 4, 1908

tion of the second aircraft of the Association was begun.

The motor used in the *Red Wing* was the original airplane engine developed by Glenn Curtiss. It had eight air-cooled cylinders, designed to develop 40 h.p. at 1,800 r.p.m., but during the actual flight test only 25 h.p. was required to lift the *Red Wing* into the air. The cylinders had a bore of 3⅝ inches each, with a stroke of 3¼ inches. Eight small carburetors served to supply fuel, one to each cylinder. In later experiments, however, the number was cut down to two, each to serve four cylinders.

As for the *Red Wing* itself, the wing span was 43 feet 4 inches, with a chord (the width of the wing from front to rear) which tapered from 6 feet 3 inches at the centre to 4 feet at the tips. The lifting area of the wings was 385 square feet, and the total weight of the craft, including 150-pound Casey, was 570 pounds. The rudder was 4 feet square and the elevator in front of the aircraft 8 feet by 2; control surfaces being operated by suitable arrangements

from the pilot's seat. The *Red Wing* possessed no ailerons or lateral control of any kind, and it was this lack that helped to bring it to grief on the second flight it made, which was on March 17.

The second airplane built by the A.E.A. was credited to Baldwin; it was named the *White Wing*, and was fitted with wheels. It embodied the good points of its predecessor and proved remarkably airworthy.

Since the instability of the *Red Wing* was due to its lack of lateral control, the inventive minds of the group went into action to devise some good method of meeting the challenge. The result was a hinged, controllable arrangement of wing-tip flaps, which, when built into the *White Wing*, proved their worth from the start.

At the time of the *White Wing* flights, no airplane in the world had made lengthy controlled hops using flaps which could be freely operated upwards as well as downwards. At that date, wing warping, such as the Wrights,

THE *LOON*

Louis Blériot, and several others had on their machines, was in use. Wing flaps had already been tried, but they were of the same type as those fitted to Henri Farman's biplane—they were hinged, but hung down freely when the airplane was at rest. Only when the machine picked up speed for a take-off and during actual flight, did these flaps become horizontal with the wings, as a result of the wind blowing under their lower surfaces. When a pilot wished to make use of such lateral controls while in flight, only one flap could be used at a time. The operation was to pull down whichever flap was desired to lift its particular wing tip.

On the *White Wing*, however, there was a great improvement—the two flaps were made to work in conjunction. They were operated by a shoulder yoke into which the pilot fitted snugly. If one flap was lowered, the other raised automatically. Thus both wing tips were controlled simultaneously, making for quicker control and greater safety in the air. When McCurdy explained the operation of the flaps to the French airman Henri Farman at Coney Island Race Track in September 1908, Farman dubbed them *ailerons*, that is, "little wings."

A lengthy legal battle raged later in the United States courts between the Wright brothers and Curtiss about patent rights on the invention of wing-tip control. These lawsuits were still in progress when the United States entered World War I. As settlement then became imperative, the Wright and Curtiss companies were brought together and a settlement was reached.

In 1918, an American, Dr. W. W. Christmas, laid claim to a similar invention which he had used in a machine he had built and flown at Fairfax Courthouse, Va., on March 8, 1908. He had never come forward with such proof before but the United States government investigated the matter, and apparently were satisfied that his claim was legitimate. They settled his demand by payment of a large sum of money, so that the vital hinged lateral control design could be freely used by aircraft manufacturers the world over, without the possibility of future litigation.

7

THE *SILVER DART*

Above: McCurdy at the controls on February 23, 1909

Below: The *Silver Dart* in flight on the same day

The *White Wing* was test flown by Baldwin on May 18, and flew 279 feet. It was flown the next day by Selfridge who made a short hop of 100 feet. A second flight by him the same day covered 240 feet.

Curtiss tried his luck in the same machine on May 22, going 1,017 feet, and on May 23 McCurdy flew 600 feet. An accident in landing put the craft out of commission, as well as slightly injuring the pilot.

The third machine built by the A.E.A., the *June Bug*, was now begun. It was credited to Curtiss, and except for three short hops in it by Selfridge, was flown exclusively by Curtiss and McCurdy. Between June 21 and August 31, 1908, some remarkably fine flights were accomplished, the longest being on August 29, when McCurdy guided it through the air for over 2 miles. The *June Bug* became one of America's most famous aircraft when Curtiss flew it on July 4, 1908, to make the first official one-kilometre flight to be recorded in the western hemisphere. For this he was awarded the *Scientific American* Trophy. The actual distance flown was 5,090 feet, and the time was 1 minute, 42⅕ seconds.

In November 1908, the *June Bug* was renamed the *Loon*, and pontoons were fitted in an attempt to fly it as a hydroplane (seaplane); this effort, however, was unsuccessful.

The fourth machine, named the *Silver Dart*, was credited to McCurdy, who was the only one to fly it. Much had been learned from experiments with the first three airplanes, and this craft was one of the finest of the pioneer machines built. The wing span of the *Silver Dart* was 49 feet from tip to tip, with a centre chord of 6 feet, tapering to 4 feet, exclusive of ailerons. Its wing area was 420 square feet, and its weight, fully loaded with fuel and pilot, was 800 pounds.

McCurdy first put it through its paces at Hammondsport, N.Y., where he flew it 600 feet on December 6, 1908. He made ten more flights of varying distances before Christmas Day.

In January 1909, the *Silver Dart* was shipped to Baddeck for further flying experiments, as Dr. Bell was desirous of having one of the A.E.A. machines flown in Canada; and on February 23, Canadian history was made when McCurdy flew the *Dart* for over half a mile above the frozen surface of Bras d'Or Lake, thus achieving the first heavier-than-air machine flight in Canada. Years later, the exploit was given official recognition as having been the first controlled flight of an airplane, by a British subject, at any point in the British Commonwealth.

On March 10, 1909, McCurdy made the most impressive flight ever accomplished with an A.E.A. machine, when he flew the *Dart* on a circular course over a distance of twenty miles. In February and March of the same year experiments were also conducted at Baddeck with what was actually the fifth and final aircraft constructed by the organization. Named *Cygnet II*, and weighing 950 pounds, it was a larger construction of Dr. Bell's tetrahedral kite, *Cygnet I*, in which Selfridge had been carried aloft in 1907. The engine of the *Silver Dart* was used in testing *Cygnet II*, but on the first attempt on February 22 a propeller broke, so that little was learned of the craft's lifting ability. A further experiment was conducted on March 15, but the huge apparatus was never taxied at sufficient speed over the ice to enable it to lift free. *Cygnet II* had 3,690 separate "wing-shaped cells" built into its light but rigid framework, for lifting purposes, and in appearance it was quite a massive machine. It was never actually flown.

The engines used in all the A.E.A. aircraft were of Curtiss design and fabrication. They were the forerunners of a long line of airplane motors that helped much towards the advancement of flying during the barnstorming era in the United States and Canada. Later still, in improved design, thousands were manufactured and used in the Curtiss JN4 biplanes, the Jennies in which so many pilots received their training in Canada and the United States during World War I.

The engine used in the *Silver Dart* was compact and reliable, with eight cylinders, arranged in V-type formation. Instead of being air-cooled as were the engines used in the previous machines, it was water-cooled, and actually was the first successful aircraft motor of this type in the world. Developing 35 h.p. at 1,000 r.p.m., it turned a propeller which was carved from a solid block of wood. The drive was not direct from the crankshaft, but through suitable pulleys and a belt, which cut the prop speed down to the ratio of 18/24 of the engine speed.

Having accomplished their whole purpose, which was to make a heavier-than-air machine that would sustain a man in controlled flight,

CYGNET II

the A.E.A. was disbanded on March 31, 1909, and the participants went their various ways. Baldwin and McCurdy remained partners for a considerable time and built two additional aircraft, *Baddeck I* and *Baddeck II*, which were flown with great success. Curtiss became world-famous as a pilot, designer, and builder of airplanes. Lieutenant Selfridge had been recalled by his government late in the summer of 1908 to act as a military observer in connection with the flight tests being made by the Wright brothers for the United States Army. During a flight at Fort Myer, Va., on September 17, 1908, a propeller drive chain snapped, fouling a propeller. In the resulting crash, young Selfridge was carried to his death, and Orville Wright, who was pilot at the time, was seriously injured.

In July 1909, McCurdy and Baldwin prevailed upon the government authorities at Ottawa to observe tests of their aircraft, hoping to have its military value recognized. Eventually the *Silver Dart* was shipped from Nova Scotia to Petawawa military camp, northwest of Ottawa.

Few in high places at the capital were sympathetic towards, or even interested in, the demonstrations, for the general opinion at the time was that airplanes could be of little practical use in actual warfare. One official who thought otherwise, and said so, was Major Maunsell, Director of Engineering Services in Canada, and it was mainly through his influence that plans came to a head.

By flying at Hammondsport and Baddeck, McCurdy and Baldwin had proved their ability beyond all doubt, but the sandy terrain at Petawawa was quite unfit for take-offs and landings for machines of that early vintage. The diameter of the tires of the three bicycle wheels which formed the landing gear was only about 2 inches, and great difficulty was experienced in getting away from the ground. It was particularly tough going since both Baldwin and McCurdy were aboard on each flight.

In spite of these handicaps, four successful flights were accomplished on August 2, 1909, just seven days after Louis Blériot had made the first flight across the English Channel, at the Strait of Dover.

On the fifth flight of the day, McCurdy, who was piloting the craft, was coming down for a landing with the setting sun blazing right in his eyes, when one of the wheels struck the top of a slight rise in the ground. In the resultant smash-up the right wing was torn completely off. The two airmen, although covered with débris from the wrecked machine, were uninjured, but the illustrious career of the *Silver Dart* was at an end.

In case it should be needed, another machine almost identical to the *Silver Dart* and named

Baddeck I had been shipped from Nova Scotia, and by August 12 it was fully rigged and ready for tests. On the 12th and 13th, before a large throng of officials and other interested spectators, *Baddeck I* was flown on five occasions. Bad luck still dogged the airmen, however; the machine was severely damaged on its fifth landing, and thus terminated the first military demonstration of aircraft in Canada.

The remains of the *Silver Dart* and the damaged *Baddeck I* were shipped back to Baddeck. The faithful engine of the *Dart* later found service in a motor launch, being used to drive the dynamo. Eventually the vessel sank in shallow water, and the hull and engine lay beneath the surface for over two years before they were salvaged. Years later, the historical value of the old motor was realized. It was then cleaned up and put into condition to be sent to Ottawa, where it now holds a prominent place in the Aeronautical Museum of the National Research Council.

Throughout 1909 and 1910, McCurdy and Baldwin continued their experiments and made many successful flights. On March 7, 1910, McCurdy flew *Baddeck II* a distance of over 20 miles in 16 minutes. In that same year he won the world's biplane speed record at the second International Aviation Meet, at Belmont Park, New York. During a flight at Sheepshead Bay, New York, on August 27, McCurdy transmitted the first wireless message from an airplane in flight to a ground operator. The man on the ground was Clair Horton; the simple apparatus was of the telegraph key and relay recording type.

In the same year that Baldwin dropped out of active flying to concentrate on the theoretical aspects of flight, McCurdy made a most daring venture. On January 30, 1911, he set up his machine—a regular Curtiss plane—on a beach at Key West, Fla., proposing to fly the 95 shark-infested miles that separated the Florida Keys from Havana, Cuba. It had been the airman's intention to make a short test hop, then return to the beach and make the big hop later in the day. But he was scarcely in the air before the beach swarmed with spectators, making it impossible to land. Swinging southward, McCurdy opened the throttle and was on his way.

All went well until he was within ten miles of his destination. Then, glancing at the oil pressure meter, he saw the needle sink slowly down to zero. Anxiously he watched the coast of Cuba, listening to the first overtones of a seizing motor. A mile off shore it happened; and in silence, except for the wind whistling past the wires, McCurdy made his landing in the Gulf of Mexico. With typical foresight he had had two metal tanks fitted to the underside of the lower wings of his wheel-equipped biplane, and these proved to be his salvation. By the time his escort vessel, U.S.S. *Paulding*, came alongside, a circle of sharks had gathered round, eying the marooned airman with hungry relish as he perched on the upper wing, just beyond their reach.

At a huge demonstration and banquet held later in Havana, the airman was presented by the President of Cuba with a silver cup and a beautifully ribbon-bedecked envelope. The envelope was supposed to contain the prize of $10,000 for the first airman to fly from the United States to Cuba, but when McCurdy opened it in his hotel room shortly after, he found the contents to be only a bundle of pages

THE *BADDECK I*

Being pushed into position for a take-off at Petawawa, August 1909

AERO CLUB OF AMERICA LICENCE
Issued to J. A. D. McCurdy on October 5, 1910

from a Havana newspaper. Evidently his landing a mile off shore disqualified him for the record, but the officials had gone through all the motions of awarding the prize anyway. In the excitement of the moment, the airman had overlooked taking the silver cup along with him. When he returned later to pick it up, it, too, had vanished. Only his memories of the flight remained, but they were real enough.

In 1912 McCurdy made exhibition flights at Mexico City, where, as the first airman seen in that vicinity, he was widely acclaimed. He was one of the British Empire's earliest pilots to obtain a flying licence from the Fédération Aéronautique Internationale of Paris, doing so through the affiliated Aero Club of America. He passed his tests at Chicago, Ill., on a Curtiss biplane, October 5, 1910, and was awarded a licence which carried the low number of 18.

Glenn Curtiss had previously obtained No. 1, while Orville and Wilbur Wright earned Nos. 4 and 5 respectively, so it will be realized how early McCurdy gained honours for himself and Canada in what was then a very risky business. It was not until 1916, when he developed a

slight defect in vision, that he ceased active flying.

From 1911 onward, Baldwin devoted his attention and the fruits of his experience largely to the development of hydrofoils, particularly in their application to waterborne aircraft; and over the years he contributed a vast amount of scientific data, which were to become so vital to Canada's bush flyers some twenty years later.

As we have already seen, Turnbull of Rothesay had been actively experimenting in aeronautical theory since 1902. Until 1906, while he was testing his theories about dihedrals in the wind tunnel, he also experimented with waterborne hydroplanes, propelled by motor-driven air-screws. For motive power he imported a two-cylinder Duryea aero engine. The first internal combustion engine specifically designed for the air to be brought into Canada, this motor is now on display in the Aeronautical Museum of the National Research Council at Ottawa.

His initial design had a scow-shaped body, with hydroplanes attached beneath, which were supposed to lift the main part of the apparatus clear of the water surface if sufficient speed could be maintained. The engine supplied power to two air-screws, set out on each side of the craft on outrigger framework, and driven by pulleys and belts.

The second "hydro" Turnbull built embodied the better points of the original, and was much lighter. It consisted of tandem floats connected by a light framework on which the operator sat and to which the engine was bolted. A single and more efficient propeller was fitted slightly to the rear of the pilot's seat. Turnbull's intention was to fit suitable wings to these craft if they showed any degree of speed, but the engine proved to be unsatisfactory and its uneven torque proved destructive to air-screws. However, the inventor learned much from these experiments.

Turnbull then turned to the construction and testing of different types of air-screws. To facilitate the work, he built a 30-inch-gauge track, some 300 feet in length, on which a wheeled truck could be operated back and forth.

Instruments to record propeller thrust, revolutions per minute, and forward speed of the

W. R. TURNBULL

BADDECK I

Second from left: Professor Charles Manley, who worked with Professor Langley and also with the Aerial Experiment Association. *Third from left*: John A. D. McCurdy. *Second from right*: F. W. (Casey) Baldwin. *Courtesy Public Archives of Canada*

TURNBULL'S GREAT INVENTION

The controllable pitch propeller in its first completed form
June 6, 1927

England by the British government to under-take numerous scientific tasks, one of which was the development of nets and screens to guard warships against torpedo attacks.

At the end of 1918 Turnbull returned to Rothesay to continue experiments with the controllable pitch propeller, an idea he had been developing since 1916. In a controllable pitch propeller the blades can be adjusted at different angles, or pitch, during full flight, an adjustment which is under the complete control of the pilot at all times. Such adjustable propellers are much more efficient during take-offs and in flight than an ordinary "fixed blade" propeller. After innumerable tests with an apparatus which embodied a rotating or "whirling" electric motor, Turnbull gradually evolved a controllable pitch propeller. It was given ground tests in 1923 on an Avro aircraft, under government and Air Force supervision at Camp Borden, Ont.

Considerable research had yet to be accomplished before such a revolutionary type of air-screw could be tested in the air, but the eventful day came on June 6, 1927, again at Camp Borden. Turnbull's product was fitted to another Avro biplane, piloted by Flight Lieutenant G. G. Brookes. The control of the pitch (blades) was by a small electric motor, mounted on the hub in front of the propeller. So encouraging were the air tests that a major development of the invention was launched without delay.

In December 1929 the inventor disposed of his patents, the American rights going to the Curtiss Wright Corporation, and the English rights to the Bristol Aeroplane Company.

Thus it is that, thanks to Turnbull of Rothesay, Canada is recognized as the birthplace of the controllable pitch propeller, the develop-

apparatus, were an integral part of the experiments. An immense amount of valuable data was thus obtained relating to the thrust and pitch values of air-screws. The diameters of the propellers varied from 1½ to 3½ feet, and designs of many shapes were formed and thoroughly tested to learn their efficiencies.

Turnbull wrote a treatise on the subject in three articles. The first, "The Efficiency of Aerial Propellers," was printed in the *Scientific American* at the early date of April 3, 1909. The two others, entitled "Laws of Air-screws," were published in the October 1910 and January 1911 issues of the *Aeronautical Journal*. By this time the value of Turnbull's research was recognized, and he was awarded the bronze medal of the Royal Aeronautical Society and elected a Fellow. By 1914 Turnbull had published a large number of scientific articles and had become one of the world's authorities on the subject of aerodynamics.

During World War I, he was invited to

TURNBULL'S APPARATUS

Used in experimenting with a rapidly revolving electric motor

PROPELLER MODELS

Right: Models used by Turnbull in his experiments. *Below*: One of Turnbull's controllable pitch propeller models

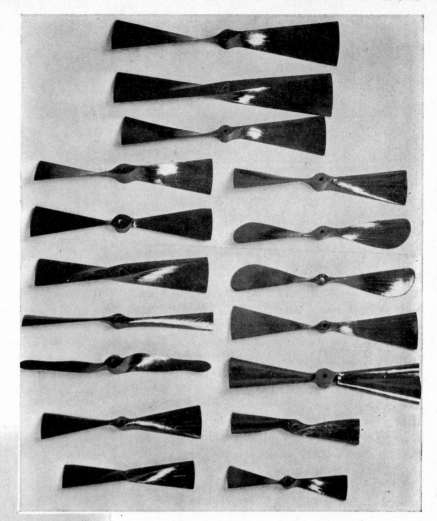

ment of which has played a tremendous part in the advancement of aviation throughout the world.

The role of Bell's Aerial Experiment Association in world aviation is well known; Turnbull, too, received world recognition for his pioneer investigations and experiments. Here we are in the main stream of aviation history.

But little has been told, and less has been recognized, of the pioneer work that was proceeding unobtrusively in the Canadian West, where unique experimental craft—even an aero engine—were not only being built but also flown, under financial handicaps that would have daunted even so stout a heart as Dr. Alexander Graham Bell's.

It is not too surprising that the world's first flights were made in eastern North America. From the 1850's on, one invention after another appeared in the region between the Great Lakes and the Atlantic Coast. It is small wonder that the Canadian Maritime Provinces which were adjacent to this hive of inventive activity should have become involved in the first experiments at flight.

In 1908, when Baldwin became the first Canadian to fly a heavier-than-air machine, Alberta and Saskatchewan had been provinces for only three years, with a population of barely half a million between them. British Columbia, although established much earlier, was no better off for population. The largest city west of Winnipeg was Vancouver, just topping the 100,000 mark; western universities were in their infancy; scientific societies and reference libraries conspicuous in most communities for their absence. There was no industrial tradition, no body of skilled mechanics, no motor engineering firms—a would-be inventor of flying machines would be tempted to add, "No nothing!"

Krugerville, Alta., no longer exists, even as the local designation for a crossroads near Stettler where a well-patronized blacksmith once had his house and shop, and where the district post office was located. Today there is nothing to mark the whereabouts of the old Underwood farm a mile from Krugerville, where, in the years 1907 and 1908, the first aircraft experiments on the mainland of Canada were conducted.

It would be well, however, in order to get the full flavour of this story, to recall that in the very early flying era the problem of obtaining the power to sustain a man in even a light machine baffled many inventors. Light motors of adequate power were next to impossible to obtain at that time, and even when they were available, or built to order, they cost thousands of dollars. The only bright side to a situation which very effectively prevented many a clever but hard-up inventor from ever attempting a full-fledged flight was this, that more of them lived to tell the tale.

In 1872 an American-born farmer of Alberta, John K. Underwood, set about to find a better means of breaking his wide acres than by the old-style ploughshare. After many tests with various designs, he evolved and in that year took out the original patent on the disc plough which, incidentally, still bears his name. Today it is one of the most popular farm implements throughout the North American continent. Underwood was the proud father not only of the Underwood Disc Plough, but also of several fine sons, who also became tillers of farm lands bordering the Krugerville road, several miles from the thriving northern Alberta town of Stettler.

The boys all inherited their father's inventive ability, and when in 1903 and 1904 the news spread that the Wright brothers had made a number of successful flights, the young Underwoods—George, John, and Elmer—began to think how wonderful it must be to fly. During the next few years, as flights in Europe also became news, they mustered their thoughts into action.

Their efforts to obtain information met with

THE UNDERWOOD MACHINE
Being towed behind a wagon and team along the main street of Stettler, Alta., 1907

dismal failure. Publications on the subject could not be procured, but, nothing daunted, they set to to design a craft according to their own ideas. At that time nothing was known about aerodynamics, nor the lifting effect on a wing caused by the vacuum on the upper surface, but the young men managed to pick up a treatise on the subject of flight, entirely theoretical, written by an eminent Nova Scotia-born mathematician, Simon Newcomb. Newcomb proved that it would be impossible, in theory, to build a plane large enough to carry a man in flight because the surface would have to increase as the square of the dimension, while its weight would increase as the cube of the dimension. Apparently the article had been written before the Wrights' achievements, and it read logically enough. Newcomb stated that a small kite could undoubtedly be made which would still be strong enough to lift 25 pounds or so, but that would about be the limit! Probably many would-be inventors, studying the laws as outlined by Newcomb, gave up in despair at once, but the treatise had the opposite effect on the three persistent Albertans. They cast about for ways and means of constructing a machine which would have strength and lightness, and yet could be kept within the laws and limits described by Newcomb.

They began their experiments on May 14, 1907, with tests of a rectangular, tailless kite, 8 feet across. A few days later they were out in the wind again, testing one of similar type, but with a 20-foot span. The kites they designed were actually flying wings, and each possessed a canvas jib which projected vertically from front to rear in the centre of the kite. Under actual test, the jib or fin gave the kite great lateral stability, and George Underwood admits frankly that they cribbed the idea from a brother-in-law, Charles R. Haas, who had experimented with a similar idea when conducting kite-flying tests several years previously.

The Underwoods then made models to determine the most suitable design. These models were flown for distances of up to 150 yards with the use of elastic bands, or a weight which, unwinding by gravity, turned the axle to which the propeller was attached. The model, fitted with a weight for propeller power, was first sent aloft as a kite and then, as it was cut loose, the falling weight swiftly rotated the propeller. Some very creditable flights were achieved, which thrilled the young men greatly and urged them on towards the construction of their full-sized machine.

Great ingenuity went into the design and building of the big craft, and occasionally the

THE UNDERWOOD BROTHERS
Left to right: John, George, and Elmer

inventor of the disc plough would lend a hand when some particularly knotty problem was involved. Today the plane would be classed as a flying wing, since it had no fuselage or outrigging of any kind. When completed it weighed only 450 pounds, without man and engine, and had a lifting surface of 900 square feet. The wing framework was elliptical in shape, 42 feet from tip to tip, with a fore and aft span at the centre of 26 feet. The outer circle of the wing was fashioned from long, laminated strips of fir, the finished size being 5 inches by 1¼ inches in cross-section. From a centre post, or hub, wire spokes went out to the laminated rim, after the style of a bicycle wheel. Turnbuckles on each spoke made possible accurate adjustments, allowing for perfect alignment of the entire wing. A man was able to stand on the outer wooden rim at any point, and jump up and down upon it without damaging it in any way. The hub was fashioned from a 12-foot length of cedar 4 inches by 4 inches, and a small platform for the operator and the engine was built to the lower section of the hub-post.

Below the platform, attached to a rigid axle, were two sturdy motorcycle wheels. At the rear of the laminated rim, towards each wing tip, bicycle wheels were suitably attached to turn freely on a castor-designed fitting, giving great handling ability on the ground. The fin, or jib, previously mentioned as a part of the kites, was quite a high affair on the full-sized machine,

10 feet above the wing at its centre, but later experiments proved that it was far larger than necessary, and it was cut down to a height of 4 feet.

A rudder and stabilizer framework were hinged direct to the fin and wing rim respectively at the rear, and two small, hinged frameworks projected in front of the wing, close to either side of its centre. John Underwood, Jr., recently explained to me that the two front control surfaces could be operated singly or together and he thought they could be designated as the first known ailerons. (But such a claim might revive an old controversy which has had flying circles in a dither many a time.)

ALBERTA'S FLYING SAUCER AT REST

The movable surfaces on the front of the Under-wood machine were in reality too close to the wing centre to be classed as wing tip controls, or ailerons, since their position would serve purely as a longitudinal control. However, Canada's right to claim the invention of the aileron was established by the Aerial Experiment Association in 1908, as related in chapter I.

All lifting and control surfaces were covered with a light-weight canvas, which was held firmly in position by many lengths of cord. These were first sewn firmly to the correct point on the canvas; later, they were tied tightly to their positions on the framework. When the craft was being handled between their barn and the testing area, the canvas surfaces were "un-tied" and furled on the framework, just as sails are furled on the spars of a ship.

By the middle of June 1907, the main work on the big machine was almost completed, and the directors of the Stettler Exhibition, hearing of its construction, invited the builders to display their handiwork at the fair-grounds. The Underwoods accepted the offer.

The craft had not then been fitted with its wheels but, nothing daunted, the trio placed its centre on a wooden stoneboat, and, lashing the whole thing firmly behind a horse-drawn farm wagon, set off over the 10 miles of bumpy roads to Stettler, where their invention went on public display beginning July 1, 1907. It proved to be quite an attraction, and, strangely enough, received much favourable comment.

As a matter of fact, the news spread rapidly. Even in far-away Toronto, an item appeared in the *Globe*, dated August 6, 1907, stating that the Krugerville *balloon* was completed, and that it had a "gas bag 300 feet long"! A Manitoba newspaper evidently had the "gas bag length" a bit jumbled up with engine power. Not to be outdone by the Toronto report, it announced that "the Underwood machine was equipped with 500 horse-power"! Apparently they were under the impression that it was a flying loco-motive or steamship.

The newspapers closer to home were better informed and therefore comparatively conservative. The Stettler *Independent* rightfully head-lined it as "Stettler's Aeroplane," and the Edmonton papers were only slightly off when they described it as "Alberta's Airship."

Enthusiastic promoters, dreaming of easy money, even began to circularize the suggestion that special trains might be run to Stettler from prairie points, so that the public might be on hand to witness the trial flights! During the exhibition, a criticism was made by one of Alberta's highest government officials. After examining the plane's "skeleton," he remarked to John Underwood, "Young man, you have missed one very important principle in your airship. You know, a bird's feathers have hollow quills, and when the bird's quills warm with vigorous flapping, the air they contain becomes lighter, and thus helps to buoy them up."

When the exhibition was over, the tedious job of hauling the framework back over the rutted roads to their own pastures was accomplished, and the Underwoods went to work to develop their brain child further. Because they had no engine, they were at first obliged to test the "flying wing" as a kite. This was done by laying out 700 feet of quarter-inch rope along the ground, fastening the "ground end" to a stout fence-post, and the other end to the nose of their machine. The first such test with the man-sized machine took place on August 10, 1907, and results were most encouraging. The stability of the "wing" was remarkable, and the high degree of lift it exerted in a 20-mile breeze was amazing. Five sacks of wheat were placed on the "cockpit platform" and they were lifted with ease. The exact weight is not known, but it was in the vicinity of 350 pounds.

Then John, who was 22 years of age at the time, asked if he could make the hundred-foot flight in place of the bags of grain. This his brothers flatly refused, thinking more of John's neck than he did himself, but at last he prevailed upon them to use a shortened rope so that the "kite" would rise only a few feet from the earth. This was done, and taking his place on the platform, John was lifted gently some 10 feet in the air, where he stayed 15 minutes, swaying smoothly before the apparatus was again brought down to earth. Although no claims for such a record have ever been made, this was undoubtedly the first time in Canada that a man had been lifted into the air by a kite.

As harvest and threshing operations began to crowd upon the brothers, further experiments had to be put aside until the following spring. By that time they had procured a 7 h.p. motor-cycle engine on which they based their hopes of flying. After some fancy figuring, they coupled it up with a belt to drive a four-bladed

VIEW FROM ABOVE

RUDDER — ELEVATOR

FIN

WING

26 Ft.

AILERONS

|← 42 Ft. →|

←FIN

WING ↘ PROP.

MOTOR CYCLE WHEELS
BICYCLE WHEELS

FRONT VIEW

|← 26 Ft →|

ORIGINAL HEIGHT 10 Ft.

FIN

LATER HEIGHT 4 Ft.

RUDDER

PROP.→

WING ↘

ELEVATOR

MOTOR CYCLE WHEELS

BICYCLE WHEELS

SIDE VIEW

DIAGRAM OF FLYING SAUCER

AUTHOR'S MODEL OF FLYING SAUCER
Showing details of construction of the machine

smoothed and rolled the area to the last degree. They had little difficulty in taxiing about at speed, but success in the air with that engine was not to be theirs.

Their "air field" is interesting. They laid out a regular "air strip," 20 rods wide and 80 rods long (330 feet by 1,320 feet), along a field of summer fallow, close by the side of the Kruger-ville road. The surface was as smooth as that of any strip today. Unfortunately, no aircraft ever took off or landed on its hard-packed surface, but unless use is made a criterion, it can fairly rate as Canada's first air strip.

When the experimenters found their engine had insufficient power they tried to procure a better one. Inquiries were sent to various points in the United States, and one, sent to the Curtiss Motor Cycle shop at Hammondsport, N.Y., asking for quotations on an eight-cylinder engine of around 40 h.p., brought back the answer that the cost would be $1,300.

Interest in the Underwoods' efforts ran high in Alberta. The local member for the federal Parliament actually took up the matter in Ottawa, suggesting that a bill be introduced to allow any motor they purchased in the United States to be imported into Canada duty-free. It is regrettable that the measure never got any further than a suggestion in the House, for it might have set a precedent which would have helped Canadian aviation enormously in later years.

The cost of the engine was not excessive as things were in those days, but it was enough to make the Underwoods think matters over very seriously. They reached the conclusion at last that as many of the best prizes offered for various flying achievements were being snapped up, the expense of an engine would not be

bamboo-and-canvas propeller, 10 feet in diameter, which was placed in the centre, and in front of the wing. The propeller was inefficient, and with the low power available the wing could not be induced to leave the ground. The boys had a site prepared for these endeavours, having

worth while. They decided to just go on flying their craft as a kite. During the spring and summer of 1908, their huge bird was seen high aloft at the end of its tether by many farmers of the district. Even at night, when the winds were right, the boys would send it up, with a lantern swinging beneath, and the weird light bobbing about in the sky caused much comment, and even some alarm, in the Krugerville area.

As time passed, the young men became a little careless in handling the big apparatus. One day, in a particularly boisterous wind, the barbed wire they were using in lieu of a rope became tangled. As the kite lifted and the pull came on the wire, a kink snapped; the kite fell to the ground, rolled over several times, and was quite badly smashed. The brothers could have repaired it, but apparently the novelty of flying it as a kite had palled, and the possibility of ever trying it out with a larger engine having evaporated, their ambitions also faded out, and they just piled the remains beside the barn.

The Underwoods never worked secretly, as many aircraft inventors found it wise to do owing to public ridicule, but they never went out of their way to seek notoriety. Had it not been for their one splurge at the Stettler Exhibition, which enabled many thousands of persons to see their handiwork, their efforts might well have vanished without a trace.

The fact that their machine never made a controlled, free flight under its own power should not be allowed to reduce the credit due them for the pioneer work they accomplished. In 1907, flight by man in heavier-than-air machines was in its infancy. No one besides the Wright brothers had accomplished officially recognized controlled flight at any point in the western hemisphere, while in the eastern hemisphere, only in the year before, 1906, had the first official airplane flight been recorded. The flying models made by the Underwoods proved beyond all doubt that their design was airworthy, and later the stability and powerful lift which the full-sized machine exerted was amply demonstrated in the many tests they conducted when flying the apparatus as a kite.

It is frustrating to reflect that another thousand dollars might have put the Underwood brothers in the front ranks of the world's aerial pioneers. However, in the same year that the Underwoods were abandoning their dreams of flight, a Saskatchewan storekeeper and contractor, who had moved to the Pacific Coast to take a fling at gold-mining, sold his claim and began work on the first airplane motor ever to be fabricated in Canada.

3. Gibson's twin and multi-planes, 1910-1911

ON a day in 1883, a small boy from a farm near the Cree Indian Reserve of Piapot, 25 miles northeast of Regina, in what was then the Northwest Territories, found himself face to face with the great Chief Piapot in person. What was more, the great man (with whose grandson he often played) actually condescended to address him. "You, too," he said, "will one day become a great chief." Sixty-five years later the prophecy was literally fulfilled when W. W. Gibson, a San Francisco manufacturer of mining machinery, became the fifth white man to be adopted chief of the Plains Cree Indians, under the name, which had been Piapot's own, of Kisikaw Wawasam—freely translated into English as "The-Man-That-Flashes-in-the-Sky."

Almost any day in the middle eighties the same lad might have been seen lying on the sun-dried grass of the Assiniboian prairie, his eyes riveted—except for an occasional swift glance in the direction of his father's herd of cattle—on the kite in the sky above him, soaring and diving in the westerly wind, as he held or slackened the taut string. He became very adept at making the kites, changing the design with almost every one he built, and he learned the art of keeping them aloft when scarcely a breath of wind was blowing. On calm days, which were a rarity on the widespread plains, he would gallop his pony for miles over the open land, towing his kite behind. It was then that the thought came to him that if his pony could supply the energy to keep a kite aloft, some way could surely be devised of applying power directly to the kite itself.

William learned quickly that a kite with a backward sweep on either side of its centre stick would fly with much more stability than one with a rigid flat surface. So for a number of years the boy experimented, making innumerable kites and pondering their principles as he worked.

In early manhood Gibson moved to what was then the tiny hamlet of Balgonie, 15 miles east of Regina, in the District of Assiniboia. By the year the province of Saskatchewan was created —in 1905—he was established as a hardware merchant; but in his spare time he immersed himself in experiments with model airplanes of his own design, stimulated—like the Underwood brothers of Stettler—by the newspaper accounts of what the Wrights had accomplished.

The stiff-paper models he made during the winter of 1903–1904 were based entirely upon ideas gleaned from his kite-flying days, for it was not until three years later that Gibson first saw a picture of an airplane taken close up. The wings of Gibson's models were given a sharp upward sweep towards their wing tips, similar to one of the best-designed kites he built. The front wing was adjustable, so that the degree of lift could be regulated in order to control the model in flight.

In his search for motive power to drive the propellers for his models, Gibson displayed great ingenuity. He used the spring end of a window-blind roller, cutting it off three-quarters of an inch beyond the spring; into the wooden end he drilled a hole, inserting a length of bamboo to form the rest of the "fuselage." The wooden part of the roller around the spring was lightened by whittling it down to a thinness of less than one-sixteenth of an inch. The metal part which pro-truded was also filed down to reduce weight, and a suitable screw was soldered to the end of it, to which was attached a pusher propeller, carved from Spanish mahogany. The "prop" was fitted at the very rear of the model, rotating between two rudders.

To launch his models the inventor devised a chute fashioned from a 9-foot length of board, 10 inches in width. Along the centre of the chute was a guide. Skids on the model served to keep the propeller clear in height, and also acted as a landing-gear at the termination of each flight. The surface of the chute was var-nished and highly polished, and it was sup-ported on four legs, the two at the front being 3 feet and the two at the rear 12 inches high.

In addition to the store at Balgonie, Gibson had acquired by this time a hardware store at Craven, and another branch at Cupar. As a respected citizen of these communities, he real-ized that if the residents in general, and his banker in particular, learned of his experiments, his credit rating might be endangered. He deemed it best to conduct his experiments as quietly as possible.

At dawn on a Sunday in the middle of June 1904, Gibson made the initial test of his first powered model. His store at Balgonie had a flat-topped roof, and here Gibson set up the launching device. Then, having wound up the spring of his model, he placed it on the chute, and let go. It shot into the air when only half way up the polished surface of the board, and in good, stable flight sped fast across the street. There it struck the top of a box-car standing on a siding, damaging one wing: but Gibson was elated, for he had satisfied himself that his model could fly. So he went on to make many other models of similar type and flew them with varying success from the roof of the store, but always during the dawning hours, to avoid pub-licity. Gibson kept his secret well—only one or two people saw the models in flight.

Dr. Kaulbfleisch, returning home very early one morning after attending a patient in the country, came into the village unnoticed by the inventor. Later in the day he called in at the store, and said: "Say, Billy, what kind of a funny looking bird was that you were trying to catch on your roof this morning? Never saw anything like it—flew right over my buggy, and lit on the grass over by the station."

GIBSON'S FIRST ENGINE

Jimmy Hicks, who was an early riser and lived close by, was another witness. One day he was heard telling a neighbour that he had seen a queer-winged bird fly off a roof, and that he thought maybe it had a nest up there. But Gibson did not enlighten them.

Encouraged by his experiments, Gibson de-cided to design and build a man-sized craft. To ensure privacy, he planned to do the work on his farm some four miles south of the village. The first step was to build a four-cylinder, four-cycle, air-cooled engine of his own design, but when it was only partly constructed the rail-road boom reached that vicinity.

With the prospect of making money, Gibson decided to contract to build a 20-mile stretch of right of way for the Grand Trunk Pacific, which was then being pushed through about one hundred miles north of Balgonie. Later he took another contract for an additional 22 miles, west of the Touchwood Hills. Through one cause or another, the two contracts, instead of returning a profit, cost him $40,000. He was 27 at the time, and, in his own words: "When I had the banks cleaned up, or rather they had cleaned me, I had no stores and no farm, so with what capital I had left, I decided to go to the Coast, and start anew."

Gibson arrived in Victoria, B.C., in the

autumn of 1906, with his personal effects and his partially completed engine.

In the spring of 1907 he met a prospector named Locky Grant, who owned a very good claim at Elk River, up Muchalat Arm on the west coast of Vancouver Island. But Locky Grant was without means to work the claim and offered the mine to Gibson for $500. On the day Grant sailed for Clayoquot in the steamer *Tees*, Gibson decided to look the property over.

Having lived all his life on the prairies, Gibson knew nothing about boats and still less of the Pacific Ocean, but being an adventurous soul he scraped together $300, bought a 17-foot launch, provisioned it, and set off northward without batting an eye. He had studied the course of his water route on a map and expected no difficulties: in his own words, "It looked so calm and peaceful." But he was eight days reaching his destination, and lost 25 pounds on the way! The Pacific off the west coast of Vancouver Island is noted for its extreme violence on occasion, and apparently it resented a landlubber's intrusion on that particular trip.

Gibson met Locky and was shown the mine. It was the first "hole in a rock" he had ever seen. When some of the weathered decomposed quartz was gathered up and washed in a pan, a 3-inch "string" of gold was among the showings. That was enough for Gibson; he made a deal forthwith. He gave Locky his launch, a camera, his field glasses, a rifle, and $100. The two men shook hands in solemn agreement, standing there beside the Elk River, and thus the mine came to its new owner.

Gibson returned to Victoria on the *Tees*, procured a small stamp mill and a water wheel to run it, and hurried back. Locky stayed and worked with him until they brought out a gold brick worth $1,200. Gibson is of the opinion that it may have been the first of its kind mined from quartz deposit on the west coast of Vancouver Island. With such proof of the mine's value, Gibson had no difficulty in disposing of the property for $10,000. Now that he had cash in hand he immediately made plans to resume construction of his engine and airplane.

The work began in 1908. It was a heartbreaking struggle, and Gibson freely admits that if he had not learned the blacksmith's trade as a young man he could never have accomplished what he did, for he had to make every part himself, by hand. He had formed a partnership with another man whom I am allowed to refer to only as "Dave." With this aid and his own capital, Gibson felt he was assured of relief from financial worry, but his partner got into difficulties in other connections and in the end was able to contribute only $500 towards the undertaking.

In 1908, flying was still so little believed in, chiefly as a hangover from Langley's fiasco, that the very idea was ridiculed, and most inventors tried to keep their work secret. Gibson tried his best, but the news travelled around Victoria that he was building an airplane, and reporters soon were on his trail. He worked eighteen hours out of every twenty-four, week after week, and many people who knew him ridiculed his obsession openly. Passing him on a down-town street, they would hold out their arms and go through the motions of flapping. Even the minister of his church tried several times to dissuade Gibson from going on with his work.

At that time Victoria was just about the last place in which one might expect to find many of the materials required to build an airplane, nor was there anyone to whom Gibson could turn for good advice. Aviation publications were not available, and he had to start from scratch, gaining knowledge as he went from one experiment to another. Delays were frequent and long, and his capital diminished at a frightening rate.

Many tests were conducted with quite large models, usually from the hilltop in Beacon Hill Park. The four-cylinder engine to which Gibson had fastened his hopes was found, when completed, to be of little use. The long 6-inch stroke caused far too much vibration, and in Gibson's own words, "When the motor was running, it jumped around like a chicken with its head chopped off." So he set to work to design and build a better engine.

Having decided to construct a six-cylinder, air-cooled motor, he drew up the plans, then went to Hutchinson Brothers, machinists of Victoria, and asked them for a price on it. Dan Hutchinson, the engineer in charge, looked the drawings over, and never having seen anything like them before, expressed the opinion that the engine would probably be a complete failure; six cylinders could not be expected to work as a two-cycle engine, and an air-cooled motor would be sure to overheat. Gibson stuck to his

GIBSON'S SUCCESSFUL ENGINE

W. W. GIBSON

guns, insisting that was the way he wanted it, so the firm undertook to help make the motor.

The cylinders had a bore of 4½ inches, with a stroke of 4¼ inches. Ignition was by battery, through a suitable coil and distributor. As with modern outboard engines, oil was mixed with the gasoline for lubricating the engine's internal moving parts, but in addition, self-feeding pressure grease-cups were fitted to aid in lubricating the crankshaft and the connecting rod bearings.

Great originality was displayed by Gibson, both in the design of the motor and in the method adopted to operate the air-screws. Two were fitted, one at the front of the engine and one at the rear, a radical departure from any other airplane ever built up to that date. Both the original air-screws were two-bladed, 6 feet in length. The front screw, which had a 6-foot pitch, was coupled directly to the crankshaft, but the one at the rear, of 4-foot pitch, was geared to rotate at twice the engine speed, and in the opposite direction. Gibson's object in designing and fitting the air-screws in this manner was to overcome engine torque, which persisted for many years as a real bugbear to designers in the building of single-engined, single-propellered aircraft. In tackling the problem Gibson showed skill and insight, and it is certain that he can claim the distinction of having first made use of the idea of overcoming engine torque in a fullsized machine by using two air-screws re-

volving one directly behind the other in opposite directions. A number of the world's most modern aircraft have had this principle incorporated in their design.

The crankshaft was made by Gill Brothers of New York, and the aluminum crankcase was moulded in Seattle. The cylinders and all other parts were fashioned in Victoria and assembled there. When completed in March 1910, it was the first successful aircraft engine to be produced in Canada.

When tested, the engine fulfilled the inventor's greatest expectations. It seemed to run as smoothly as an electric motor, developing between 40 and 60 h.p. depending on its revs. Complete, it weighed 210 pounds.

The airplane itself, which its builder named the *Gibson Twin-plane*, had many unique features. It had two complete wings, one well behind the other, each measuring 20 feet from tip to tip, and 8 feet at the widest point. The framework was of spruce, covered with a pale blue waterproof silk material obtained from Jeune Brothers, tent manufacturers of Victoria. Steel sockets were attached to the framework of the wings at their point of contact with the "fuselage" of the machine, and clamps, applied at these points, enabled the wings as a unit to be moved backwards or forwards to obtain a correct centre of gravity in relation to lift.

The design of the wings, which Gibson originated, gave great automatic lateral stability.

THE *GIBSON TWIN-PLANE*
Showing clearly its unorthodox design

A basically similar design, in use on some aircraft of modern make, is known as the "gull wing" type. Numerous wing ribs were fitted between the wing spars to keep them rigid and their covering taut, and a large main spar, 14 inches in width and 2 inches in depth, extended from wing tip to wing tip above the wings. The spar was carefully streamlined, a matter to which Gibson—in contrast with many designers of that age—gave much thought.

In the damp climate of the British Columbia coast it was essential to keep the wing covering taut at all times, and Gibson met the need by another ingenious device. Each wing rib was encased in a loose pocket of material which was sewn to the wing covering. Each rib had a metal tube at its centre, designed so that it could be used as a turnbuckle. By this method the lifting surface of the wings could be kept tight during flying tests, but could be slacked off when desired, to reduce strain on the wing framework and the fabric covering.

Two streamlined metal fuel tanks, each with a capacity of ten gallons, were placed one on either side of the engine, and well above it to ensure a good gravity feed of fuel to the carburetors. The tanks were fitted with baffle plates inside to prevent undue surging of the gasoline,

a clever forethought of Gibson's, used in all aircraft fuel tanks today. The engine was placed in the exact centre of the machine.

The undercarriage, then termed the running gear, was fashioned from metal tubing attached to the main frame-bearers. There were four wheels, one at either end of each bearer. Although well braced with piano wires, they proved to be one of the weak spots of the machine as they had no resilience other than that offered by the small bicycle tires with which the wheels were fitted.

The "fuselage" consisted of two frame-bearers, built eight feet apart, each approximately 35 feet in length. They were of Douglas fir, and on each were fitted 25 cast aluminum "collars," with steel arms protruding from each of their four sides. Through small holes at the outer ends of each steel arm a wire ran from end to end of the bearer. A strong turnbuckle attached to each wire kept the entire structure very rigid, and it had great strength.

A pointed front elevator of laminated cedar, 8 feet by 4 feet, was hinged to the front of the bearers, and was controlled by a lever in the hand of the pilot. Two rudders at the rear were manipulated by means of control lines attached to a shoulder yoke worn by Gibson. There were no ailerons or other means of lateral control.

At last the momentous day arrived when the *Twin-plane* was to be tested. It was dismantled at the workshop in Victoria and moved as secretly as possible by horse and wagon to a large grassy meadow on the farm of Mr. Dean, near Mount Tolmie, several miles north of the city. (Fittingly, though accidentally, the meadow is now a part of the Lansdowne Airport. Perhaps some day it will be renamed "Gibson Airport.") There, with two helpers, Gibson erected his craft, and on the early morning of September 8, 1910, made a successful short flight. Unfortunately, the running gear suffered damage in landing, necessitating a two weeks' delay before new parts could be secured and repairs made.

Once again, this time on the 24th, everything was in readiness. At 5 o'clock in the morning the machine was wheeled from its shed and the engine set in motion. Gibson took his seat, and after the motor had warmed up he signalled his helpers to let go. A light cross-breeze was blowing, and aided by a slight incline down the field

the *Twin-plane* lifted quickly into the air, after a run of only 50 feet. As it picked up speed and soared away from the ground the craft began to drift sideways, and Gibson, unfamiliar with the working of the controls in the air, made an unfortunate error. He leaned to the wrong side in an effort to overcome the drift with the rudders. This of course caused the machine to swing farther around and head directly for a stand of oak trees, well down the field. The pilot immediately shut off the engine and made a level landing, after a flight estimated at close to 200 feet. The momentum of the airplane carried it on with considerable force, and it struck a sturdy oak. Gibson was hurled from his seat but sustained only minor scratches and a severe shaking. Unfortunately, the machine suffered serious damage. As winter was coming on, further experiments were abandoned for that year.

It should be emphasized that, in those days, a flight of even a hundred feet with a machine of entirely new design was a momentous achievement. Short as the flight was, it qualifies

GIBSON AT THE CONTROLS OF THE *TWIN-PLANE*

as the first free flight to be made by a Canadian-built airplane. The initial hop by Wilbur Wright on his first powered flight at Kitty Hawk, N.C., was only 120 feet, and the earliest made by A. V. Roe in England in June 1908 was much shorter, only about 100 feet. The original flight of the Canadian Aerial Experiment Association, in which Casey Baldwin flew the *Red Wing* at Lake Keuka, N.Y., in March 1908, was 319 feet, and Santos Dumont flew less than 200 feet in 1906, when he received world-wide acclaim by making the first officially credited flight of a heavier-than-air machine in Europe.

Compared with these and similar early flying endeavours by other pilots who later became internationally famous, the initial flights of William Wallace Gibson have a place in the forefront of pioneer efforts. It is greatly to be regretted that they did not receive early official recognition.

In the fall of 1910 Gibson obtained a copy of *Artificial and Natural Flight* by Sir Hiram Maxim, published in London and New York in 1908 and 1909. This was the first authoritative work on flight that Gibson had ever read. Sir Hiram was a firm believer in the possibilities of using wings with a narrow chord rather than one of greater width, claiming that wings of such design could lift more per square foot of wing surface. The book included many cuts and drawings illustrating the different wing structures with which he had experimented.

Gibson decided to re-design completely the wings of his twin-plane, and he chose the shape which Maxim claimed gave the most efficient lift. By the spring of 1911 the new craft, now named the *Gibson Multi-plane* had taken shape. In place of the original wide silk-covered wings were a number of narrow ones made entirely of thin spruce, properly cambered, and beautifully finished.

It is worth noting that wherever Gibson used wooden cross-members as braces, they were carefully designed to give added lift. They were also properly cambered, and wooden strips, which Gibson termed "webs," were glued to their undersides at frequent intervals to help preserve their correct shape.

At this time financial difficulties threatened again; the last of the $10,000 was fast being gobbled up, and Gibson had a wife and three children to support. However, like most inventors, he had unlimited faith in the ultimate success of his efforts, so he sold his home in Victoria, and devoted the money from the sale to completion of the second machine.

The *Multi-plane* had the same frame beams that had been used on the *Twin-plane*, and the same running gear, but the latter had been strengthened and made more rigid. Gibson had learned, too, that it was necessary to have some sort of workable lateral control, and ailerons were made and fitted to the machine, the hinges being simply tough strips of leather. The ailerons, or wing flaps, in keeping with the rest of the machine, were made of spruce, and were operated by a wheel in the hands of the pilot. This wheel was fitted with a groove at its centre, which slid backwards and forwards on a keyed shaft at the pilot's will, operating front and rear elevators in conjunction with one another. Like the rest of the lifting surfaces, they were of spruce. The single large rudder at the rear of the craft was made from laminated cedar and was operated by a rope controlled by the airman's feet.

The engine was placed in the same position as before, but in the *Multi-plane* Gibson eliminated the front air-screw, using instead a new eight-foot pusher propeller, with a six-foot pitch, which he later found to be much more efficient than the two originally fitted.

Gibson confessed that since his experience at the Dean Farm he "was anxious to get away from the oak trees," and Thomas Paterson, then Lieutenant-Governor of British Columbia, granted him permission to carry on experiments at his farm at Ladner, several miles south of the Port of Vancouver. Gibson spent six weeks on the Paterson ranch in April and May, 1911, testing and adjusting his machine, but the weather was most unseasonable, and it rained almost continuously. He managed to get the craft in perfect balance, but the muddy, soggy ground, the small area, and the incessant downpour prevented concerted effort. No actual hops were made, although there were several attempts.

In the month of June a "Made-in-Canada" fair was held at Hastings Park in Vancouver, and the committee invited Gibson to place his aircraft on display. Hundreds of people saw the *Multi-plane* and it became the centre of interest during the week of the fair.

The continuous rain at Ladner was having a detrimental effect on the many glued parts of the craft, and as the wet weather had every

THE *GIBSON MULTI-PLANE*

Right: Gibson in the pilot's seat

Below: Twelve men stand on the machine to demonstrate its strength. Gibson standing at right foreground

THE *GIBSON MULTI-PLANE*

appearance of continuing, Gibson decided to transfer his experiments to sunnier skies. He went first to Kamloops, B.C., shipping the machine by freight. He set up his plane in the exhibition grounds, in preparation for tests, but was pestered by a crooked promoter who tried to inveigle him into advertising the flights. Gibson left Kamloops, taking his machine with him, before any tests had been attempted. Being strictly an inventor, and determined to avoid further publicity, Gibson arranged to conduct work on the farm of a friend on the outskirts of the city of Calgary, Alta. The ground altitude in that part of the country is over 3,000 feet, but that did not daunt Gibson and he persevered, although many adjustments had to be made. Among other things, the engine had to be fitted with three carburetors, instead of one as originally used at the coast, a change which markedly increased the motor's efficiency.

By this time, Mrs. Gibson had become alarmed at the risks her husband was taking, and since she had to make a visit to eastern Canada, she persuaded him not to try to fly the machine himself during her absence. Gibson

accordingly obtained the assistance of a friend, Alex Jaap, a native of Forest, Ont. Jaap had previously served as a helper, and as a "test pilot" he proved himself fully worthy of the trust.

Several good hops were made but the best flight of all, which also proved to be the last, took place on August 12, 1911. On that date Jaap took the plane up for a test, and unfortunately smashed it up badly when making a landing. The accident and the flight were reported in the Calgary and Regina newspapers. On August 12, 1911, the Calgary *Herald* carried the following account:

. . . When attempting to make a landing from a height of over 100 feet in the air during a flight, the airman working with Gibson to make tests, the latter being the inventor of the multiplane, crashed down into a swampy coulee near this city and narrowly escaped death.

Aviator Jaap had flown about a mile and miscalculated the ground on which he proposed to descend and when within 50 feet of the ground noticed the turf honeycombed with hundreds of badger holes. He attempted to ascend and elevated the machine, but too late. His engine stopped and refused to start again. The wheels struck the uneven surface of the ground, and were torn off, and the machine was literally torn to pieces.

For six weeks back, Gibson and Jaap have been experimenting with their machine at a ranch a few miles from this city, making short flights with great success. Gibson intends rebuilding his damaged machine with pressed steel planes and intends going to Toronto, where they have factories for that sort of work. . . .

The *Multi-plane* was very badly wrecked and the season was getting late. Most serious of all, the inventor's funds were again at a low ebb. He had come to the crossroads where he had to decide again just what to do. He was now forced to abandon further work on his ideas and to go back to making money. From the time he began experiments at Balgonie to the date of the final crack-up at Calgary, $20,000 had been expended.

Had Gibson been able to continue he might well have become one of the outstanding designers of the day, but as he himself remarked, "I might just as easily have gone to an early grave." In later years he often expressed the wish that he had continued with his original twin-plane design, since its basic idea was very sound, and if it had been developed further, far-reaching success might well have been achieved.

I endeavoured to trace Gibson during the 1930's and it took me several years to discover where the inventor had finally settled down. I was twice informed of his demise, and was once told of the cemetery in which he was buried! When I finally ran him to earth in Oakland, Calif., he was able to give me a lead to the whereabouts of the historic six-cylinder engine, which eventually turned up at Victoria, B.C., in the possession of a nephew. Today the first successful aero engine fabricated in Canada, the engine that powered the first free flight to be made by a Canadian-built airplane, has a place of honour in the National Air Museum at Ottawa.

4. More wings in the West, 1909-1915

THE TEMPLETON-MCMULLEN TRACTOR BIPLANE, 1909-1911

In the *Colonist*, the daily newspaper of Victoria, B.C., there appeared on September 24, 1910, an item which any air-minded reader must have read with attention. It stated that the Western Motor and Supply Company of Victoria had received an order to supply an engine for an airplane under construction in Vancouver. The news note went on to state that the engine desired was a three-cylinder, air-cooled, English-made Humber motor, of the Anzani type, especially designed for use in aircraft, and that the order had already gone forward to the British firm for shipping and delivery as quickly as possible.

Other details recorded in the report read as follows: "The machine built by the Vancouver men is of the Glen Curtis [*sic*] type, with some additional features added by the builders. It is a biplane with planes of 28 feet (tip-to-tip) and 5 feet wide, and each plane will be covered with a special rubberized silk. The operator's seat will be below the level of the lower plane which is different from that of the Curtis machine. With the motor installed the aeroplane will weigh 400 pounds."

The ambitious young constructors of the craft were the brothers William and Winston Templeton, with their cousin, William McMullen, and the machine embodied a number of ideas of their own. Plans of the machine were first formulated early in 1909, and by the end of that year all the drafting had been completed, and

the would-be aviators were ready to begin actual construction. What was of more importance, the three of them had saved sufficient money to go ahead with their project. Painstakingly, all the many pieces and parts were made, the entire undertaking being accomplished in the basement of the McMullen residence in Vancouver. By early spring of 1911, their ambitions were realized in a full-sized airplane.

During this initial construction, the machine was never completely assembled, for the smallness of the basement workshop made this impossible. It was not until the inventors had transported all the parts to Minoru Park race track, on Lulu Island, near Vancouver, that they were able to put it together on April 23, 1911.

The design of the machine was remarkable, for it was a tractor-biplane, that is, with the air-screw in front of the wings instead of at the rear, as was then the general practice with all craft of biplane design. Only monoplanes at that early date were supposed to be suited to that type of air-screw. It is also worthy of note that the engine, delivered f.o.b. Vancouver, cost the tidy sum of $1,200, and the entire undertaking set its builders back $5,000.

The inventors incorporated the best known points of several noted aircraft into their design as well as a number of original ideas. Ailerons, similar to those used in the Curtiss machines of the time, provided lateral control, being placed between the upper and lower wing. On either side of the machine, canvas "curtains" were fitted, as they were then thought to be a great

THE TEMPLETON-McMULLEN BIPLANE
With McMullen aboard, at the Minoru Race Track, Vancouver, in April 1911

help in overcoming the possibility of side-slipping. This danger was very general in most of the earliest airplanes, because of the lack of dihedral, which results when a wing is built straight from tip to tip, without any angle from the centre outwards. The curtain idea was borrowed from the Voisin biplane, which Santos Dumont first flew in Europe. Longitudinal control was worked by a hand lever, operating a box-type front elevator. A large tail surface was also fitted as a part of the machine but it was of rigid construction. A square rudder, also worked by a hand lever, completed the controlling surfaces.

Again and again each of the budding aviators made promising hops, but it soon became apparent that the 35 h.p. developed by the engine was not quite sufficient to keep the machine aloft in sustained flight. The most impressive hop was over a distance of 260 feet, and shortly after that the craft ran afoul of the race-track fence, suffering serious, though not irreparable, damage. In spite of this bad luck the local newspapers—The *World* and the *Daily Province*—gave the tests a favourable write-up. However, the inevitable problem reared its ugly head: the inventors' bank accounts were dwindling.

The machine was dismantled and the parts were taken to a boat-works on the Vancouver waterfront. The young men intended later to construct a pontoon and fit it to the repaired craft, thus turning it into a seaplane. Unfortunately, fire broke out in the shed where the machine was stored, completely destroying it; and the high flying aspirations of the three young men disappeared in the black pall of smoke that drifted lazily over the waters of Burrard Inlet and Vancouver's waterfront.

Although these three young Canadians only managed to place their feet upon the lower rungs of the ladder of pioneer flying fame, it should be borne in mind that they designed, built, and made experimental hops with a type of aircraft which had many original features. With the low power available, the performance of the machine was in every way equal to initial tests of many other aircraft. It is only too typical of the period that, through no fault of their own, their high hopes came to a sudden end.

THE PEPPER BROTHERS OF SASKATCHEWAN, 1910–1911

Back in the nineties, and on the other side of Canada on a farm near Addison, Ont., boys were born to the Pepper family with monotonous regularity. When the Peppers moved west to farm the prairies, few of the eight brothers took kindly to the new life; most of them roved away in their late teens to seek their fortunes elsewhere.

In 1907, young George, then seventeen, launched on a wandering career of sign painting which eventually brought him to San Francisco in 1909. Here he read a good deal about the

THE PEPPER BROTHERS' BIPLANE
George Pepper at the propeller and "Ace" in the centre, July 14, 1911

successful flying which the Wright brothers and the Aerial Experiment Association had already conducted, and when he returned to his home at Davidson, Sask., he brought with him a strong determination to fly and a generous sheaf of news clippings. With the aid of these he soon had persuaded his brother "Ace" to join him in conducting experiments of their own. Early in 1910 the brothers began by making models, all with propellers driven by elastic bands. Their experiments were not confined to any one type, for they made monoplanes, biplanes, and triplanes, all of which were built and tested at Regina, where "Ace" then made his home.

As they progressed with their research, the boys finally settled on a biplane type for their forthcoming man-sized machine, for they learned that although monoplanes had better manoeuvrability, biplanes stayed aloft longer with the same power. This discovery accorded with their belief that a biplane could lift more weight than a monoplane equipped with an engine of equal horsepower. Another of their conclusions—shared by few airplane designers of that date—was that an air-screw in front gave better performance than a pusher propeller. The Peppers' final model had a wing span of four feet. It had adjustable wing flaps, rudder, and elevators, and the small craft behaved remarkably well on the short flights it made, answering to various adjustments in a very promising manner.

In the summer of 1910, George and "Ace"

moved to Davidson—half way along the railway line between Regina and Saskatoon—to reside with their parents, while they undertook to build their full-sized machine. They spent almost a year on the work. The wing framework was fabricated with glued laminations of opposite-grained, kiln-dried spruce, a wood which was not easily obtainable on the prairies at that time. The ribs were fashioned of bamboo, bent to the correct camber after being immersed in boiling water, and then lashed firmly in place with heavy waxed cord.

The undercarriage and engine mounting were formed from steel tubing, which was also hard to procure, and the builders experienced great difficulty in welding the various sections properly together, as there were neither craftsmen nor facilities for such work in their vicinity. They had several mechanics from near-by towns try their hand at brazing the tubes together and as a result a considerable amount of tubing was burned beyond further use. Eventually the Peppers did the job themselves by using an ordinary blowtorch, utilizing a brick furnace to pre-heat the lengths of tubing while the brazing was being done.

Another trouble was that they could not obtain the right type of wheels and were forced to resort to bicycle wheels, which were not strong enough for such experiments, as many other early airplane designers found out to their sorrow.

As their craft neared completion the Peppers

had to face the problem of obtaining an engine. As a result of wide interest in their project, a small, private, and unregistered company was formed, the chief subscribers being the late Doctor Craig and the late Mr. Rutlege, barrister, both residents of Davidson. With the money thus provided, an engine designed specially for airplanes was purchased. It was air-cooled, with twin, horizontally opposed cylinders, rated at 30 h.p. Although its trade name is known to have been *Miss Detroit*, its source in the United States is now unknown.

After the motor had been placed in their machine, the Peppers ran into trouble trying to make it run properly. They knew nothing of the intricate workings of internal combustion engines, and they had no one to turn to for advice. At that date, engines for no apparent reason would act up and behave with incredible obstinacy. Eventually the Peppers overcame *Miss Detroit*'s tantrums, and got her working at full horsepower.

Early in July 1911, they began experiments by taxiing their machine about on a near-by pasture. On July 14, George took his place in the pilot's seat, which was in the centre, at the rear of the lower wing. Pushing the throttle wide open, the budding airman essayed his initial attempt at a solo flight. But it was not to be. One of the bicycle wheels, coming into contact with a hefty grass tussock, could not stand the strain. It buckled, and the machine swerved, wiping off the undercarriage and snapping the precious propeller.

Undaunted, the brothers immediately set about repairing the damage, and their constructive ability came to the fore. They made a new propeller by gluing a number of opposite-grained spruce boards together, and then laboriously fashioning it by hand. With everything once again shipshape, they trundled the biplane to the near-by race track, which was considerably smoother than the pasture they had previously used, and on August 1 George again took his place in the machine, giving the engine full throttle for another attempt.

To begin with, success crowned his effort; the craft quickly picked up speed and lifted lightly into the air. But like many novice aviators, the Peppers had very limited knowledge of actual flight conditions, and they had completely overlooked what might happen when the plane got into the fresh crosswind that was blowing at the time. The biplane reached the end of the race

AUTHOR'S MODEL OF PEPPER BIPLANE
Showing the interesting layout of the landing-gear

track and was flying beautifully at a height of some thirty feet, when a strong gust caught the left wing and tipped the machine alarmingly. Either George was not speedy enough, or the wing flaps were not sensitive enough to correct the trouble quickly. No one seemed to know for certain, but there was no mistaking what happened. The machine settled, and the right wing, brushing the earth, swung the craft into violent contact with the ground. George was only shaken up in the smash, but the wreck of the machine, at a time when funds were exhausted, brought a complete and bitter ending to the hopes of the pioneer airmen. Having suffered two severe reverses in rapid succession, the Pepper brothers were forced to throw in the sponge, and postpone further experiments.

One year slipped into another, and then came World War I. With the great advancement in aircraft which followed, the Peppers gave up all further thoughts of building aircraft and never again attempted to master the skies of Saskatchewan.

STRAITH: MANITOBA'S DESIGNER-AIRMAN, 1908–1914

Bill Straith was born in North Keppel, Ont., and his first experiments with heavier-than-air machines began in the Ontario town of Owen Sound. In the face of mild parental warnings, he constructed a glider during the year 1908,

WILLIAM STRAITH
At the controls of his dual-operated craft

in collaboration with a friend, Reginald Hay. Using the latter's motorcycle as a towing medium, young Straith made a number of hops in their home-built glider.

In 1910, Straith moved to Brandon, Man., later going to Winnipeg to join the mechanical engineering firm of Maclean and Company. He began the construction of powered aircraft as a hobby, and in Brandon in 1912 he completed his first machine, a Curtiss-type biplane. Un-

AUTHOR'S MODEL OF STRAITH BIPLANE

fortunately, he could do nothing but taxi the machine about, as it lacked sufficient power for flight. Later, in Winnipeg, he installed an engine of greater power and also constructed a pontoon, which he fitted to the machine in the hope of turning it into a hydroplane. It still lacked sufficient power to enable it to fly, but the budding inventor learned many things from its construction. Attempts to fly the craft were made on the Assiniboine River, at a point near the site of the old Granite Curling Rink at Winnipeg. No doubt citizens looked with astonishment at the spectacle of one of their well-known business men sitting in the flimsy craft, doing his best to urge it aloft for his first solo flight.

In 1913 Straith became interested in the construction of an aircraft of the Blériot type, a monoplane, and was able to make a number of short hops, but he never considered it a successful machine. During 1914 and 1915 he built his most pretentious pioneer aircraft, a dual-controlled, two-seater biplane fashioned along Farman lines, 42 feet from wing tip to wing tip, with a height of 9 feet and a length of 22 feet.

It was engined with a 65 h.p., water-cooled, Emerson motor, which drove an eight-foot pusher propeller, giving the biplane an approximate air speed of 65 m.p.h. Fuel tank capacity was eight gallons.

After days of taxiing and short hops, Straith began making numerous flights of between five and ten minutes' duration, and rapidly gained skill. His trials took place on the outskirts of Winnipeg in the neighbourhood of a power line on which he kept a wary eye. Eventually the inevitable occurred—Bill and the high tension wires made contact. The machine was completely demolished and Straith became a patient in the Winnipeg General Hospital, where he remained until his fractured hip and other injuries were healed.

Straith was destined to become a skilled and experienced pilot, with over 15,000 logged hours in the air to his credit, an undoubted leader in flying hours among Canadian pilots.

William Straith, the Pepper and Templeton brothers, and William McMullen all rightfully belong to the small group of inventors and aerial pioneers of the early flying era. Knowledge of their work never became widespread, and history cannot rank it as successful. Yet, as the contrasting fates of Langley and the Wrights at the very beginning had demonstrated, the balance between failure and fame was exceedingly delicate. The least we can accord these western Canadian pioneers is the honour and recognition justly due a noble failure.

5. A close-up of two early birds

It would be a simple matter, and more in accord with my ordinary way of thinking, to dispose of my partnership with Tom Blakely in Calgary and our brief Alberta adventures in a page or two. And yet, in a way, I would feel that something had been left out of the total story of Canadian flying, something that ought to be told. There aren't too many of us left from the pioneer flying days; each year there are fewer, and our memories, perhaps, are not as keen. Some of the best are gone, and have taken with them the memory of a host of little incidents— none of them much in themselves but, added together, telling a story that mere facts just can't convey. There's a special feel to every period that comes out best in the intimate close-ups of a person's experience, the significant little things that a factual history rejects.

A five-line ad, for instance, in a Calgary daily paper:

> PARTNER WANTED—Would like to meet someone interested in aeroplanes, willing to become partner and share expenses in building a machine at Calgary.   Box 87, News-Telegram.

That was my first intimation that there was another air-minded individual in the city. I'll never forget the day when I walked into the Calgary branch office of The Locators, a Winnipeg real estate firm, and asked for Mr. Blakely, the manager. The man who came out was a tall rangy redhead, with a mass of curly hair, a stiff red moustache, and a warm smile that was as genuine as the freckles that spattered his face. I liked him at once, and never had any reason to change my mind later. Walking down 8th Avenue West, after that interview, was like walking on air. Not only did Blakely have a large number of parts of a Curtiss biplane— there was an actual motor as well. All this he had acquired for a mere $200. At last I was to have a part in the actual building of a full-sized, man-carrying airplane, with every expectation of flying it when that job was done.

We rented a large shed near the then un-completed Lancaster Building situated at 8th Avenue and 2nd Street West, which later housed Calgary's General Post Office. The various parts of the machine together with the two partially constructed wings had been stored for over two years in a barn, where they had been very useful, as anyone could see at a glance, to a great many chickens. Our first task was a messy one.

But before we go into details of our venture, it might be as well to look back for a moment. Tom Blakely was born on a Manitoba farm of Irish parents, who gave him a pretty thorough religious background. They moved to Winnipeg eventually, where Tom grew up, and where in 1910 he saw his first plane, piloted by Eugene Ely, the early American barnstormer. About the time that Straith was trying out his machine on the Assiniboine River, Tom landed his Calgary job.

My own background was English. All through my teens I had been air-crazy, filling my room with home-built models of every known experimental craft, and going to every air meet I could wangle my way to. In 1913 (just before I came to Canada) I won a prize with my model of a Blériot which had a four-foot wing span and could fly 300 feet with ease. I was working then in the Office of the Raleigh Cycle Company and taking an elementary course in internal combustion engines. My one great ambition was to become a pilot, or, if that was not practicable, at least to get into the mechanical end of the motorcycle business. The average Englishman at that time knew practically nothing of Canada or the United States, and I was no exception. However, there was a business slump in England, and the C.P.R. was flooding Europe with glowing accounts of Canada. My mother, who was a determined woman, made up her mind to emigrate. She sold her millinery business in Nottingham and came out to Alberta, settling on a farm which she bought near Stanmore. I had no definite prospects in England so I tagged along. I was nineteen at the time and took a job with the Hudson's Bay Company department store in Calgary. It was 1913, but even so most people were completely unaware of the imminence of World War I; there were no radios in those days, and even the newspapers didn't use much space in reporting the daily war temperature of the world.

Now to get on with the story of the *West Wind*.

In spite of the mess that the parts of our machine were in we went to work with a will, and I must say that from start to finish I never had a job to do before or since that I worked at with half the pleasure and anticipation.

The machine had originally been flown in Canada at Moose Jaw, Sask., by an American barnstormer, and had been cracked up against a wagon on its first flight. The owner thereupon sold it to a concession owner who had shipped it to Calgary, where the chickens enjoyed its roosting facilities undisturbed until our appearance. Our biggest job was to restore the wings and undercarriage, which had been pretty thoroughly mangled. Every cracked wooden rib, strut, or spar had to be replaced, the ribs with laminated wood which we glued ourselves and then shaped to exact size. In the earliest machines built, considerable use was made of lengths of bamboo for outrigger spars, but in our machine bamboo had no part: the main beam of the landing-gear (which we had to replace) was fashioned of fir, all other wooden parts of spruce.

A MODEL BLÉRIOT
The author with his model Blériot-type monoplane in 1913

Back in 1910 and 1911 designers had learned that rounding off the portions of an airplane that faced into the wind when in flight was a sure method of decreasing head resistance, which consequently greatly aided the machine in flight because less power was required to drive it through the air. This is mentioned because many of the parts of our machine were rounded and all of the eight struts which served to keep the lower and upper wings separated were very well designed from the standpoint of streamlining, although the whole was tightly bound together as a unit.

All of the metal fittings and wires and much of the metal tubing showed evidence of expert workmanship, and were still serviceable, though a good deal had to be hammered back into shape. The sockets into which the strut ends fitted were simply small sections of suitably sized tubing, brazed to strips of metal.

The early machines are often referred to as "crates," and by comparison with modern planes, perhaps they were just that. But the word is not an entirely fair description of machines which, in the hands of a skilled pilot, not only flew, but flew well. They were built of wood and wire because these were the materials which could be relied upon to withstand the innumer-

THE *WEST WIND*
At Shouldice Park. Blakely (*left*) and author (*right*)

able strains and stresses involved in taking off, flying, and landing.

The wires that braced the wings and other parts of the machine, to keep it rigged rigidly and true, were of two types. For the wing bracing, a steel wire approximately an eighth of an inch in diameter was used, woven from many fine steel wires to form a tough whole which could stand tremendous strain. This type of wire was used to operate all the control surfaces, which were the front and rear elevators, ailerons, and rudder. Most of the other bracing wires were of heavy-gauge piano wire, which also had a breaking strain of many thousands of pounds. Wood and wire—yes: but in flying trim the machines of that day were dainty and graceful to the eye—far more beautiful, to my way of thinking, than the massive metal monsters of the present.

The Ellis-Blakely machine was just as fragile-looking as any of its race, but it was much sturdier than its appearance divulged. Even so, no matter how well-built the pioneer machines may or may not have been, their ability to lift from the ground and accomplish full flight depended entirely on the ability of the engine to produce sufficient power to be imparted to the propeller which, in turn, gave the entire machine ability to fly, thrusting it forward if a pusher propeller or dragging it forward if a "tractor."

In our own machine, we had a very reliable engine in a 45 h.p. Maximotor. However, its power output did not give us too much leeway for safety in the air. The carburetor adjustment of most early aviation engines was a constant

source of trouble, since no variation in the air and fuel supply could be made manually during flight.

That the carburetor was extremely primitive probably had a good deal to do with the trouble. It was simply a bowl, with an ordinary cork

IN THE PILOT'S SEAT
The author in the pilot's seat of the *West Wind*

float that regulated the flow of the fuel to the carburetor from the fuel tank through a suitable needle valve, which operated open or shut depending on the level of the float in the bowl. That is all there was to the actual carburetor. A simple flat metal valve in the intake manifold, which was operated at the will of the pilot through suitable wires and a foot lever, was all that was needed to complete the entire fuel intake system. The ignition system was very dependable, being a well-made German Bosch magneto, and I can state that we were happily free of ignition troubles, except with spark plugs, which frequently cracked through excessive heat; this was a problem we were never able to overcome. Even though plugs were then priced at approximately 75 cents each, to poor and striving airmen they were quite an item. The copper-built oblong-shaped fuel tank our machine was fitted with held approximately four gallons of gasoline, not sufficient for a record-making flight attempt to be sure, but quite enough for general purposes when one is learning to fly, with quarter-mile hops the order of the day. The fuel capacity of the Ellis-Blakely machine enabled it to stay aloft about 15 minutes; but, as none of the individual flights we ever made exceeded 10 minutes in time, a shortage of fuel in the air was never one of our worries.

I said earlier that our engine was reliable, and so it was—relatively speaking. In modern terms, however, the earliest engines were of course most unreliable pieces of mechanism, in the main because although man had invented them, he had not then solved all the minor or major difficulties which kept them from running smoothly and reliably. In the pioneer internal combustion engines, there wasn't a part which could invariably be relied upon, and spark plugs, magnetos, carburetors, and the like, would give up their usefulness without warning.

Our Curtiss was equipped with a tricycle landing-gear with the single wheel forward. The main purpose of the beam already mentioned was to connect the forward wheel with the framework of tubing that held the rear wheels in place. Like other craft of its day ours had no device for softening the landing impact after the flight—a hangover, perhaps, from the days when the main problem was almost exclusively one of getting *off* the ground. One look at our tires would have scared the pants off

a modern pilot—or any sane person for that matter. They were heavy two-ply rubber tire and tube combined, into which was set a regulation valve, and there was no beading of any kind to help bind the tire to the wheel. The tread was perfectly smooth, except for the odd patch of aging rubber—and anything but suitable for runs on stony prairie. One feature of the landing-gear I should mention here was known as the "drag." At the rear of the main wooden beam that connected the front wheel of the tricycle landing-gear to the tubing, which was fashioned to hold the rear wheels in place, was attached a double 8-inch length of flat iron which could be controlled at the hand of the pilot through a length of flexible wire and a lever. By pulling it downwards to dig into the ground the pilot could apply a crude but effective brake. We had little occasion to make use of the one fitted to our machine as we had ample space in which to manœuvre on the ground, but many early barnstormers, coming in to land in small fields or race tracks, found the drag on their Curtiss machines a life-saver indeed.

The great day came at last. There she stood, inside the shed, as trim a little Curtiss-type biplane as one could wish to see. Later we named her the *West Wind*, but to Tom and me she was always the "Old Girl" or sometimes the "Old Bus." For an airplane of that day, she was certainly getting old, as we discovered later.

As it turned out, we needn't have been so secretive about our final operations, but even as late as 1914 an aviator had the feeling that he was generally regarded as a bit of a crackpot at least, and more probably a suicidal maniac. Without blowing any bugles, therefore, we hired a team and a large dray, packed the dismantled machine aboard, and at dawn on a summer morning set off for our chosen testing

THE MAXIMOTOR

grounds some six miles to the west of Calgary. There on the level prairie out towards Bowness Park, we soon had her rigged.

There is no denying the fact that once we had our craft set up, standing there on the wide prairie, Blakely and I were very proud young men indeed. When we first sat in the seat, with the engine running smoothly behind us, it was a tremendous thrill to place a foot on the throttle pedal, and pressing down, feel what we considered a tremendous surge of power from the sudden roar from the engine.

We soon discovered, however, that the roar of our 45 h.p. Maximotor was not everything. For a good many days we alternated our time between taxiing back and forth and patching the tires, which kept blowing out until we had them pretty thoroughly rebuilt with patches. We took it all in our stride and came back for more, until June 25, when Blakely flew her off the ground for the first time. My own first solo came the morning after Dominion Day, on July 2, 1914. By the end of that month we had become quite proficient in straight hops, continually lengthening our individual flights and becoming more confident and expert with each.

TESTING THE MOTOR

An outdoor test of the motor. Tom Blakely on left; a friend, Walter Scott, on right.

Our greatest difficulty in lifting clear from the ground was the long runs needed. At full output, the Maximotor developed only 45 h.p., so no stretch of imagination could alter the fact that our Curtiss was underpowered.

The ground level in the part of Alberta where we were operating is over three thousand feet above sea level, and that fact, coupled with our meagre power output, was a very real handicap, because the air is much thinner the higher the altitude. It does not possess the same "lifting power" as denser air at sea level, and also the thinner air radically affects fuel adjustments. Therefore we had to make very long runs on the ground in order to get into the air. Another thing, too, was the fact that the majority of our early attempts were made just after dawn when the air was calm, so that there was no wind to help "in giving our plane a lift."

All through the early tests, the first short hops, and above all, on the longer ones, one experienced the same feelings. First, the thrill of acceleration—the feeling that there's power enough behind you to thrust your machine up into the blue and out on the other side of it. Then you are concentrating: watching the ground to gauge your speed—for there was then no other method—listening continually to the rhythm of the engine, hoping not to detect a warning cough, your tension mounting as the ground goes by at a faster clip. She's lifting, you think—yes!—no!—yes, she's up! And so you are concentrating for your very life as the ground falls away and you try to remember simultaneously everything you've ever learned about what to do in the air—that's high enough—don't push your luck!—level off.

You look down now at the prairie below, suddenly aware of the height you've gained, and how flimsy your perch. Then a gentle forward pressure on the controls, and you're coming down, easing your foot off the throttle—down—down, take it easy—the ground gets closer—closer—you level off, touch, and bounce a bit too much perhaps, touch and bounce again, a wing comes up for a split second of panic, then the wheels are on the ground together, the front one settles too, and you cut the throttle completely, by pulling the ignition switch, and roll to a standstill. It's all over!

There is no experience in the world like the surge of relief that comes over you then. It's the way you might feel on an ordinary day perhaps, busy with your familiar daily routine, if everything started to slow down till it stopped—even the water dripping from the tap suspended in the air. All of a sudden the flow of life goes back to normal, and you feel a relief and a terrific exhilaration. That sounds pretty strange, maybe, but it's as close as I can get to the sensations of those first flights.

I suppose we were flying for the thrill of it, too, but neither Tom nor I was the daredevil type. We certainly never took anything but a calculated risk, although we had our off-guard moments, as will appear. And yet, if we were cautious, it was not out of consideration for our own necks. I honestly believe that our chief, and probably our only, concern was for the safety of our beloved machine. We were young then: death seemed but a vague spectre, scarcely to be considered. But the Old Bus was very real and dear to us; and her loss or damage an unthinkable catastrophe to be averted at all costs. Perhaps the daring young pilots of this fast flying age feel the same close tie with their thundering jets. But, somehow—well, best leave it unsaid.

During our 1914 experiments, in order to be up and doing at the crack of dawn, when the air was still, we slept in a kind of pup tent; if you can dignify with even that label a square of canvas stretched over a rope between two four-foot poles and open at the ends. It was a bit drafty, but it kept off the heavy dew—and it certainly encouraged us to get up in the mornings. In spite of sleeping in our clothes we usually woke up cold. We both held jobs in the city, which was awkward at times. Tom was his own boss, but I had some explaining to do from time to time when we got absorbed in what we were doing. As a matter of fact I was fired three times, but was always reinstated through the intervention of the general manager, who seemed to have a soft spot for me and even dispatched his assistant all the way out to the Park one morning, to inform me that "all is forgiven—now get back on the double and get to work before we change our minds!"

It was just as well that our beards were still rudimentary, as all our water had to be fetched from the Bow River—a one-mile walk—and by the time we had filled the cooling system of our engine we were lucky to have any left over for personal requirements. I can't think of a better instance of Tom's sterling good humour than the fact that he invariably grinned when he lost the toss to see who would get the next pail.

On August 4, war broke out with Germany. Everything was in such a turmoil that we decided to call a halt to our activities. So we took our plane apart and hauled it back to the shed near the Lancaster Building. Then we wrote offering our services as "pilots," to the Canadian, English, and French governments. The letters we received from Ottawa apologized for refusing our offer of service as airmen, explaining, in part, that "it was not anticipated the military authorities would have any requirements for the service of aviators." Our letters to England and France were equally futile. One came back from the Royal Flying Corps stating briefly that if we wished to present ourselves at the airship shed at Wormwood Scrubs, our applications would be duly considered. From France came a similar reply: if we desired to enter the French Air Force it would be necessary for us to apply personally at their headquarters at Rouen!

During the winter we overhauled the engine and christened our machine the *West Wind*.

In June 1915, again using a horse-drawn dray, we trundled our craft to the hills above Shouldice Park, several miles west of Calgary, in the direction of Bowness as before, but this time not quite so far out from the city. According to war regulations, a government permit to operate the machine was required, and this was granted by the Officer Commanding the Royal North-West Mounted Police at the Calgary barracks.

The area we chose this time was much more suitable than the one used the previous year. It was a quarter-section of land owned by Mr. Shouldice, once farmed, but now gone back to weeds. It was ideal for our purpose. Tire troubles were also overcome. Early that spring we were able to get inner tubes for our tires, which we split around the inside so as to insert the tubes. Holes were drilled very close together along the splits, and the tire then laced up with strong twine, the whole then being replaced on the wheels, to be taped tightly in place for good measure. From then on, flats were quite a rarity.

By now we had become bold enough to make flights in a fairish breeze. During one of Blakely's earliest efforts in 1915, he made a fine take-off in a stiff crosswind, but coming in for a

THE *WEST WIND* CRASHES
After Blakely attempted his first landing in a crosswind

landing under the same conditions he ran into trouble. He noticed that the machine was drifting sideways as he settled down, but not knowing how to overcome it on the spur of the moment, he took no preventive action, and the results were startling indeed. After the dust had cleared, watchers arrived breathless on the scene to find the plane badly twisted, with broken spars and wires sticking out here and there. Blakely, standing near by, was ruefully rubbing his chest: he had landed on it when he left the seat very suddenly. He was not aware that a long jagged piece of wood was sticking out half way up the inside of his trouser leg! Luckily for him it was his clothing, and not his flesh, which had been pierced so neatly. Ten days of work were required to repair the damaged machine, and as we were desperately hard up (I had just been fired again) we had to economize severely in order to purchase the necessary replacement material. For those two weeks we lived on boiled turnips, bread, and tea. If we seemed to thrive on this diet it might have been because the particular variety of "turnips" which we had purloined from a farmer's field was later identified for us as mangel-wurzels—grown exclusively for cattle fodder.

Our early attempts at secrecy had long been abandoned. As a matter of fact, what with the oil boom and outbreak of war in the previous year, we could have sailed over the centre of Calgary, without its citizens lifting much more than an eyebrow. On occasion, trolley cars on the near-by tracks would pull to a stop, and passengers and crew would alight to watch one or the other of us winging by overhead.

But our real gallery were the Shouldice boys, sons of the owner of adjacent farmlands—and Jackie, a ten-year-old who lived with his widowed mother and a younger brother in the only house near by. Bright and early on one Sunday morning he was there as usual, but Tom had not shown up from his boarding-house in Calgary. The time dragged by, and still no Tom. By nine o'clock there were signs of a breeze springing up, and the best flying hours of the day were dwindling. I made some such remark to Jackie, and we looked at each other: one thought unspoken in both our minds.

"Listen, Jackie—how would you like to hold the throttle down while I spin the prop to start her up?"

Jackie knew the routine by heart, having watched it any number of times, but I carefully showed him how to straddle the two wooden bracing diagonals that ran down to the front wheel and the way to hold down the throttle lever. Then I went round to the rear of the machine to swing the prop.

One good yank, and the engine broke into

song. I stood back waiting for Jackie to release his hold to let the motor idle.

"Hey, Jackie!"

But the roar of the open exhausts drowned out my voice, the plane quivered, and began to move.

I ducked under the rear spar, and whipped around the end of the right wing to get to the lad and release the throttle. By then the *West Wind* was doing some whipping along on its own account, and going faster every second. I was not a good runner even in my prime, and I had only made about half the distance to the boy when the leading edge of the wing caught me in the rump and down I went. Fortunately I was watching it, and was able to grasp part of the landing-gear bracing as I fell, and so we set off across the waving grass, the boy still jamming the throttle wide open, myself being dragged along, and the machine driving along almost at flying speed.

I hadn't realized it, in the excitement, but our machine was heading straight for a section of the prairie that we normally gave a wide berth, criss-crossed as it was by Buffalo trails of a bygone era. As I tried desperately to drag myself forward to reach the now thoroughly frightened lad and knock his hands away from what we later called his "death grip" the machine hit the first of the depressions with a sickening jar. Jackie's hands were jerked free, and the engine, dropping back to normal idling revs, brought the craft to a halt; but not before the two front

spars to the elevator had been broken and the main landing-beam cracked.

When Tom arrived, he fortunately saw the funny side of it, and as for Jackie—he disappeared from the scene of our operations for many days.

By this time we had bought a proper tent, and had established a cosy little camp on the "Isle of Birds." We never knew who gave it this poetic name, but it was appropriate enough: the island was a veritable bird sanctuary, probably because of the trees. We made our initial invasion by building a raft upstream to float ourselves and equipment down. But the current of the Bow past the island was swift, and we missed our landings on two occasions. Finally, with the third raft we built, we made the island. To cut a long story short, we eventually rigged a wire cable from shore to island, and ferried ourselves back and forth in a seat on pulleys, always with one eye cocked warily on the deep fast-flowing current swirling fifteen feet below.

During 1915 we evolved a take-off technique which was most unorthodox: To meet the difficulty of gaining height, we would put the machine in a lengthy run along the level area on the hills above Shouldice; then when we had risen a few feet off the ground we would keep going, flying right out over the lip of the cut-banks, two hundred feet above the Bow River and the valley below. We would then swing around in a wide circuit, without gaining more than a few extra feet of altitude, and back we would come a-roaring to the hilltop, to make a landing at practically the same height at which we had taken off.

It worked like a charm, but looking back, I shiver a bit, for we knew little or nothing of the dangers of vicious up-and-down drafts which might be lying in wait at the brink of the hills. Often we experienced a severe "bump" as we shot out over the valley, but our plan always worked. No doubt we were lucky.

On the eastern side of another area which we used, the poles and high tension lines from Kananaskis to Calgary should have proved a hazardous obstacle. In modern flying they would, but their proximity did not worry us one whit. Whenever the wind was from the west, as it was most of the time, we were obliged to approach for landings from the power line side.

PREPARING FOR AN EARLY START
At Shouldice Park in 1915

MACKENZIE WITH HIS BLÉRIOT-TYPE MONOPLANE
Tom Blakely on the wing, Frank Ellis at right

We came in low, flying *below* the high tension wires, and *between* two of the high, widely spaced poles. The devil, they say, looks after his own.

So the summer slipped by. Meanwhile, unknown to us, exposure to the sun and weather for two seasons, coupled with age, had gradually weakened the canvas covering of the wings, particularly the upper one. One day I was coming in for a landing, and when a few feet from the ground, I heard a resounding "plop." The single surface of the upper right wing had given way along one of the seams. The wing dipped instantly and the machine struck the ground quite heavily, damaging the undergear and various other parts. I was fortunate: if the machine had been at any height when the covering ripped, a crash would have been inevitable. Many airmen lost their lives through similar trouble in the early days of flight.

It would have been possible to repair the damage, but a short time afterwards a terrific windstorm swept across southern Alberta. Although our craft was well staked, the pegs and ropes proved no match against such hurricane force, and when we found our cherished *West Wind*, a quarter of a mile away from where she had been tethered, she was a rolled-up, jumbled heap of wreckage. With sad hearts we took away what was worth salvaging, mainly the engine and propeller, leaving the shattered remains where they lay. They lingered on as a stark sort of landmark for several years, until souvenir collectors and the passage of time gradually obliterated all trace.

Even now I catch myself hoping that one of these days someone living in one of the neat homes that now stand where we first conducted our flying will turn up a part of the *West Wind*, and recognize it for a part of Calgary's pioneer airplane, that challenged and was finally destroyed by the strength of her boisterous namesake.

So ends the story that must stand for all the untold stories that are gone now, and beyond recall. I believe that my fifteen-odd years of delving have turned up most of what is left; though it has amounted in some cases to nothing but undated newspaper clippings.

News items from old papers, for instance, relate that as early as 1910 a J. Watts was building a Blériot monoplane at Victoria, B.C., and had ordered an engine from England. In Winnipeg—out Norwood way—another Blériot was being built between 1911 and 1913 by a certain Albert Contant, one report stating that it was almost completed. In the same period Winnipeg had its Boswell brothers building their own machine. In Alberta, in the same year that Tom and I were making our trial runs, a Mr. Mackenzie constructed a Blériot-type monoplane which he stored in his barn for two years before he was able to procure an engine. He had scarcely got the assembled craft warmed up for its first test when the motor's crankshaft snapped. The Blériot went back to the barn and waited in vain for another trial.

In 1912, Billy Stark, a famous barnstormer whom I later got to know, witnessed an amazing attempt to fly on the Vancouver waterfront, by two now unknown youths. Without previous tests, training, or taxiing, they set their Blériot up at the landward end of a long wooden ramp which ended over the waters of Burrard Inlet, and one took his seat in the machine and prepared to take off. He intended flying out over the Inlet when he reached the end of the runway. An accident may well have saved his life, or at least favoured him, for as he was revving up the engine the propeller flew apart, damaging the front of the monoplane and scattering pieces far and wide. Stark, who had tried his best to stop the foolhardy attempt, almost became a casualty himself when a piece of the shattered prop ripped into a wooden wall beside him.

Foolhardy youth, we who are getting along are apt to say. And yet—but I must not get ahead of the story. Let it suffice now to say that this ends the flying period of Canada's own pioneers in the West, and we are ready to go east again, where there were youngsters just as venturesome.

FLYING in those days was obviously a pretty risky business, and for those concerned, a deadly serious one. And yet, looking back from this distance there is something just a bit on the comical side about the combination of an over-whelming urge to get up with an utter disregard of the manner in which one got down again. In the course of reading everything I could lay my hands on that had anything to do with flying I ran across an old rhymed couplet from a poem with the quaint title of *Scribleriad*, written by Richard Cambridge and published in 1751, which somehow conveys the mood of the air-minded youth nearly two centuries later.

> Let brisker youths their active nerves prepare,
> Fit their light silken wings and skim the buxom air.

The fact is that to skim the buxom air, many young men of active nerves—or as we would say now, with "real" nerve—fitted themselves into wings so light and flimsy that today we look at their pictures and shudder.

Young Larry Lesh for instance, whose story will appear in due course, built, and flew, in Montreal in 1907, a glider made of bamboo and muslin, with a wing span of 16 feet and a lifting area of about 125 square feet—with a total weight of 25 pounds. Frail wings, these, to trust one's life to—even if you were small, and not yet 15 years old!

But before we go into the story of gliding in Canada, leaving the pioneer pilots behind, two eastern "starters" deserve a brief, but honour-able, mention.

Today, at Dayton, Ohio, in the centre of a park dedicated to the memory of the Wright brothers, a slim shaft of stone rises on the crest of Wright Brothers Hill. Inscribed on the shaft in alphabetical order are the names of the 119 pupils who learned to fly at the Dayton school, and at the head of the list is the name "Alex-ander, J. M." In the record of the school is listed a John M. Alexander who earned his American Aero Club licence, No. 335, in 1915.

At Shawville, in the province of Quebec, an early air enthusiast constructed a plane and—according to a number of his former schoolmates who saw the test in 1914—got it sufficiently under way to demolish the machine pretty thoroughly on its first trial. The enthusiast's name was John M. Alexander. The old adage, "He made of failure his stepping-stone," seems to fit this man, if, as seems likely, the two are the same person.

The second "starter" was Jack Pickels of An-napolis Royal—a west-shore town which looks out from Nova Scotia across the beautiful Bay of Fundy. Pickels was only fifteen when he began to build a Curtiss-type biplane, in 1914. The youngster was clever with his hands, and made all the wooden parts of the biplane him-self, fashioning them mainly from spruce, and fabricating also many of the metal fittings. The metal parts and all the bracing wires were pur-chased from the Curtiss Company, from whom came also the plans of the machine's construc-tion and dimensions. The work progressed slowly until the summer of 1915, when a two-cylinder, air-cooled, four-cycle Kemp engine was purchased—doubtless with the help of his

LARRY LESH
(Age 14.) On his horse-drawn glider
in 1907

parents—a motor designed specially for use in an airplane. Complete, less propeller, it weighed only 100 pounds.

Apparently there was no suitable area in the vicinity of Annapolis Royal on which to conduct experiments, and when the great day arrived for the actual tests, the only available place was a very confined and rocky spot. Nothing daunted, the young man had a go at it. Taxiing at speed, he came to grief on the rough terrain. The remains were toted back to the Pickels' home, and there the pieces lay. So ended another dream.

But long before this and only three years after Kitty Hawk, a thirteen-year-old protégé of the renowned glider experimenter Octave Chanute, Larry J. Lesh, performed a number of towed glider flights at Chicago. The glider he used was a Chanute design, but was built by the boy himself. An automobile driven by C. A. Coey towed the glider in a number of 250-foot flights. Early in the following year Larry came with his family to Montreal, where he attended the Peel Street School and where he continued to live and experiment with glider flights for a good many years.

The size and weight of the glider he brought with him from Chicago have already been given. In it he made numerous towed flights during the summer of 1907 on a farm near Dominion Park. Motive power for the towing was supplied by a horse, which was mounted by a farm boy, a friend of young Lesh's. The average length of each flight was 250 feet, and the heights were up to 100 feet. Some were made by the daring

youth in winds blowing at 25 miles per hour.

In the rented shed at Dominion Park which Larry used as a workshop, the lad had built another glider by midsummer. Like his former craft, it was a Chanute biplane but slightly larger, and was fitted with small, movable flaps to aid in lateral stability. It was somewhat more pretentious in having a single spar boom tail, to which was rigidly attached a small stabilizing surface and rudder. The method of control was to move the boom up or down or to either side with suitable ropes.

Numerous flights were made with this machine, but one of outstanding interest was made in August 1907. Octave Chanute, who was still sponsoring the experiments of young Lesh and helping to finance them, nevertheless strongly objected to the flight in question.

The new glider was attached by a 300-foot tow-line to a motorboat on the St. Lawrence River. The flight began from a concrete dock. As the boat sped away and the tow-line tightened, Larry ran swiftly along the dock, holding tightly to his glider, and was quickly launched into space. He faced a 10–15 mile wind as he rode triumphantly downstream, perched on a trapeze-like seat, with control well in hand. At times the tiny craft was rocked by wind turbulences as it carried the young aeronaut swiftly past the docks along the Montreal waterfront, and on past Dominion Park. For seven miles the flight continued before Larry gave the signal for the motorboat to slow down opposite Pointe aux Trembles. He had planned to cast off the rope with a device he had worked out, and then to glide to land or shallow water, but his signal was misunderstood. The boat turned in a sharp circle and its engine stopped, leaving the glider flat-foot in mid-air with the rope still attached. Larry had to drop from the craft when about twenty-five feet from the water. A good swimmer, he received no hurt, but the glider was wrecked.

In the summer of 1908 Larry rented a small workshop on St. Francis Xavier Street and built a third glider, of radically different design from the Chanute types. During the first Aeronautical Society meet held at Morris Park, New York, in the fall of 1908, he demonstrated his machine. Towed by an automobile, he made several successful hops but a final flight proved disastrous when the motorist overspeeded his car. The strain on the glider was too great; it collapsed in mid-air and catapulted to the ground in a tangled heap. The young pilot suffered a broken leg and other injuries, and was lucky to escape with his life. Thereafter Larry Lesh's story belongs to the American side of the border, and the remaining "brisk youths" of the early days might have had nerves that were just as "active" as young Lesh's, but they didn't do very much "skimming" in the "buxom air" of Canada's eastern skies.

We now come to the slighter records of a few Canadians whose first efforts to fly did not meet with sufficient success to justify their continuing.

A news item in the Toronto *Globe* of August 11, 1909, reported the successful first flight of an aircraft on the previous evening. "The airship, which was watched with interest by many in the vicinity, rose to a height of a hundred feet, and remained in the air for a considerable time, appearing to be under perfect control." The "airship" was a glider, designed and built by F. B. Fetherstonhaugh, who lived on the Lake Shore Road, towards Mimico, Ont. No further reports were published about it; apparently secrecy was maintained by its builder during the entire time of its creation and existence. A Toronto teacher of mathematics, A. Dalmero, helped in the construction of the machine. The Toronto paper stated definitely that the plane flew, but the owner of the craft affirmed, during my brief correspondence with him, that it did not fly—"it was only towed aloft."

The glider was a biplane and certainly underwent tests on Lake Ontario, off shore from Mimico. On the final test, according to Mr. Fetherstonhaugh, it was towed by a motorboat, and was pulled along the surface on a kind of pontoon rigged out of two canoes. Heavy waves swamped the canoes, the biplane slipped into the water and was wrecked, and the pilot, believed to be Mr. Dalmero, was almost drowned.

In 1911, an ambitious young fellow, J. H. Parkin of Toronto, built a Chanute-type glider, fashioning the framework from bamboo and the ribs from spruce, covering the surfaces with a light-weight factory cotton fabric. The ribs were steamed and bent to a form which had been advocated by Turnbull in various treatises. The craft was placed on view at a boys' handicraft exhibition in Toronto, and the workmanship of the glider received much praise from the press.

Young Parkin was never permitted to test his craft in actual flight. This was perhaps just as well in view of the risks involved and the fact that he was destined to serve his country later in aeronautical research.

During 1912 and 1913, interest in gliders was greatly stimulated by efforts and tests conducted in Europe. A large prize was offered in France for anyone who could make a flight in a propeller-equipped glider by manpower alone. Although the distance was to be only 30 metres, the prize was never won, despite repeated attempts, and gradually it became apparent that a man could not lift himself and a small machine by his own muscular efforts.

Young Jack Burton of Hamilton, Ont., was not attempting to qualify for any prize when he evolved a glider by fitting short monoplane wings to a bicycle, with a small propeller which rotated through suitable sprockets and chains and was operated by the foot pedals. One day— date unknown—he took himself and his small craft to the hillside near Ferguson Avenue in Hamilton. After assembling it he took his seat, and pushed off, pedalled furiously down the incline.

Success was short but emphatic. After a brief but speedy run, he soared into the air and there control apparently ended, for man and glider crashed from a height of some twenty-five feet. When the inventor was extricated from the tangled framework, he was found to be suffering from a dislocated shoulder and a broken leg.

The last heard of Jack, many years ago, was that he had taken a job in China and was living in Shanghai.

The one remaining instance of early gliding takes us back to the West.

As with other lads all over the world, news of flying was an ever-intriguing thought in the minds of three young high-school students of Kelowna, B.C.: Ralph Bulman, and George and Stanley Silke. The idea of building a glider took form when they were out camping on the Bulmans' cattle range during the summer holidays of 1912.

Some distance from their camping spot, the remains of a dead steer had attracted a number of turkey buzzards, which circled close enough to the boys for them to observe the skill of these birds in gliding. The boys then trapped one of the birds and tried without success to form the skelton into a kite, as a model from which to make further calculations. The actual making of a glider from these plans never advanced beyond the collecting of a quantity of bamboo poles and silk thread, and a mass of goose feathers!

On the advice of their mathematics teacher, they purchased a set of biplane glider plans from the *Scientific American*. Considerable difficulty was experienced in following the complicated details of construction; but with the aid of their teacher and the encouragement of their parents, they managed to assemble the apparatus. Meanwhile they studied the articles on wind resistance, wire and wood stresses, and so on, in the *Encyclopedia Americana*.

Ralph's father let them use a large barn to work in, and in the summer of 1914 the glider took final shape. When it was found that it could not be angled around to be taken out by the barn door, Mr. Bulman, Sr., had part of the side of the barn removed, so that the glider might emerge intact.

At the time the war broke out, interest in flying was at a high pitch, particularly in their locality, as an American airman, Weldon Cooke, had exhibited and flown his hydroplane at Kelowna in August 1914. The young students' craft was exhibited at the Kelowna Fall Fair, and attracted all kinds of attention.

Several times after the exhibition the boys took their glider to a down-hill area near the Bulman home which was free of trees, bush, and cattle. On each occasion lots were drawn to determine who should be the one to make the first attempt. Time and again they were obliged to give up because wind conditions were not exactly right. The glider was quite light, but had no running gear of its own, this being supplied by the aeronaut's own legs. A helper at either end steadied the wings and assisted in carrying the weight as all three ran forward into the wind, till the craft gained sufficient momentum to lift its own weight and that of its pilot.

The control system was most elementary. Lateral stability was provided by warping the wings through cords attached to a strap around the pilot's shoulders. Longitudinal control was provided simply by thrusting the legs forward or backward to change the centre of gravity, a crude but serviceable method which glider experts had used for many years.

One autumn day, with conditions just right, Ralph's turn came to take the controls. After a surprisingly short run with the craft, it lifted easily into the air, and the astonished Ralph, hanging grimly to its middle, found himself speeding out from the crest of the hill. With the fear of stalling, the young pilot jerked his legs forward in an endeavour to bring the machine down. He overdid the manœuvre however, and came down—alack—in a nosedive. Before he could correct it, or perhaps before he realized

LARRY LESH IN MID-AIR
On his glider, towed by a galloping horse

what was taking place, Ralph and the glider struck near the base of the hill. Damage to the plane was confined to the lower wing, but the youngster suffered a badly twisted leg and two broken ribs.

It was the only flight made by that glider, and although the top wing was later attached to a bicycle in the hope that hops could be made, no success was achieved. Its career ended finally in a dispute of right of way involving the glider, a horse, and a fence-post. The fence-post won.

During the winter of 1915, the Silke brothers built another glider on lines similar to the first, but with improved controls, which George handled with success on the one flight it ever made. Shortly after, it was blown from its moorings by a sudden windstorm, and ended as a wreck at the bottom of Creek Gulch, near Kelowna.

Perhaps the best way of summing up the story of early gliding in Canada is to enumerate the casualties: 1 ducking, 1 near drowning, 1 dislocated shoulder, 2 cracked ribs, 1 badly twisted and two broken legs, 1 dead turkey buzzard, 1 machine collapsed in the air, 1 destroyed on the ground by wind, 2 destroyed in the water, 2 smashed in ground impacts, and 1 operation to remove the side of a barn.

It will almost sadden the reader to learn that many of these brisk youths, who were so anxious to get up that they never sufficiently considered how they would get safely down again, are today solid respectable citizens of Canada and the United States. Larry Lesh, for example, went into the radio field and eventually established himself as a Florida business man; Ralph Bulman became the president of a substantial Okanagan Valley fruit and vegetable firm. No early Canadian flyer, to my knowledge, has ever robbed a bank, or committed suicide in an orthodox way.

But the story of the beginning of Canadian gliding should not end without recalling the coincidence that in the first such towed flights in 1907, both Larry Lesh and Lieutenant Selfridge received duckings at the end of their remarkable aerial trips, and both machines, one of very light build, and the other of quite substantial construction, came to battered endings in their respective localities.

By 1914 the era of practical flight was established, and the story of its immediate precursors —the barnstormers—must now be told.

In Canada the outbreak of World War I in 1914 is generally accepted as the end of the pioneer flying period. Strangely enough, no strictly experimental flights from first to last resulted in the death of any person participating. The first flying fatality in Canada came early enough; but it came in association with another variety of the species *homo aeris*, popularly known throughout the continent as the barnstormers, who had already begun to usher in the new age of practical flight in heavier-than-air machines.

PART 2

THE BARNSTORMERS, 1906-1914

7. Balloons and airships, 1906-1914

A MAJORITY of the flights, or attempts at flight, that I have described up to this point were basically private and experimental. In some instances the machine was exhibited at a fair, where frequently there was a good crowd around to watch proceedings, and some tests were even publicly announced and witnessed by crowds; but essentially the experimenter was wrapped up in his own efforts and probably expected to lose money rather than earn it.

The barnstormers, on the other hand, were men—and women too, as we shall see—who had achieved flight either through their own experiments or through having been taught to fly, and were primarily interested in making a living through flying. The only way to do this was to make a cash arrangement with a city's representatives or an exhibition committee, usually at a summer or fall fair. Air meets, too, were organized with increasing frequency, but these were not strictly paying affairs. The barnstormer was originally a vaudeville performer who, singly, or with a troupe of actors, roamed the country playing one-night stands at every populated point along the way. The barnstorming aeronaut or aviator who managed to make a precarious livelihood out of demonstration flights was a direct pioneer of practical flying. Whether the craft operated was lighter or heavier than air, he, or she, became a public performer whose scant financial rewards were augmented by the glory of public acclaim.

As early as 1879, on July 31, the ascent of a Canadian-built balloon is recorded, occurring at Montreal. The balloon was constructed by two Montreal men, Richard Cowan and Charles Page, assisted by an American friend, Charles Grimley of New York. When finished, the huge bag had to be inflated with gas drawn from the city mains, a job which required almost 24 hours to complete. When a test was made for buoyancy, the balloon being anchored to the ground at the end of a long rope, its lifting ability was found to be too great. According to *La Minerve*, the gas capacity was 70,000 cubic feet. So the balloon was hauled down, more ballast was added, and the crew and a passenger, Mr. Page, climbed aboard. The valiant aeronauts then cast off and soared to a considerable height. Their balloon, which was named *Canada*, had no mechanical means of propulsion, and the oar-like paddles attached to the basket, with which they had hoped to steer, were at once found to be useless. Drifting before a slight wind, the aeronauts covered a distance of fifty miles from Montreal before descending at their own will.

Cowan and Page continued work of this nature for some time, developing an elongated bag instead of the spherical type in general use, but apparently they got no further than constructional stages. However, all the available evidence points to them as the first Canadian aeronauts, regardless of whether they made any more ascents.

Although balloon ascents continued to be made at fairs and exhibitions until about 1920, the appearance of airships stole much of their glory from 1900 onward. Airships, also known as dirigibles (from the French *diriger*, to direct), were the next development. Their lifting-power was still the balloon, elongated instead of spherical, and they were motivated by a propeller (run by a motorcycle engine). Navigation was by rudder and elevators.

The method of inflating the airships was simple, but ingenious. A supply of iron filings was usually obtained locally. The filings were placed in a large wooden barrel which could be tightly sealed, with a suitable length of piping protruding from the top. A quantity of hydrochloric acid was then poured on to the filings until the chemical action set up had generated sufficient hydrogen to inflate the envelope of the lighter-than-air machine. The majority of these one-man dirigibles had a gas capacity of about 6,500 cubic feet. They were approximately 50 feet in length and 16 feet in diameter, and about 350 yards of Japanese silk were required to make one. The net spread over the gas bag, to which the framework was attached, was of linen seine twine and weighed only 6 pounds. Some 5 gallons of acid and about a ton of iron filings were required to form sufficient gas to inflate the bag to capacity. Obviously, to transport the cumbersome outfit from city to city and set it up ready to fly was not only an arduous job, but an expensive one as well.

The elements were none too kindly disposed towards these fat "flying sausages." No record has come to light proving loss of life in Canada through misadventure with airships, but it is a wonder, for many of the bloated bags bit the dust of exhibition grounds or tangled with the roof-tops they had come to conquer.

The first recorded and advertised exhibition flight of a power-driven airship in Canada took place in Montreal on July 12, 1906, when an American pilot, Lincoln Beachey, rose from the grounds of Dominion Park in a machine of his own construction. Beachey and his assistant, Charles Earl Hess, had built the craft in Toledo, Ohio, during the previous winter. Following the design of the Knabenshue airship, a well-known type of that date, the bag was fabricated of Japanese silk, and varnished with boiled linseed oil. Motive power to drive the cloth and wood propeller was a V-type Indian motorcycle engine. A light framework of spruce was suspended

below the bag by means of a net and ropes.

The operator of the craft stood on a thin plank at the lower part of the framework. In addition to running the engine and the controls, he had to shuttle back and forth along the plank to help the ship to rise or descend by changing its centre of gravity along the framework. These aerial experiences developed in Beachey the qualities which in later years made him the most famous of the barnstorming aviators.

After Beachey's visit, a year passed before another airship rose again in Canadian skies, this time at the Toronto Exhibition. As the affair was well advertised, much was expected—altogether too much. The airship was owned by Roy Knabenshue, a "dirigible man" who really did know his business, but who was not thereby immune from trouble.

The first ascent was billed for August 27, 1907, but at the last minute it was discovered that rain had seriously affected the motor's ignition system, and the flight had to be cancelled. The next day there was too much wind. It was blowing from the north, out across Lake Ontario, and Knabenshue showed good judgment in not venturing aloft. Visitors to the fair, however, were disappointed and disgruntled. The newspapers too, were not very polite, as witness the following: "The practical demonstration by the Knabenshue airship will apparently remain unsolved, and what was expected to prove one of the sensations of the fair will not materialize, and yesterday afternoon the directors of the fair decided to request that the Knabenshue airship be removed from the grounds."

At the very moment the directors were wrangling about how to get out of paying for the airship's appearance, the pilot of the ship, Carl Robinson, made an attempt to fly over Toronto, but before he was even well away his motor failed and the wind had him at its mercy. Drifting low out of the grounds, the airship settled fast, dropping almost to the railroad tracks in front of a fast-moving freight train. Throwing out ballast, Robinson avoided that danger, only to drift on to the roof of a school a few moments later. Here the airship got tangled with the cupola, became deflated, and was badly damaged. The pilot was unhurt, and remarked later that his better judgment had told him not to take the ship up, but that

JACK DALLAS WITH HIS BALLOON
At the Calgary Exhibition Grounds, July 1908

after his courage and the airship's abilities had been called in question, he had decided to go up if it was the last thing he ever did.

A lawsuit developed over the affair, the airship men seeking payment, the Exhibition Board maintaining that the demonstration was unsuccessful. The outcome of this, the first case in Canada connected with aviation, was that half the fees due were paid.

In September 1907, Lincoln Beachey was back in Canada with his airship, this time at Sherbrooke, Que. His ascents were the great attraction of the exhibition there. He had good weather and was able to fly as scheduled. His first ascent was at 5:00 P.M. on the 4th, the opening day of the fair, and consisted of a circuit of the grounds. The next day he flew over the city, almost to the centre of the town, while thousands of citizens stood cheering in

the streets. Numerous other flights were made, all highly successful. The Sherbrooke press stated these were the first flights in Canada, but aparently they had not interviewed Beachey on the subject.

Airships were never numerous in Canada. During 1908 they were particularly scarce. The only flight that has come to light took place at the Calgary, Alta., exhibition.

This machine was in charge of Jack Dallas, although owned by an American compatriot named Strobel. On the opening day of the exhibition, July 1, a stiff wind was blowing, but the aviator got away from the grounds safely. Gaining a good altitude, he set off towards the city and, spotting the parade, circled above it for some 15 minutes before heading back to the take-off point, where a landing was made without mishap. On the four days following, success-

THE BALLOON AGE

Left: A hot-air balloon being inflated at the Calgary Exhibition, 1906   *Right*: An early parachutist jumps from his hot-air balloon at Montreal, 1910

ful ascents were made; then came misfortune. Dallas and his assistant, Bert Hall, were filling the bag with hydrogen. Suddenly the wind slapped the bag against a high bracing pole, ripping the silk covering. In some manner the escaping gas caught fire, the airship was totally destroyed, and the two men slightly burned, thus terminating Calgary's first airship flights.

In September 1909, another American-owned airship, piloted by a man named Nassr, was slated as the principal attraction at the exhibition at Ottawa. Excitement was provided a-plenty, but by a chapter of accidents. The airman started away from the grounds, but before he could gain height he ran foul of some electric wires, and the propeller became tangled in them. One end of the framework dropped at a precarious angle, and spectators ran forward to steady the ship. Some of them grasped a steel rod beneath the structure, receiving a terrific jolt as the electricity from the power lines short-circuited through them to the ground. One man later died as the result of the shock and burns he suffered.

Later in the day the airship made a short but successful flight, but on account of a strong wind was unable to return to its starting point, being forced to run before the breeze until a landing was made in Ottawa south. There the bag was deflated safely and transported back to the grounds on a large, horse-drawn dray. Once back, arrangements were immediately begun to inflate it for another flight the same day. By late afternoon the work was accomplished, the daring Mr. Nassr took his place on the framework, and soared off again to brave the elements, which were still in hostile mood. No sooner was the airship clear of the ground than it struck some high tension wires near the entrance to the park. The bag was ripped open, the escaping gas immediately caught fire, and the whole thing blew up in a mass of flames. Nassr was uninjured but the airship, valued at several thousand dollars, was a complete loss.

The first airship to be seen in British Columbia made its appearance at Victoria, in September 1909. Owned by two Americans, J. Strobel and J. C. Mars, it was flown exclusively by the latter, who had begun flying hot-air balloons in 1892 and had become one of the country's outstanding balloon-parachutists and airship pilots.

Plans at Victoria called for flights beginning on September 20, but the winds proved too strong and no attempt was made. On the 21st they were blowing even more strongly, and Mars's good judgment overcame the insulting insinuations which were beginning to reach his ears; he stayed on the ground. Fortunately the wind moderated somewhat on the morrow and preparations were made to fly in the evening, when the wind usually dropped. Trouble with lights prolonged the take-off until 10:00 P.M. Darkness had fallen when the airship made a circuit of the grounds and the searchlight which was to keep it in view did only a partial job, as its operator had great difficulty in keeping the beam on the craft. But this was the first time that Canadians had ever seen a powered craft flying over them at night.

On the 22nd, the flight scheduled for 2:30 P.M. was held up until 7:00 P.M.; many people were disappointed and left the grounds before it was made. The third and final flight made during the exhibition took place on September 23, when over 12,000 people were on hand to watch Mars as he ascended at 6:00 P.M. Everyone was happy, and there were no more disparaging remarks.

Another visiting airship came to British

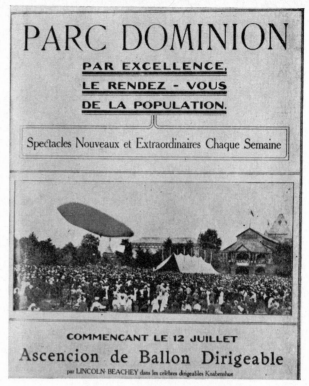

LINCOLN BEACHEY'S AIRSHIP
Advertised in Montreal, 1906

Columbia in 1909; it was owned by an American named Harry Ginter, who was hired to fly during the Provincial Exhibition held October 12 to 16 at New Westminster. A number of ascents were made, but before the exhibition came to an end, the airship had already done so. A strong wind carried it across the Fraser River, and Ginter was forced to bring the craft down on the bushy surface of a peat bog, where it was badly broken up by the wind before it could be deflated.

This was the hey-day of airships in Canada and thereafter interest faded fast. One further item, however, can be included.

During the great air meet at Montreal in June 1910, a young American named Cromwell Dixon had an airship on display, along with a second airship owned by another American, Fred L. Owens. It was the only occasion in our early flying history when two airships performed at one exhibition, although they were never in the air at the same time. In fact they were not to be together very long in any connection. On the first day, Dixon was taking off when the motive power gave out. Being only a few feet

CROMWELL DIXON

from the ground, he hopped overboard, intending to grasp a rope and tether the craft. Released of his weight, the airship shot skywards before he could tether it, rising quickly to some 3,000 feet where the silk envelope burst under the strain of sudden expansion. As the gas rushed from the ripped bag, the dirigible plummeted to the ground—a total loss. Hundreds of people in a special train near by, waiting to return to Montreal, went back to their own city believing that the young pilot was in the ship when it crashed.

Dixon was then only 17 years of age. His first airship, constructed by himself and his mother in 1907, was fairly successful. Mrs. Dixon was responsible for the fabrication of the envelope, the entire job being done on her small sewing machine. Cromwell Dixon was one of several young men who switched from airships to become full-fledged airplane pilots. In 1911 he received his Aero Club of America licence, No. 43, and a month later, on September 30, he made the first airplane flight over the Rocky Mountains, flying from Helena, Mont., to Blossburg, Mont. Two days after, he crashed to his death in a take-off at Spokane.

One of the earliest known balloon flights in Canada also had an international aspect. Two American aeronauts, John LaMountain, and John Haddock, set off in a large balloon, the *Atlantic*, from St. Louis, Missouri, on September 22, 1859, intending to make a cross-country flight during the night and to land somewhere the following morning. A south wind of tremendous velocity caught them, and at an estimated speed of 100 miles an hour they were swept across the border into Canada. It was an astounding trip for a balloon in that era; at times their altitude was as high as three miles. After a wild ride over a vast wilderness of trees, they valved away sufficient gas to allow them to make a rough landing in the forest. They knew they were in Canadian territory when the flight ended, but had no idea where. Finally, after wandering about in the bush wet and hungry, they came across a trapper's cabin, inhabited by a Scotchman named Angus Cameron, who fed them and was able to tell them they were 150 miles north of Ottawa; they had travelled over 300 miles in the few hours they had been aloft.

They had been given up for lost in the United States, where a reward of $1,000 was offered for knowledge of them alive, or $500 if they were dead. They received a great ovation upon their eventual return to the United States.

In August 1883 a small balloon was used to take aerial photographs at Halifax, Nova Scotia, and as they were undoubtedly the first to be taken in Canada in this way, the event is worthy of record. The camera was attached below a balloon with a gas capacity of 1,000 cubic feet which was tethered to the ground by a rope and a suitable windlass. The mechanism of the camera was operated by clockwork, as the balloon was too small to lift the weight of a man. The instigator of the experiments was Captain H. Elsdale, a member of the British Imperial Army Ballooning Establishment, Woolwich, England, who was on a tour of duty in Canada at the time, attached to the Royal Engineers at Halifax. He conducted the tests at his own expense. His first photographs were very clear and showed the north portion of the barracks at Halifax. They were taken from a height of 1,450 feet, but some later pictures were taken from altitudes varying up to 3,000 feet.

One other famous balloon incident with an international flavour involved the Gordon Bennett balloon trophy race of 1910. Starting from St. Louis, two American airmen, Allan R. Hawley and Augustus Post, established a remarkable non-stop flight record of 1,172 miles. After leaving St. Louis on October 17, they vanished for ten days, and their long silence convinced everyone they had perished. Then the news came from Canada that they were both alive, and had ended their long drift in the wilderness near James Bay, at the southern end of Hudson Bay. Like their compatriots back in 1859, they were extremely fortunate, having come across an Indian trapper who brought them to civilization by canoe along Ontario's northern waterways.

When airplanes came into the news, the public everywhere clamoured to see them at exhibitions and fairs. The one-man dirigibles quickly passed into the discard, and by the end of 1910, in Canada at least, they and their doughty pilots had become mere memories.

8. Airplane barnstorming begins, 1909-1910

It will be quite as surprising to Canadians as to their neighbours in the United States to learn that the first of a long line of American barnstormers who flew heavier-than-air craft made his début not in the United States but in Canada —at Toronto—in the year 1909.

As a matter of fact nothing was of less consequence to the early flyers than the imaginary line that separated one country from the other. It is true that there were regulations on both sides respecting the entry of persons and goods, but there was no practical way of applying them to flying machines or their contents, and duty that applied to imported machinery was enforced chiefly on dismantled machines coming in by rail. It was not until 1919 that specific regulations covering air flight and transport in and out of Canada were drawn up, and a year later before they were at all effective.

To the Canadian for whom the made-in-Canada label is important the story of the barnstormers will be disappointing. The fact is that, with two exceptions, *all* the barnstormers who appeared over Canada between 1909 and 1914 were nationals of other countries. The disappointed reader will, however, derive what comfort he may from the knowledge that a Canadian, John McCurdy, was the first barnstormer to appear over many an American town —particularly in the South—and that British Columbia's Billy Stark was not unknown in the American Far West.

The reason for this American predominance is not difficult to explain. At an early date the Wright brothers and Glenn Curtiss, together with many others, had established flying schools in the United States where ambitious young men could learn to fly, but in Canada there was no organized flight instruction before 1915.

WILLARD, 1909

The American who made his barnstorming début at Toronto was Charles F. Willard. How he came to fly at all requires a few words of explanation.

In 1908 the Aeronautic Society of New York decided that experimentation had gone far enough to warrant public demonstrations of flying. They first tried to persuade a French pilot to come over with his plane, but failing in this, they prevailed on Glenn Curtiss to build a plane and train a pilot for them, for a total sum of $5,000.

By July 17, 1909, the plane was delivered to the Society—the first airplane to bear the now renowned name of Curtiss.

It was christened the *Golden Flyer* because of the yellow, rubberized silk fabric with which the lifting and control surfaces were covered, and the orange shellac finish of its struts. The outriggers to the front and rear surfaces were of bamboo, braced with light, nine-strand wire cable. The main wings were braced with heavy gauge steel wires, threaded at their ends to enable them to be screwed into motorcycle spoke nipples, held in place by suitable fittings.

The front elevator was of "box-kit" type, and of large size, which assisted in keeping the machine aloft in flight, for, as regards longitudinal balance, the *Golden Flyer* was nose-heavy. An aileron attached to each outer *front* strut supposedly gave the craft full lateral

CHARLES F. WILLARD

ment Association, also made numerous short flights in the *Golden Flyer*, and was so successful with the machine that he had a duplicate craft built. He went to France with it and on August 29, 1909, he rose to fame by winning the first Gordon Bennett Trophy race.)

Willard became the first barnstorming pilot in the western hemisphere. The Aeronautic Society contracted to demonstrate their *Golden Flyer* in Canada, and the machine was shipped by train from New York to Toronto, arriving on August 28, 1909. It was assembled at the old amusement park at Scarborough Beach, east of the city.

The exhibition committee, completely ignorant of the requirements for a suitable take-off, had chosen a confined area between several buildings, a spot just large enough to set up the tent in which the plane was housed, with only sufficient space outside for it to be wheeled out into the open. Apparently the committee expected it to take off like a bird, or "swoosh" up into the sky like a rocket. As Willard naïvely explained to me, the "only escape" was between the buildings, towards the near-by lake, a space so narrow that it allowed clearance of only six feet at either wing tip! Not to be beaten, the airman sized up the situation and arranged for a wooden trough to be constructed down the centre of the "alley," to act as a guide-track for the front wheel of his machine. When completed, it ended atop a three-foot breakwater, just above the surface of Lake Ontario. An inauspicious sort of runway, if ever there was one!

Rain delayed flight attempts until September 2, when, under a lowering sky, with evening fast drawing in, Willard seated himself in his machine, warmed up the engine, and gave the signal to let go. With a roar, the *Golden Flyer* rushed down the track towards the breakwater, and from there shot out into space.

The flight was billed as the grand finale of the day, but the "grand" could well have been omitted. When he was scarcely airborne, the pilot nosed the craft down a few feet in the hope of picking up speed. The engine just wasn't doing its best, however. After sagging along for some 300 feet, man and machine smacked into the water, and night had fallen by the time Willard was rescued from his cold and wet surroundings.

The plane was not damaged; it had settled on the shallow bottom, its front wheel resting on the sand, the tail sticking high out of the

control, but Willard later confided to me that the ailerons were practically useless. Every time he operated them, they greatly retarded the speed of the machine, evidently acting like flaps on modern ships.

The motive power was the creation of Curtiss, a four-cylinder engine of vertical type, with 3¾-inch bore and a 4-inch stroke. At full throttle it turned over at 1300–1400 r.p.m. Flights at that date were of short duration, the capacity of the fuel tank being exactly one gallon!

The plane could be quickly assembled or taken apart and crated for shipment, a very necessary requirement.

The training of young Willard—chosen by Curtiss to be his first pupil—had begun in the spring of 1909 at Mineola, Long Island. Direct instruction was impossible, because the machine could accommodate only one person at a time. Willard thus practically taught himself to fly, and at last became quite proficient. (Incidentally, Curtiss, who was already an established pilot through his flying with the Aerial Experi-

THE *GOLDEN FLYER*

Curtiss' first plane undergoing tests early in 1909

water. Willard and his helpers worked like slaves to put it in condition again, and five days later it was completely restored and the intrepid pilot was ready to have another go at it. This time, small air bags were fitted to the undersides of the lower wings, in the hope that they would keep the biplane afloat should it again end up in the water.

On the evening of the 7th, when Willard tried again, his efforts were crowned with success. He got his machine off the end of the breakwater at full flying speed, and soared gracefully over the lake at a low altitude. He made a wide circular flight over the water, covering a total distance of some 5 miles, and lasting for 5 minutes.

His intention had been to alight on a strip of level sand along the exhibition waterfront, but to his dismay he discovered that the beach was now packed with spectators, so once again he was forced to take a dip in the lake. The air bags proved useless and Willard was found seated in the machine, with the waters of Lake Ontario rippling around his neck.

Again, the airman and his *Golden Flyer* were rescued undamaged, and four days after, both were dried out, fit as ever, and ready for the fray.

On the evening of September 11, Willard got his ship off to a good start, but the magneto went dead when he was only fifty feet off shore, and quite low in height. So down went Charles and his Curtiss into the lake for the third and final count, and as the propeller was slightly damaged on this occasion, further attempts were called off. Thus the first advertised exhibition flights on the American continent came to a harmless but water-soaked conclusion.

At the time of Willard's really sensational attempts at Toronto, newspapers throughout the land were acclaiming Peary's discovery of the North Pole, and scant space was left for reports of the flying done at Scarborough Beach. So the young man's efforts slipped by almost without comment in the press. The Toronto papers apparently were about the only ones that referred to it at all, and they were far from complimentary. Dry captions such as: "Airship went up, airship came down," "Willard's plane prefers the water," did not worthily publicize the young man's daring and determination.

HAMILTON, 1910

In 1910, within a year of Willard's Toronto exhibition, flights had taken place which were the first of their kind at other widely spaced points such as Vancouver, B.C., Winnipeg, Man., and Montreal, Que.

THE *GOLDEN FLYER*
Flown at Toronto by Charles Willard in September 1909

Aviation exploits in Canada in 1910 began in British Columbia when the American airman, Charles K. Hamilton, arrived there from Seattle, Wash., with his Curtiss-type pusher biplane. He had constructed the machine in California and had taught himself to fly during the latter part of 1909.

A crowd of 3,500 British Columbians was on hand at Minoru Park race track on Lulu Island, near Vancouver, on March 25, 1910, to witness the first airplane flight in western Canada. After being in the air 10 minutes, Hamilton returned to the oval in a safe landing. A newspaper reporter stated in part: "There was no engine trouble, but a sudden landing was made." No doubt the landing seemed sudden only to the newsman, who probably had never seen an airplane before.

On a second flight later in the day, the pilot circled the track. In an attempt to make comparisons of speed, an automobile raced around the oval, but Hamilton, unaware of the test, flew in too wide a circle, and the automobile won. The first day's flying was a great success —everyone who had witnessed it was fully satisfied. The Vancouver *Daily Province* published a fine picture of the machine flying past the grandstand at 55 miles an hour, and the caption —"Hamilton's Aerial Clipper"—is proof that his

plane pre-dated the present-day air clippers in name, if not in airworthiness, by almost thirty years.

The flying took place during the Easter weekend, and on Saturday the 26th Hamilton rose into the air on three separate occasions. His third flight was outstanding for its distance and for the additional reason that it was the first cross-country flight record to be established in Canada.

After taking off from the race course, Hamilton followed the north arm of the Fraser for about 20 miles up river. The spectators on Lulu Island had no sight of him for a considerable time, and they cheered him to the echo upon his return. The report in the *Daily Province* described his return thus: "Everyone crowded about him, asking where he had been. 'New Westminster,' came the reply. Mr. Hamilton was shaking with cold and was immediately supplied with a stimulant. . . . He described how he had followed the winding course of the North Arm of the river, mounting to 2,500 feet; then finding the temperature too chilly, had descended to a lower altitude, which he maintained until he reached New Westminster, where he sank to within nearly 100 feet of the ground. His arrival there caused considerable interest, the street cars stopping while the occupants watched his

CHARLES K. HAMILTON'S BIPLANE
At Lulu Island, Vancouver, March 25, 1910

evolutions. Turning just west of the bridge he began his homeward flight. . . . His speed averaged about forty miles per hour, but during the course of his flight, he must at times have attained a rate of 50 miles an hour." That was fast flying once!

Easter Monday, the 28th, dawned cold, with a strong westerly blowing, but Hamilton was not deterred, although many an able airman would have stayed on the ground. Several flights were made during this last day of his visit to Vancouver. The highlight, probably unique in the annals of aviation, and certainly in Canada, was a contest with a racehorse. The

horse was Prince Brutus, ridden by Curley Lewis. The track was a mile, and Hamilton conceded his opponent three-eighths of the course, a handicap which proved just too much. The horse and rider made the five-eighths they had to cover in the fast time of 1 minute and 7 seconds, against the biplane's time of 1 minute and 17 seconds, but Hamilton was overhauling them as they passed the post.

At the conclusion of his flight, the pilot gave the crowd a real thrill. He went up to a considerable height, then cut his engine. With the propeller barely turning over, he put the nose of the machine down in a very steep dive, holding it thus until almost at the ground, then flattening out just in time to make a perfect landing on the infield. This feat shows him to be a master airman, especially when it is recalled that he had taught himself to fly only four months earlier.

Hamilton lived only two years longer, and then, unlike many of the Early Birds, he died in his bed. He was anything but robust, and to be perched on a seat wide open to the rushing winds was more than his frail body could stand. After reaching the heights of fame as one of North America's outstanding pilots, he died from pneumonia during the winter of 1911–1912.

CHARLES K. HAMILTON (centre) WITH
MECHANIC AND MANAGER

OFFICIAL SOUVENIR PROGRAMME
Issued for Hamilton's flights

Ely, 1910

In the summer following Hamilton's exploits at Vancouver, a series of exhibitions of flying were staged at Winnipeg, Man. An American named Whipple Hall had been billed to give the exhibitions, but he became ill and was re-

placed by another American pilot, Eugene B. Ely.

The machine Ely then owned was the third one built by Glenn Curtiss; the maker had sold it to an advertising firm in San Francisco, supposedly for display purposes only. The biplane was on view in a department store in Portland, Ore., when Ely spotted it, and his offer of $5,000 was accepted. He immediately began learning to fly, and later set off on a barnstorming tour.

Ely reached Winnipeg the second week in July 1910. His first flights were advertised for the 14th, but a boisterous prairie wind was blowing that day and he wisely stayed on the ground. As usual, people began to murmur acrimonious things against flying machines in general, and against one in particular, so on the 15th, although the wind was still blowing at some 30 to 40 miles per hour, Ely decided to make the attempt.

"I am not going to disappoint the crowd any longer," he said. "I'll go up if I break my neck in the attempt. I have heard people say that this airplane stunt is a bluff and that we are making this wind a reason for keeping to the ground. Well, I'll show them."

Perhaps being thus "forced" to fly had its advantages, for pilots certainly learned to navigate the air under tougher conditions than they might otherwise have known.

Although he was pitched about like a cork on rough water, Ely made a good flight, and after a wide circle in the air came down to a good landing, to receive many congratulations. A second flight shortly after terminated unfortunately. A strong gust caught one wing just as the wheels were touching down. Striking the ground with force, the wing collapsed, and the machine was quite badly damaged.

In a crack-up, those early machines, with their many wires and wooden outriggers, often looked much more battered than they actually were. People were astonished that Ely emerged from the tangle with only a cut on the leg.

No further flights were made in Winnipeg by Ely, although the president of the T. Eaton Co. posted a $1,000 prize to be awarded to the airman if he would fly to Portage la Prairie, 60 miles to the west, and a total prize of $5,000 for a flight there and back. But the prizes went begging.

From Winnipeg, Ely went to Minneapolis,

EUGENE B. ELY
At the controls of his pusher biplane

Minn. This was a turning-point in his career. He ran into Glenn Curtiss who was barnstorming with Charles Willard at that city, and when Curtiss saw Ely and his machine in flight, he secured an injunction against its being flown again, as he had originally sold it only for display purposes. This led Ely to sign up with Curtiss to barnstorm for the Curtiss Exhibition Company, and he continued through the eastern states, rising to fame as one of North America's foremost airmen.

His special claim to glory is the fact that he was the first airman to fly an airplane off the deck of a ship. This event took place at Hampton Roads, Va., on November 14, 1910, from the fore-deck of the anchored U.S.S. *Birmingham*. The airman landed on a near-by beach.

When Curtiss realized he had such a fine airman at his command, he supplied him with a brand-new machine with a more powerful engine, and Ely flew to further fame by landing on and taking off from a warship, U.S.S. *Pennsylvania*, anchored in San Francisco Bay, January 18, 1911.

Ely was the man who said, "Give me enough power and I'll fly a barn door." It is a pity he did not live to see the great aircraft carriers, but he came to his death in an airplane crash at Macon, Ga., October 19, 1911.

Through Willard in Toronto, Hamilton in Vancouver, and Ely in Winnipeg, thousands of Canadians had their first glimpses of an airworthy plane handled by a skilled pilot, and in a few short years it became apparent even to the most sceptical that the flight of heavier-than-air machines was not an impractical dream, but an accomplished fact with rapidly expanding possibilities. A new enthusiasm swept the continent.

THE sudden surge of public interest in flying that took place at the end of the first decade of this century came as a delayed reaction to the early achievements of the Wright brothers in 1903 and after. Earlier, the public had built its hopes on a more immediate and spectacular conquest of the air, hopes that were thoroughly dampened by the plunge of Langley's "aerodrome" into the Potomac. The simple, unspectacular first flights at Kitty Hawk, even when they became recognized, hardly compared in lay eyes with the long-distance flights of drifting balloons and the impressive bulk of gasoline-powered airships with which the public were already familiar. But by 1910 the heavier-than-air machine had demonstrated its superiority over the airships.

This was true not only in America, but—as I well remember—across the Atlantic, too. I think the greatest thrill of my youth in England was the occasion when I encountered the famous French airman, Count Jacques de Lesseps. A friend and I were on our way home from a cycling jaunt in the outskirts of Nottingham, near Colwick Park, when we heard a hum which grew quickly louder and louder, and suddenly just over the treetops loomed a flying machine, which to our startled eyes seemed enormous. Swiftly circling almost over our heads, it settled gently to a landing on one of the great grassy stretches contained within the park area. Excitedly we mounted our bikes, and pedalled furiously to where the landing had taken place, arriving breathlessly just as the pilot was climbing out of the cockpit of his Blériot monoplane.

The minute he pushed up his goggles we recognized him as the famous Count de Lesseps, hero of a dozen air meets in Europe. He grinned at us, lit a cigarette, and said something in French. We tried our scanty French on him, but he must have thought it was English, so we just grinned all round until the park officials arrived and whisked him away, and also posted guards around his machine.

THE MONTREAL AVIATION MEET

The youngsters who attended the great air meets at Montreal, Que., and Toronto, Ont., in 1910 are now well along in years, but any who read this account will thrill again as they recall the excitement of those early air shows.

The first air meets held at Montreal and Toronto during 1910 were organized and directed by E. M. Wilcox of Toronto who foresaw at that early date the great possibilities of aviation. Associated with him as a syndicate were Duncan MacDonald, president of the Royal Automobile Club of Canada, and William Carruthers, of Montreal Tramways.

The Montreal aviation meet was the most pretentious undertaking of its kind to be held on this side of the Atlantic up to that date, for although other meets had previously been held in the United States, none had brought together so many airplanes at one time.

The Montreal newspaper representatives began early to publicize the event and chief among these was a young reporter named John Bassett, now very prominent in the publication of the Montreal *Gazette*. As a result of the press build-up, public interest ran high. City authorities and outstanding citizens helped with donations toward cash prizes and expenses.

The site of the "air field" chosen was on the level farm land to the north of the railway tracks near Lakeside, about 10 miles west of Montreal. The owners of five farms of the long narrow strip type, so often seen in Quebec, were approached and after considerable hard bargaining work was undertaken to fill in the ditches at the farm boundaries and to level off fences and other obstructions to make a flying landing-field!

A grandstand to seat 10,000 spectators was built of lumber and additional ground space was arranged to accommodate another 10,000. Then, to ensure that as many as possible entered the airport by the "pay as you enter route," an arrangement was made with the personnel of a militia unit to patrol the boundary on air-meet days!

The prices of admission varied from 50 cents to $2, and twenty special trains were scheduled to run daily on both the Canadian railroads between Montreal and Lakeside. The meet was to have lasted ten days, from June 25 to July 4, but eventually was shortened to eight days.

Some of the world's most distinguished pilots and planes eventually attended the meet. Contracts had been arranged by personal contact by Gordon McGarry, representing the syndicate,

WALTER R. BROOKINS

with Count Jacques de Lesseps at Reims, France, to bring over two of his "flying machines" for the sum of $10,000! Another contract was also made with the Wright brothers in the United States for a similar sum; and as a result the Wright organization sent five famous flyers, Walter Brookins, holder of the then world's altitude record of 5,460 feet, Frank Coffyn, Ralph Johnstone, Duval Lachapelle, and Paul Miltgen.

The eminent French airman, Count de Lesseps, arrived at New York on June 24, 1910, accompanied by a sister, a brother, and his personal physician. His two mechanics, M. Vanoni and Henri Steiner, had arrived a few days before, bringing with them de Lesseps' two Blériot monoplanes. One was the famous *La Scarabée*, second airplane to fly the English Channel, a two-seater with a 50 h.p. Gnome engine; the other was a single-seater fitted with a 35 h.p. Anzani motor.

Certainly the most interesting personality of all those participating in the 1910 air meet was Count de Lesseps. Born in Paris in 1883, a son of

COUNT JACQUES DE LESSEPS

LA SCARABÉE

the renowned Count Ferdinand de Lesseps, builder of the Suez Canal, young Jacques de Lesseps had seen the start and growth of flight in France. He had closely followed the experiments of Farman, Blériot, and other Frenchmen, which were climaxed with Blériot's sensational flight across the Strait of Dover on July 25, 1909. He purchased a machine of his own and mastered the art of flying under the direction of Louis Blériot himself. Then he proceeded to follow his famous tutor's example in flying from France to England, and on May 23, 1910, he became the second airman to negotiate this crossing. The exploit won him immediate fame.

Another Blériot monoplane was purchased by the syndicate mentioned above but, as will be seen, it seemed to be a "hard luck" machine.

The meet was officially opened on June 27. The first flight was attempted by Paul Miltgen in the syndicate's Blériot machine but unfortunately something went wrong just after take-off and he was forced to land. Walter Brookins then took off and made a successful flight of 7 minutes, thereby becoming the first pilot to fly in the province of Quebec.

The best flight on the opening day was Johnstone, who stayed aloft for 35 minutes. In the afternoon Brookins went up again and established the first official Canadian altitude record of 1,650 feet. Count de Lesseps also made a short flight of 4 minutes.

From that day until the meet concluded, the various pilots made numerous flights daily. Interest was always keen but it was increased on the 28th when two Canadian flying records were established in the presence of 5,000 people. While both Brookins and Johnstone were in the air, flying separate Wright machines, another machine took off, and gaping Canadians witnessed the unprecedented sight of three airplanes in the air at once. The second sensational event was Johnstone's flight up to 2,000 feet, the highest attained in Canada to that date.

During the afternoon Brookins took off with Frank Coffyn, one of his flying mates, as passenger. This was heralded by Montreal newsmen as the first passenger flight in Canada, but the statement was incorrect, as almost a year before McCurdy and Baldwin had flown together in their machine at Petawawa military camp.

On the 29th Brookins took Count de Lesseps up as a passenger in the Wright machine—the French airman's first trip in a biplane. From a considerable altitude Brookins made a spiral glide with the engine throttled down, demonstrating to the Count the accuracy with which the manoeuvre could be accomplished. Although de Lesseps could not speak English and Brookins was equally tongue-tied in French, when they landed the former showed his delighted appreciation by a number of friendly slaps on the American airman's back. He enjoyed the flip so much that the next day he had his brother taken up and Brookins circled the grounds for 15 minutes with Bertrand de Lesseps aboard.

On this day, too, three airplanes were again seen in the air at the same time. But the big sensation of the day was the arrival at the grounds of William Jennings Bryan and his family, who had come all the way from Havana, Cuba, to witness the flying.

On July 1 a capacity crowd held its breath as Johnstone, after an informal performance in the air, swooped down to the field for a landing. A hush of anxiety fell on the crowd as a wing tip struck the ground and the biplane was completely demolished. When the airman emerged without a scratch, the crowd went wild. Later, as compensation, a number of very fine flights were made by the other airmen.

On Saturday the 2nd, a huge throng of 20,000 people were on hand. To mark the day, Brookins established a new Canadian altitude record of 3,510 feet, a record which remained unchallenged for two years.

The big event of the entire meet was a spectacular cross-country flight by de Lesseps in his famous *La Scarabée*, late on Saturday afternoon. He left the grounds at 6:15 P.M., and, climbing to over 2,000 feet, made a 49-minute, 35-mile flight in a wide circuit over the city of Montreal and its environs. This was the first airplane flight over any Canadian city, as well as a record in Canada for distance and time. The famous flyer was hailed as the hero of the hour by the crowd and his fellow airmen alike.

July 3 was a Sunday, a non-flying day. On the 4th, under lowering skies, only de Lesseps and Brookins accomplished short flights. This brought an official ending to the great affair, but the Count made two short flights on the 5th to satisfy a large number of people who turned up on the odd chance of seeing an additional flight.

A secondary attraction at the meet was a contest of airplane models, sent in by boys from all parts of Canada under the sponsorship of a Montreal newspaper, the now extinct *Daily Witness*. Described as the Boys' Model Aeroplane Competition, it was the first of its kind to be held in Canada, and the workmanship displayed in most of the models was stated by the judges to be excellent. Wright biplanes and Blériot monoplanes predominated, as might be expected, but there were a number of others built from designs developed by the inventive young competitors. The selection of the best model took place on July 4. The judges were chosen mainly from among the airmen taking part in the meet. The grand prize of $50 was awarded to a young Torontonian named J. H. Parkin, for a carefully constructed Blériot XI replica.

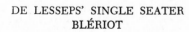

DE LESSEPS' SINGLE SEATER BLÉRIOT

Mechanic Maurice Vanoni at left

COUNT DE LESSEPS

Judging the winning entry in the first Model Airplane competition held in Canada, July 1910

Parkin in 1929 became Director of the Mechanical Division of the National Research Council of Canada in which are located the wind tunnels and aeronautical research facilities. He is also interested in carrying on the collection of Canadian aviation relics, a project which was begun many years ago but which was given impetus in 1937, when the Associate Committee of Aeronautical Research sponsored the setting up of the Aeronautical Museum in one of the exhibition halls of the National Research Council building on Sussex Street, Ottawa (see photograph of plaque, p. 370).

THE TORONTO MEET

After their highly popular visit to Montreal, most of the aviators went on to a similar meet which took place at the Trethewey Farm at Weston, northwest of Toronto. The programme, which began on July 9, 1910, was only a little less impressive than that of the Montreal meet. Five machines and five airmen were present: Johnstone, Brookins, and Coffyn with two Wright biplanes; de Lesseps and his two Blériots; and an American airman named John G. Stratton, who had been engaged by the syndicate to pilot the Blériot monoplane which had failed to perform well at the Montreal meet.

A blustery wind made the opening day bad for flying, but Johnstone did some risky tricks, diving down steeply several times towards the ground, to straighten out when only a few feet separated him from what seemed a certain crash. He was the first one up, and he flew for 18 minutes. Next, Stratton attempted to get aloft, but was unsuccessful.

By 7:00 P.M. the wind dropped a little, and de Lesseps flew his single-seater for 6 minutes at a low altitude. Another flight by Johnstone for 10 minutes to a height of 800 feet, and a single circuit of the grounds by de Lesseps in *La Scarabée* concluded the first day's performance.

On the 12th the weather was again rough. This time de Lesseps saved the day with two short flights. Johnstone's biplane developed engine trouble and he was unable to get up.

Stratton finally managed to get into the air in the Blériot owned by the syndicate. He climbed to 600 feet and was doing nicely until a control wire snapped, and the aircraft became almost unmanageable. He shut off the engine and attempted to glide back to the flying field but three pine-trees barred his path. He managed to avoid two of them but not the third. Thirty feet above the ground he piled into its outstretched branches, shearing off the Blériot's port wing, breaking the prop, and snapping the fuselage in two just behind the pilot's seat. The broken monoplane and its pilot lodged firmly in the tree, prompting Ralph Johnstone to remark, "Jack's a real bird man; he can fly, and he can light in a tree and stay there!"

The bad performance of the syndicate craft in the air was the result largely of the damage it had suffered when being shipped from Montreal to Toronto aboard a railway flat-car, having been out in the open the whole time and exposed to a drenching rain during transit, so that it was badly warped when it reached Toronto. The damaged condition of the plane and the snapping of its control wire might well have caused disaster, and it was probably just as well for Stratton that he had ended his flight in the comparative safety of a tree.

On the evening of the 13th de Lesseps took off in *La Scarabée*, and, climbing to an altitude of 2,000 feet, headed directly for the city of Toronto. Torontonians may now have become a bit *blasé* about airplane flights over their roofs, but the Count's was the first, and thousands of people gazed aloft in amazement as the trim monoplane hummed overhead at some 40 miles per hour.

Upon his return to the Trethewey field, the French airman was wildly acclaimed. Cheer upon cheer went up as the crowd surged around him, and he was carried shoulder-high to the grandstand to receive congratulations from the officials and his fellow airmen.

There were some interesting sidelights during the flights. Mr. Wilcox had arranged with Lieutenant Colonel S. P. Biggs, Commanding Officer of the 2nd Field Company, Canadian Engineers, for personnel of the Field Company to help in various ways, one of which was to simulate a bombing attack on an earthwork fort in front of the grandstand by exploding dynamite charges planted in the ground as an aircraft flew over the field and dropped sand bags on the fort. This shows the vision of these gentlemen who foresaw at that early date the military possibilities of the airplane in warfare.

For some curious reason, people in those days were often more interested in how high aircraft could fly than in anything else. Major T. C. Irving, Jr., after talking with Count de Lesseps about the accuracy of altitude measurement, also arranged for a check to be made on July 13 by ground measurement so that announcements could be made from time to time as to the altitude at which the aircraft was flying. The help of two enthusiasts, A. MacAllister and T. R. Loudon, was enlisted to run a transit at each end of a measured base line. The idea was that sights would be taken on the aircraft and the altitudes worked out from the measured angles with the help of a pre-arranged table of results. No one had any real idea of the small rate of climb and speed of these aircraft with the result that de Lesseps took off to the west and practically disappeared before turning around to head for the city. Neither of the instrument men even got a chance to centre the aircraft in the telescopes, let alone get co-ordinated measurements. Major Irving, when he realized what had happened, turned to the grandstand and announced through a megaphone, "He is up pretty high, about 1,000 feet!" And the audience which had been eagerly awaiting the results clapped!

Between his visits to Montreal and Toronto,

A LETTER OF DE LESSEPS
Recommending his mechanic, Maurice Vanoni

Brookins had gone to Atlantic City, and on July 7 he had established a world altitude record of 6,175 feet in a Wright biplane. On July 14 he reached Toronto by train, and during the day went up in one of the Wright machines in an attempt to better his own record. Unfortunately, engine trouble developed when he had gained only a few hundred feet, and the endeavour was abandoned. The same evening de Lesseps made another sensational flight of 20 miles at some 1,500 feet, and Coffyn also put on a good exhibition.

The final day of the meet brought forth a number of fine flights by de Lesseps, Coffyn, and Johnstone, the latter making the record for the entire meet by going up to 3,000 feet on a flight which occupied 25 minutes. For the first time in Toronto, two airplanes were seen in the air simultaneously, when de Lesseps and Johnstone were on the wing at the same time.

Thrills there were a-plenty for all who were present, and no casualties marred either of the great air meets of Canada's two major cities in the memorable summer of 1910.

At the conclusion of the Toronto meet, de Lesseps shipped his Blériots back to Montreal, and after having *La Scarabée* assembled at King Edward Park, some ten miles down the St. Lawrence River, he set off on what was to become the most sensational flight in Canada up to that date.

Leaving the park grounds at 6:51 P.M. on July 26, 1910, the airman flew off towards Montreal, and after making a 40-mile circuit of the city, headed directly over the down-town area, circling above at 3,000 feet for some 16 minutes before he flew off to circle low above St. Helen's Island. The Count was 81 minutes in the air before returning to King Edward Park, and the event created widespread comment in both the Canadian and the American press.

Before leaving Montreal, de Lesseps sold his single-seater Blériot to the proprietor of the King Edward Park amusement concessions, where it went on display. The Count did not sell *La Scarabée* though, and she carried him

GRACE MACKENZIE
About to take off with Count de Lesseps

in numerous outstanding flights in the United States and in Europe.

Among the people whom the Count met in Canada was Thérèse Grace Mackenzie, a daughter of Sir William Mackenzie, president of the Canadian Northern Railway. The Count soon became a guest at the Mackenzie home at Toronto, and when, in October 1910, the airman shipped *La Scarabée* to Belmont Park near New York, to take part in a great flying meet, Grace and her two sisters also took train to New York to witness the affair.

De Lesseps was one of only three contestants who successfully negotiated a flight from Belmont Park to the Statue of Liberty and back. Although an airman named Johnny Moisant made better time, he was disqualified for flying a plane other than the one he had entered in the race. The award went to de Lesseps, who achieved the next best time.

On October 25, 1910, at the Belmont meet, Count de Lesseps asked Grace Mackenzie to accompany him as a passenger in *La Scarabée*. The Count also took her sisters, Kathleen and Ethel, on short flights the same day, and the American newspapers ran large headlines about the event. One of these caused consternation in the Mackenzie home at Toronto, and a telegram soon reached the girls from their father: "Come home at once!" No matter, they had flown, and had earned a place in history as the first Canadian women to fly.

Shortly after the Belmont meet, the engagement of the Count and Grace Mackenzie was announced in the Toronto papers. Their marriage took place in London, England, in 1911, and after their honeymoon they established residence in Paris. From then until the outbreak of World War I, de Lesseps added still more lustre to his name by taking many prizes at air meets in England and on the continent.

But the Canadian countess was to return to her native land again, and her famous husband was to play a further part in the history of Canadian aviation.

10. Exhibition flying, 1911

THE glowing press reports about the 1910 air-meets at Montreal and Toronto, together with items of flying news from American and European points, caused a great stir of interest throughout Canada. People everywhere now wanted to be able to say they had seen an airplane fly. Promoters and committees of fairs and exhibitions vied with each other in trying to obtain the best pilots available. As a consequence some very fine flying was accomplished, as well as a few fiascos.

And it was not only the youngsters who caught the enthusiasm. For instance, in England, the *Daily Mail* offered a prize of £10,000 to the airplane pilot, irrespective of nationality, who won their round-Britain race in July 1911.

The contestants were to pass over a point several miles east of Nottingham; that was a short hike for an air-crazy seventeen-year-old like me, and for my two sisters, but it was quite a jaunt for my mother, who was just as determined as I was not to miss the air spectacle. The four of us set out long before dawn on that memorable July day, and, although we missed seeing the only two who finished the course, Graham-White and Paulhan, we were in time to see Vezines in his trim Blériot, and "Colonel" Cody in his great lumbering biplane (nicknamed "Cody's Cathedral"), winging past not far above the trees. That was my first close-up of aircraft in flight, and it was not till later in 1911 that I attended my first air meet at Burton upon Trent.

THE MANNING BROTHERS AND DE PRIES: VANCOUVER

In April 1911, half-page advertisements in the Vancouver newspapers announced a three-day flying event during the Easter week-end. Jack De Pries and the Manning brothers were to be on hand to perform the "spiral glide," the "zig-zag whirl," and other thrilling stunts. The meet was to be at the Minoru Park race track, and the price of admission 50 cents.

A large and enthusiastic crowd was on hand on April 14 by 3:00 P.M., but it was beginning to get a bit apathetic when De Pries pushed his Curtiss-type biplane from its tent-hangar, one hour and forty-five minutes late. After a take-off run of some 300 feet, the pilot managed to get the craft into the air, then it swerved suddenly and rammed the fence near the end of the track. Spectators scattered and fell off their perches on the fence, but no one was injured. The biplane suffered some damage and was incapacitated for the remainder of the day. The second craft, fitted with a 40 h.p. motor, was then trundled out. This time De Pries had better luck. After a lengthy run, he hopped the high board fence surrounding the track, and sailed off at a height of about 25 feet, to alight in a near-by field after a flight of about 500 feet. This concluded the first day's "meet."

After this rather tame beginning, the affair went from bad to worse. The following day, De Pries, starting his take-off run well down the

field, was in the air by the time he was passing the grandstand. After a 2-mile flight in the direction of the small fishing village of Steveston, he landed in a field. The machine cracked up, and the pilot suffered a deep scalp wound and concussion, from which, however, he eventually recovered.

On the 18th, the performance of the Manning brothers proved that the advance advertising had greatly overrated their flying abilities. In the 40 h.p. machine, "Brownie," Manning got several feet off the ground, then landed and struck the fence. A second attempt an hour later concluded with the same result, and thus Vancouver's first flying "meet" came to an inglorious close.

Newspaper reports did not mince matters, and the airmen received a severe verbal trouncing. It is quite apparent that they were extremely amateurish aviators.

ROBINSON: EDMONTON

Better judgment on the part of promoters was shown shortly after in Alberta. Hugh A. Robinson, an American, already had a well-established reputation as a pilot when he came to Canada to perform during the Edmonton Horse Show. He arrived on April 28, along with his mechanic and his Curtiss biplane. The next day he made two flights of some 15 minutes' duration each, going up to over 400 feet. These flights constitute a record for Alberta as they were the first ever to be made in that province. On May 1, Robinson went aloft on three separate occasions

on similar flights, throwing in a few steep dips and dives for good measure.

Some years ago, he told me he received $3,000 for the five flights he made, or $600 per flight —not bad pay, he considered, for the five chances he took of breaking his neck.

Robinson qualified for his Aero Club of America licence bearing the low number of 42, at Kinloch Field, Mo., shortly after his trip to Canada. He became one of America's outstanding airmen, and later visited Europe, flying, altogether, in 11 different countries.

ST. HENRY: SASKATOON AND REGINA

Down Nebraska way, Bob St. Henry had been an actor. His real name was C. W. Shaffer, but he retained his stage name as a pilot, and few people knew that he had any other. He was first referred to as "Lucky Bob" in the newspapers of Saskatoon, where he arrived to fly at the annual exhibition in May 1911; and the nickname stuck to him throughout his flying career.

The machine he brought with him was a Curtiss, fitted with a 40 h.p. Curtiss engine, which proved to be a bit underpowered for the high altitude in that part of Saskatchewan. Bad weather had kept him grounded for two days, and the wind was still very strong on May 19 when the aviator decided to wait no longer. He managed to get off the ground, but when he was about 60 feet up a particularly hefty gust of wind struck the machine, slowing it almost to a standstill. Before the airman could do a thing about it, the nose of the craft was pointing steeply towards the ground. The plane struck the earth with considerable impact, smashing things up a bit and jarring the pilot badly. However, after a night's rest he was able to inspect the damage, and stated he would obtain a more powerful engine from the Curtiss factory at Hammondsport. He promised that after repairs had been made and a new engine installed, he would give the people of Saskatoon full satisfaction.

On May 30, the eight-cylinder, 70 h.p., motor arrived. Some fancy figuring was required before St. Henry and his mechanic were able to install it in the biplane. As it weighed considerably more than the original four-cylinder engine, the problem was how to place it on the engine-bearers so that it would not upset the balance of the machine in flight. They first established the correct centre of gravity of the biplane while it was still fitted with the original engine, by placing a length of plank, 2 feet by 12, under the wooden connecting-beam between the front and back wheels. Then, after marking the centre of gravity, they exchanged engines and bolted the new one down on the bearers at a point which kept the balance of the machine the same. Shaffer has since told me that they hit an exact balance, which was never changed during the life of that machine. Crude aerodynamics perhaps, but necessity was the true mother of invention in those days of experimental flying.

The first test with the engine was made on the evening of June 2. St. Henry went up to 250 feet, circling the field three times, before coming down to a good landing. This was not

the first flight at Saskatoon (Bob's short hop on May 19 pre-dated it), but the event on the 2nd was the first really successful flight made by an airplane in the province of Saskatchewan.

On June 3 St. Henry made another evening flight, and as it had been well advertised, plenty of people were on hand to witness it. The air was extremely rough, but the airman went up to 500 feet, and circled around for some time before landing, when he had the misfortune to damage the landing-gear. With that, his flying endeavours at Saskatoon came to a close.

Bob then went down to the United States to fill several flying contracts; but in August 1911, Saskatchewan saw him in action again, this time at the Dominion Fair in Regina.

His first flight was on Saturday evening, August 5, when he flew about 6 minutes before a great throng of people. A second flight shortly after lasted only 7 minutes, but included some daring "dips" and "swoops," which according to reports had the crowd gasping. After returning to the grounds for a short rest, he again went up, despite huge banks of black storm clouds which were welling up from the east. In the brilliant glow of a spectacular sunset, St. Henry flew off westward towards Government House, making a return journey to the grounds which covered about 8 air miles, lasted 9 minutes, and reached an altitude of 1,200 feet. On his return people were astonished to note that the airman's clothing was soaked with rain. He had fringed the storm area on his flight. On the 8th the airman made three good flights late in the day, on one of which he flew to Government House. Passing low over the grounds, he dipped the wings and waved his hand to Lieutenant-Governor Brown, who vigorously returned the salute.

On another flight he flew off to circle the Parliament Buildings, and on his return experienced the strange sight of hundreds of people waving to him from the residential areas, using white tablecloths and serviettes snatched from inside their homes. Bob was almost blue with cold after his last flight of the day, although the ground temperature was warm.

On the 9th he put on the most spectacular effort of his Regina visit. In his one flight that day, he stayed aloft for 19 minutes, reaching an altitude of 1,600 feet. Rough weather made it tough going for "Lucky" Bob, but on the day following he went into the air twice, much to

BOB ST. HENRY, SASKATOON, JUNE 2, 1911

the satisfaction of the onlookers, who had not expected him to dare the boisterous weather.

The entire exhibition put on by St. Henry did him great credit, and left nothing to be doubted as to his nerve and his flying ability. He became one of the best airmen of that era, although it was not until February 28, 1912, that he took time out to pass his tests for his Aero Club of America licence—No. 100.

WALSH: VICTORIA

While St. Henry was at Saskatoon, a flying event was in progress in British Columbia. On May 30, in the city of Victoria, there occurred the first full-fledged flights in the history of Vancouver Island.

The American airman, Charles F. Walsh, who later obtained Aero Club of America certificate No. 118, had shipped his Curtiss-Farman type biplane up from Portland, Ore., to put on a show. The band of the Fifth Regiment was on hand to enliven proceedings while Walsh made his first flight, which lasted some 15 minutes at a height of 600 feet.

On his second hop of the day, two noteworthy incidents are reported as follows: "In dodging through the trees one of his planes [wings] struck the branch of a tree and cut it off with the cleanness of a razor. He brought the twig back with him and will keep it as a souvenir. In travelling past the grandstand he took an apple from his pocket and threw it at the judge's box, striking the corner, which illustrates that the airplane might be destructively useful when employed for military purposes." The reporter for the Victoria *Colonist* of May 31, 1911, did not have to wait many years to see his prophecy fulfilled.

On May 31, the airman delighted the large crowd with two good flights, the longest for 15 minutes, and everyone went home well satisfied. Although Walsh was scheduled to fly on the day following, trouble with his engine made him decide against it. By the time the engine was repaired a howling wind had sprung up, preventing any attempt. The day had been made a school holiday so that the children might see the "flying machine," and hundreds of them were disappointed when the gale put an end to flying at Victoria. Walsh said he was deeply sorry that the flights had to be called off, and added: "Not that a man's life is worth so much, but a fellow hates to lose a good machine!"

Canadians were to see him fly again.

STROBEL AND LE VAN: CALGARY

The next flying episode occurred at Calgary, Alta. The *Golden Flyer*, which Charles Willard flew at Toronto in 1909, was later sold to a flyer named J. Strobel. That famous aircraft was the first airplane to fly successfully at the "City of the Foothills."

Strobel and his assistant, Harvard Le Van, who was also a pilot, were engaged to fly during the Calgary Exhibition which opened on July 1, 1911. Wind and rain were the bogeys which effectively prevented any flights on that day. Strobel tried nobly to lift the machine off the muddy track, but had to give up after three unavailing efforts.

Two afternoons later, after adjustment of the carburetor had increased the engine revolutions, Le Van took his place in the pilot's seat and successfully flew out of the grounds, to circle low over the widely scattered outskirts of the city to the east. He almost came to grief over the Elbow River, when a down-draft caused by high cut-banks almost stood the airplane on end.

On the 4th, Le Van ran into the fence on an attempted take-off, tearing one wing and smashing one of the wheels. The place was still soggy, making it very difficult for Le Van and Strobel, and the ground elevation at Calgary of over 3,000 feet was a tremendous handicap for the machines of those days.

After much hard work, the airmen had the craft fully repaired in time to try again by late afternoon on the 5th. This time, thanks to drier ground, Le Van made a safe departure from the area and went on to make a very presentable flight of about half a mile before returning to land on the infield. Here a wheel struck a gopher hole, and was badly buckled.

While their plane was in its tent, either being repaired or just "off duty," the airmen reaped a fair harvest by charging inquisitive citizens 10 cents a look. Since interest in airplanes was at a high pitch in the West just at that time, the idea paid good dividends.

The fair concluded on July 6. With the two final flights to their credit, Strobel and Le Van went off in search of less muddy fields to conquer.

FRANK COFFYN (*Centre*), WINNIPEG, JULY 14, 1911

COFFYN: WINNIPEG

When it was known that Frank Coffyn was to give an exhibition in Winnipeg, flying a Wright biplane, public interest was immediately aroused.

Coffyn was becoming one of the foremost American pilots. His first solo had been made in 1910 at Indianapolis, Minn., before 40,000 spectators. Recalling his début, Coffyn remarked years later: "I tell you I was scared to death, in spite of the fact that Wilbur Wright ran alongside, holding one wing to help me get off."

It was about ten days later that Coffyn was flying at the Montreal and Toronto flying meets. Winnipeggers were therefore assured of a thrilling performance when he and his plane, his mechanics and a private physician, arrived at their city on July 12, 1911.

Coffyn's flights were made during the week of the Winnipeg Exhibition, July 15 to 22, one or more times daily, except when heavy winds kept him down. Sunday, according to the stricter conventions of pre-war Canada, was, for all right-minded people, a non-flying day.

The crowds were thrilled when, in the course of a flight, the pilot made steep nosedives at a low altitude, and they enjoyed a special feature when a race was staged between the plane, a motor-car, and a motorcycle. The plane won the four laps of the race track, but only by 20 yards.

It was at this exhibition, on July 22, that the first passenger from western Canada was taken aloft. He was W. C. Power, then on the staff of McLaughlin-Buick Motors of Winnipeg.

The T. Eaton Co. offered a prize of $1,000 if Coffyn would make a cross-country flight to Portage la Prairie, 60 miles away. The Wright Company, however, considered the amount too small and since no one else volunteered to increase the sum, the venture, which would have made a long-distance record in Canada, did not take place.

Coffyn qualified for his Aero Club of America licence later in 1911. His certificate bears the low number of 26.

As a Wright pilot, Coffyn is credited with designing and using the first pontoons with which a Wright model-B biplane was flown. It is also of interest to learn that in 1912 he devised an electric movie-camera with which he took the first aerial views over New York.

Frank Coffyn maintained a continuous career in aviation. During World War II he was with the Canadian Aviation Bureau in New York, an organization which secured pilots for the Canadian and British governments in their project of ferrying aircraft across the Atlantic.

THE HAMILTON MEET, 1911

A resident of Hamilton, Ont., E. M. Wilcox, with a committee of local citizens, decided to promote a flying exhibition for that city in 1911. The event proved to be a big attraction and gave Hamiltonians their first opportunity of seeing aircraft on the wing from their own doorsteps.

The events were staged from an area known as the O'Heir Survey, near the Tuckett Farm on Beach Road, and the committee showed good judgment in selecting Charles F. Willard,

John McCurdy, and James V. Martin as the exhibition airmen. Willard had already made a name for himself in Toronto in 1909. McCurdy was also established as one of the world's foremost pilots. Martin had recently arrived from England, where he had earned his brevet, or flying licence; he brought a large Farman biplane, the only one of its type ever to be flown in Canada.

The meet began on July 27, when a total of seven good flights were made: four by Willard in his Curtiss biplane; two by McCurdy, flying in what was described as a "headless" type Baby Wright biplane; and a single flight by Martin. All of the airmen again flew on the 28th, but Martin experienced engine trouble and was forced to bring his flight to a quick conclusion.

The 29th was a windy day, and when Willard tried to fly he had to make a forced landing. The engine of Martin's machine was still unreliable, and he and his plane were forced to remain grounded. McCurdy redeemed the day with two good flights. Three days later McCurdy was the only one with an airplane serviceable enough to fly, and he raised the enthusiasm of the crowds to a high pitch with three sensational flights. Willard's Curtiss was still undergoing repairs, and Martin took no further part in the meet. The meet closed on August 1, with two more flights by McCurdy. As the weather was cold, with black, threatening skies, the attendance on the final day was very poor.

The big event of the entire "do" actually took place after the meet had concluded, when a flying race was organized, and Willard and McCurdy took off from Hamilton on the evening of August 2 on a 40-mile flight to Toronto. This was the first event of the kind to be held in Canada. Willard, with a slower machine, was started a few minutes ahead of McCurdy, but the latter was the first to reach Toronto, where he landed at a point on Toronto Island. Willard landed at the Toronto Exhibition Grounds only a short distance behind, and received a great ovation from a throng of people. (His three previous attempts to fly at Toronto, it will be remembered, had each ended with a ducking in the lake.)

The following week-end, a small aviation meet took place near Toronto, and both Willard and McCurdy made several flights. During one take-off, McCurdy had a narrow escape and his machine was completely wrecked. He was thrown clear of the craft and suffered a severe shaking-up; but, taking to the air in another machine, kept on hand for such an emergency, he amazed and thrilled the crowds by carrying on. Tough birds, those early ones!

MESTACH: QUEBEC CITY

The next flying event on Canada's 1911 aviation agenda featured a French pilot of ability, George Mestach, who had obtained his Fédération Aéronautique Internationale flying licence in France, where he had learned to fly. The setting was the city of Quebec: the date, August 30 to September 5.

He and his machine—a Blériot monoplane—were the first flying outfit to be seen at the provincial capital. The initial flight took place shortly before noon on August 30. Mestach went up over 2,000 feet, circled around for a considerable time, then returned to the grounds. Again at 5:00 P.M. he was on the wing, thrilling the spectators with his steady flying. The masterly precision with which he handled his machine proved him to be a very competent airman.

Each day of the exhibition he made one or more flights. Occasionally he included steep dives and vertical "pull-outs" which brought the crowd to their feet.

During a flight on September 1, Mestach took along a number of messages addressed to various officials of the exhibition, and dropped them as he was flying low over the grounds. They were eagerly picked up and delivered. If any of these survive, they should be valuable to collectors, for they constitute the earliest known messages to be dropped from an airplane in Canada.

Almost coincidentally, a postcard was received in Canada from England which in all probability was the first piece of air mail to be received by a resident of Canada. It was addressed to Mrs. H. E. Leveson-Gower, Traynor, Saskatchewan, Canada, and was mailed to her from her family in Bromley, Kent. The card was one of a number carried on the first airmail flight in the United Kingdom. They were issued by the Postmaster General of Great Britain as a part of the coronation ceremonies of King George V. On the two halfpenny stamps the card carries is the circular cancellation in black, "First United Kingdom Aerial Post, Sp. 12, 1911, London."

POSTCARD MAILED BY
THE FIRST U.K. AERIAL POST

The picture on the card is of a Farman biplane, flying above Windsor Castle. Above are the words: "A.D.—Coronation—1911—First U.K. Aerial Post—By sanction of H.M. Postmaster General." Below the picture are the words: "For conveyance by aeroplane from London to Windsor. No responsibility in respect of loss, damage, or delay, is undertaken by the Postmaster General." The card travelled from London to Windsor by air, then to Traynor by ship and rail, all for the stamp value of 2 cents.

Masson: Calgary

The final flying event of 1911 had its setting in Alberta, in October. Calgarians already had seen an airplane in action, when the two American pilots, Strobel and Le Van, were there during the previous July, but as only two short flights had been made, it can hardly be said that an airplane on the wing had become commonplace.

In August, the Canadian pioneer airman, William Wallace Gibson, whose varied exploits are described in chapter 3, had had his *Multiplane* in operation near Calgary where a number of very successful flights were made, but as they were conducted without publicity of any kind, few people in the area knew about them.

Flying interest throughout the country was high, and a Calgary newspaper planned and largely financed the October event, which in consequence was known as the Calgary *Herald*'s Aviation Meet.

The airman obtained by the *Herald* was Didier Masson. Of French birth, he had been a resident of Honolulu for some time. When he made up his mind to become an aviator, he journeyed from Hawaii to San Francisco, and there, with the help of two mechanics, James Archibald and Jules Brule, he constructed a Curtiss-type biplane and powered it with a rotary, air-cooled, Gnome engine. The whole thing cost him about $5,000, and it was quite a professional job. He taught himself to fly, then set off on a barnstorming career. The Calgary *Herald* heard of him and eventually he and his two mechanics arrived in Calgary.

At that time of the year, Canadian temperatures were beginning to creep down. Having come from Honolulu not long before, Masson felt the cold keenly.

The poor grade of gasoline he was obliged to use did not help matters, and as Gnome engines were very temperamental when not fed the "food" they best thrived on, engine trouble became serious. Finally Masson overcame the difficulty by taking the fuel feed line off the outside injection chamber, and inserting the pipe right inside the crankcase, through the hollow crankshaft with which all such engines were made. From then on, no further difficulty was experienced from improper combustion.

The chief flying attraction planned was a flight from Calgary to Edmonton, which at that date was a long cross-country hop and would have established a record.

Masson set up his tent on an open stretch of prairie south of Calgary beside the McLeod Trail, and made his first successful test flight on October 17. On the 18th, during another test, he was taxiing at speed when one of the wheels of his machine ran foul of a coil of bale wire, which lay hidden in the long grass. The tangled wire was whirled up into the propeller, damaging both tips and a bamboo strut in the tail, and it was not until the 20th that repairs could be completed. The wonder is that they were ever accomplished, because material for such work was unobtainable in that vicinity, and repairs had to be made with makeshifts.

Masson and his helpers did a good job, however, for on the morning of the 20th he soared aloft to some 2,000 feet, flying directly over the centre of Calgary as far as 14th Street, from which point he made a wide circle back, to land at the fair-grounds at Victoria Park. In the afternoon he made two similar flights—the first flights ever seen directly over the centre of the city of Calgary.

It was a good start, but things did not go so well from then on. The next day brought a gusty wind. In trying a take-off, the machine ran into a fence and broke a bamboo strut. Within the hour, however, it was bound and mended, and the intrepid pilot was able to make a good flight up to 2,000 feet. The strong wind made the going grim, and the flyer was bitterly cold.

The planned flight from Calgary to Edmonton was postponed from day to day, to clear up small matters, but finally arrangements were made for a special train to leave Calgary with right of way to Edmonton immediately the airman took off. To facilitate speed, the engine was to draw only one coach of 20 passengers. (The fare must have been equally sensational!) It was expected the train would be able to keep the airplane in sight as the biplane's speed was not much greater than 50 miles per hour.

The final attempt came on October 26. Masson got away to a good start from the Victoria Park Exhibition Grounds, then, just as it appeared

DIDIER MASSON, CALGARY, OCTOBER 20, 1911

that success might be in his grasp, things began to happen. First, the wires holding the fuel tank snapped. The tank fell on the airman's shoulders and lodged behind him and the seat, making control of the aircraft very difficult. Next, the broken wires tangled with the propeller and snapped off both blades. Masson was still within gliding distance of Victoria Park so he swung the machine around, stuck the nose down, and proved his airmanship by making a safe landing, although without power of any kind.

This mishap ended Masson's visit to Canada, at the beginning of one of the most distinguished flying careers in America.

11. Billy Stark, Canadian barnstormer, 1912-1915

EXHIBITION flying in the unstable "flying machines" of the early days was, at its best, a risky business. In spite of this, many pioneer men and women aviators survived that dangerous period, though few did so unscathed.

Pilots were often obliged to take off and land in areas which were little better than large back lots, because such places were chosen by some promoter or exhibition group. Rather than disappoint hundreds of spectators who had come from miles around to see a man fly—and perhaps break his neck—pilots invariably did their best, often in winds which made flying almost suicidal.

Whenever the public pays money for something, it becomes a hard master. If an airman did not get into the air on time as scheduled, cries of "faker" would begin to split the breeze. Such baiting frequently resulted in pilots attempting to fly under bad conditions, against their better judgment and the advice of their friends. As a consequence, a number met their deaths amidst the splintered wreckage of their crashed machines, and the crowd would go home awed, but happy.

As related in the story of the Aerial Experiment Association, in chapter 1, the first death anywhere in the world through an airplane accident occurred in 1908, when Lieutenant Thomas Selfridge was fatally injured at Fort Myer, Va. In 1909 there were 3 deaths, all in France. A total of 29 was the official world recording for 1910, but the number for the next year skyrocketed to 83. In 1912 the total was 122, and from then onward the yearly death-toll of barnstorming and exhibition pilots increased, until flying regulations came into force to stop daring pilots from trying to outdo each other in recklessness and glory.

Prior to the outbreak of World War I, a number of licensed airmen had flown in Canada, but the majority were from other countries. Canada still had only two licensed pilots: J. A. D. McCurdy, who possessed Aero Club of America certificate No. 18, and William M. Stark, who owned No. 110.* Although the latter did not commence flying until four years after McCurdy, he quickly made flying history in British Columbia.

A son of the late James Stark, a business man of Vancouver, Billy early showed an interest in all things mechanical, and he is credited with driving the first gas-engined automobile on the down-town streets of Vancouver, in 1901. In 1911 he became interested in the "sport" of flying, and fortune having favoured him with material means, he decided to learn to fly. Early in 1912 he journeyed south to San Diego, Calif., and there joined the Curtiss Aviation School, then operating at North Island. As an adept

---

*While it was not compulsory in the early years of flying to have a pilot's licence, membership in the Aero Club of America carried definite prestige. Tests were not severe: a pilot was required to take off and land within specified points, and had to accomplish certain basic evolutions in the air, such as circular and figure-of-eight flights. He was also expected to be conversant with engine and aircraft maintenance. A large number of pilots in the early years did not trouble to obtain licences, either because of the inconvenience of travelling some distance, or because of the fees.

WILLIAM M. STARK'S LICENCE

pupil under the watchful and expert tutelage of the famous Glenn Curtiss himself, he quickly mastered the art of flight. Graduating on March 22, 1912, as a full-fledged pilot, he was granted his Aero Club of America licence, and thus became the first Canadian subject to be taught to fly.

He at once purchased a new Curtiss exhibition-type biplane, equipped with a 75 h.p. V-type Curtiss motor, and after shipping it north in the same boat in which he himself returned to Vancouver, he prepared to go into the business of exhibition flying in British Columbia. The price of the complete machine, f.o.b. San Diego, was $5,500.

The spot chosen for preliminary Canadian trials and exhibition flights was Minoru Park race track, situated on Lulu Island, a few miles south of Vancouver. After assembling the aircraft there, Stark made his first flight in Canada on April 12, 1912. Going up to 500 feet, he flew westward over the Richmond rifle range, out over the mouth of the Fraser River, then swung back in a wide circle and made a perfect landing on the track. The flight had lasted 20 minutes. Full accounts were carried in the newspapers, but Stark's success was completely overshadowed by the early reports of the *Titanic* disaster, which occupied the major space of the same editions.

Billy's first advertised exhibitions took place at Minoru on the 20th. Wide publicity was given

in the press prior to the big event, and the British Columbia Electric Railway put on 9 extra cars to carry the crowd. Three flights were made by the airman, varying in heights from 400 to 800 feet, the longest taking 13 minutes. Numerous "stunts" which had never been seen in the West before were performed by Stark. One which caused consternation among the spectators was a steep dive with the engine throttled down, climaxed with a perfect landing in front of the grandstand without the use of the motor. It was expert flying for those days, and indeed would still be considered so.

On April 24, 1912, Stark carried as a passenger James T. Hewitt, sports editor of the Vancouver *Daily Province*. Taking off from a field near Minoru Park, of larger proportions than the race track, they flew for 8 minutes, going up to 600 feet. It was an epic aerial event for western Canada as it was the first occasion on which two persons had flown together in one plane at any point west of Winnipeg.

It must be emphasized that Stark's Curtiss was built only as a single-seater machine. To accommodate the passenger, it was necessary to fasten a wide board firmly on the lower wing to the left of the pilot, and upon that piece of plank Hewitt sat, grasping whatever was handy to prevent his being blown off when the craft was on the wing.

A lengthy account of the flight, written by Hewitt, was published in the *Daily Province* the

BILLY STARK AND JAMES HEWITT
Preparing to take off at Vancouver on April 24, 1912

following day. One of his remarks is worthy of recounting as an amusing indication that even in those days traffic hazards were beginning to be felt. "The turning of the machine gave me the feeling of sweeping around a sharp corner in an automobile but I felt satisfied because I knew there could be no collision around the turn. In fact, I felt much safer than in a Vancouver street."

Immediately after Hewitt's trip, Olive Stark, the airman's wife, took her place on the improvised passenger seat, and Billy again took off, to make a 6-minute passenger flight with as much success as the first one with the sports editor. By this short flight Mrs. Stark earned the distinction of being the first woman to fly in Canada.

As the distance between Vancouver and Minoru Park was inconveniently long for many people, Stark decided to transfer his flying activities to a place closer to the city. He chose the exhibition race track at Hastings Park as the best of several poor possibilities. No aircraft had ever flown from that spot before and it was a hazardous place for such a purpose. Huge firs and cedars hemmed it in closely on three sides; the fourth side opened directly on the

waters of Burrard Inlet. The infield of the race track was filled with large stumps. Even today, with the stumps gone, the site would be shunned by any airman with a desire to live.

On the afternoon of May 4 a large crowd was on hand to see the airman "do his stuff." His first hop lasted only 5 minutes, but a good landing was made upon his return to the infield. Shortly after, Stark again took off, flying east towards Moodyville, then circling back along the north shore. He reached a height of 1,000 feet, and made a number of figure-eights before coming down for a landing.

Incomprehensible as it may seem, a herd of cattle grazing in the infield had not been shooed out, and as the Curtiss sailed towards them, several excited animals ran directly into the path of the machine. Stark had sufficient presence of mind to yank his craft up and over them but in doing so he sacrificed most of the available landing-area. The plane was still travelling fast, but the proximity of the high trees precluded any safe attempt at getting away from the ground to try another landing, so Stark set the wheels down and headed for the fence. When about 20 feet from the inevitable impact, the airman threw himself from the seat, the wing passed above him, and an instant later the machine smacked the railings with a resounding crash. Two assistants, J. Perry and T. Watson, had actually endeavoured to slow the pace of the machine by attempting to grasp it as it rushed for the fence. The former suffered a badly cut hand, while the latter had the somewhat unique experience of being knocked down and run over by an airplane, with only bruises to show for the episode. Stark's injuries were equally slight, and damage to the biplane was not severe, but the incident ended Stark's flying at that place.

Several months before his death in March 1942, Stark in describing this affair, revealed that the risks he ran while trying to give the people their money's worth were far greater than they appear in a casual report. The lay-out of Hastings Park race track was such that, while he was able to fly out into the wind on both flights, he was forced to land *down-wind*, there being no other way to get in. As all airmen know, landing in such a manner even when ample space is available is a highly dangerous undertaking.

Stark's next contract was at Victoria, B.C., as part of a 24th of May celebration. He made a flight of 20 minutes, operating from the Oak Bay grounds. In landing, the drag brake on the front wheel refused to hold the machine on the wet grass of the enclosure, and again collision with a fence caused considerable damage and prevented further flights that day. The crowd took it in good part, however, and both the Victoria newspapers published praise-filled accounts of the flight. Repairs were effected by the afternoon of the following day, and Billy sailed aloft again for a fine flight of 15 minutes, concluding his contract with a perfect landing in the Oak Bay grounds.

Prior to this, Billy's brother, his manager, had contracted a flight for him at Armstrong, B.C., on July 1. Stark reached there on June 27, and when his helpers and his machine arrived by train the following day, the craft was assembled in the local skating rink, where many of the town's residents saw an airplane for the first time. A cold and blustery wind was blowing on Dominion Day, but Billy took to the air as scheduled, going up to 1,500 feet, then sailing off down the Otter Lake Valley, pitching and rocking alarmingly. After 2 miles, he circled around and, with the stiff wind behind him, was quickly back over the fair-grounds. Coming down low as a salute to the crowds, he went on towards Enderby, deciding to make a landing beyond the town, where, in the gusty wind he had to contend with, a greater area offered more safety. He landed safely in a large field close to the old Landsdowne cemetery and a passing motorist gave him a lift back to the fair-grounds, where the long delay had been creating anxiety among the crowd. The aircraft was a huge drawing-card for the fair, as it was the first time an airplane had visited any point in the interior of British Columbia. Over 4,000 people were on

BILLY STARK WITH MRS. STARK
Just before their flight on April 24, 1912

hand to see it fly. Stark received $1,500 for the visit—a sum justified by the high expenses and the still higher risks.

In July, Stark was asked to go to Portland, Ore., to fly the machine of another pilot there. No sooner had the machine left the ground than the engine went dead, and he was forced to crash-land in a small space amidst a number of parked autos. The plane was badly damaged and the meet was called off, but the newsmen of the Portland papers praised the airman for the quick thinking and steady nerve he showed in deliberately smashing the plane to avoid striking the spectators. He received a bad shaking up, and cuts, but no bones were broken. Family influence then prevailed upon Stark to take time off from trying to break his neck, and for the next two years he was out of aerial circulation.

By 1914, however, Stark's desire to fly again became irresistible. He had a large pontoon made by Van Dyke & Sons, a Vancouver firm of boat builders, and thus turned his Curtiss into a seaplane. This craft (at that time called a hydroplane) was the first of its kind to be developed in Canada. It was contrived with the help of a Vancouver friend, William McMullen, whose work as one of Canada's first aircraft builders is described in the Templeton-McMullen story in chapter 4.

On June 14, 1914, Billy launched his seaplane on the harbour waters of Burrard Inlet. After taxiing out beyond Brockton Point, he gave the engine full throttle and the machine rose into the air with little effort. After a practice landing near Prospect Point, Stark continued his flight, going around Stanley Park to alight on the sea close inshore to English Bay bathing-beach. Thousands of aircraft have since flown a course following along the First Narrows, the entrance to Vancouver harbour, but Billy and his hydroplane were the trail blazers.

Stark was on hand with his machine, again wheel-equipped, at the Chilliwack Fair on

BILLY STARK IN HIS SEAPLANE
June 14, 1914

Dominion Day, 1914. Practically all the residents of the Fraser Valley turned out for the event, for the airplane was a great novelty. When the airman made his first flight, a boisterous wind made things most unpleasant for him, but the crowd thoroughly enjoyed it. A reporter of the Vancouver News Advertiser wrote: "The wind was tricky, and but for the fact that the proposed flight had aroused great interest in the Fraser Valley, particularly among the old-timers, Stark would have declined to ascend. After the performance he described it as the riskiest trip he had ever made, although [as the reporter naïvely put it] it appeared quite safe from the ground." A second flight was made in the evening under calmer conditions.

With the outbreak of World War I, private and exhibition flying throughout Canada practically came to an end, but not so the story of Stark and his Curtiss. Both of them did good work in 1915 helping to train a number of young men to become pilots for war service, but an account of these events belongs to a later chapter.

IN the year 1912 three pilots who had been seen in Canadian skies met their death; two by crashing, one by pneumonia which was indirectly the result of too frequent exposure to the elements. In the same year another pilot who had flown in Canada collided in the air with a second machine whose pilot died in the resulting crash. A fifth fatality occurred when an American pilot smashed into a grandstand at an exhibition, killing one spectator and injuring 21 others, shortly after he had completed a Canadian flying tour. All of these accidents took place at American points. Ely, who flew in Canadian skies in 1910, died in an air accident across the border the next year.

It is surely a matter of sheer coincidence that from the beginning of flying even up to the end of 1912 no one had died by aerial misadventure in Canada. Certainly it cannot be said that there was any lack of opportunity. In a year that saw flying in every province save Alberta and Prince Edward Island there was more than one narrow squeak.

The first exhibition of the year was that already described, by Billy Stark at Vancouver. The second came to Manitoba while Billy was still providing excitement for the Vancouver crowds; two American "Life Defiers" and "Death Mockers" (as they were advertised) were thrilling crowds in Winnipeg and Portage la Prairie.

TICKELL AND BEACHEY: PORTAGE LA PRAIRIE AND WINNIPEG

Frank Kenworthy of St. Louis, Mo., owned a Curtiss-type pusher biplane for which he engaged Tom McGoey, of Grand Forks, N.D., as pilot. When McGoey built a similar machine for himself, Kenworthy sought and found another pilot—Sam Tickell. The two owners meanwhile had become partners.

In May, Kenworthy signed a contract to fly "several" machines at Winnipeg, although the firm possessed only two, and of these the second had never yet been flown and was not airworthy. Tickell was given the "good" machine to fly, and the other one was kept in reserve.

On May 10 Tickell raced the plane up and down the River Park race track, at Winnipeg, making short hops, while trying to get the 60 h.p. Hall-Scott engine to hit on all eight cylinders at the proper time. He was billed to make a flight in the afternoon, but told the officials that the engine was not running properly, and that it could not lift the machine over the top of the grandstand. The promoters said he must try, or the money would have to be refunded to the spectators.

So brave young Tickell climbed into the seat of the machine, and giving the engine full throttle, took off in fine style. Immediately, the engine began to falter and Sammy saw he could not clear the grandstand, but it was too late by that time to land in the infield so he tried to pull up enough to clear the grandstand roof, at the same time banking sharply to the left. The machine immediately stalled, and slipping off into a dive, dropped the forty or fifty feet it had gained, and hit the track with a resounding wallop. After the dust had subsided, a badly damaged plane was discovered with Sam pinned underneath it. Strangely enough, he was found to have no bones broken, and seemed to be suffering only from a severe shaking-up. Even his nerve was undamaged. Meanwhile, the promoters in the tent at the other end of the field were so scared that they dared not even come down to see what had happened. When Sam walked into the tent they could scarcely credit their eyes.

The Winnipeg promoter, not knowing the second machine was unfit to fly, still wanted Tickell to go aloft in it. But Sam was hustled off to his hotel under the pretence that he was badly shocked, so that he should not be forced to attempt an impossible flight. Thus ended the flying at Winnipeg for the time being.

The firm of "K. and McG.," however, had a contract to fly at Portage la Prairie, beginning on the 18th, so Tickell and McGoey set about repairing the damaged machine. Wonderful to relate, they had it in shape and shipped to Island Park fair-grounds at Portage by the 19th. Then it was that Kenworthy informed Tickell that they had procured another pilot—the well-known Hillery Beachey, a brother of the famous Lincoln Beachey. He held Aero Club licence No. 89, issued in January 1912. When Hillery saw the machine, he refused to fly it, because it had elevators at the front instead of at the rear. Kenworthy then asked Tickell if he would fly it, and got a like answer. Although he was offered $100 for a 5-minute flight, the young man remained adamant.

It was then decided to alter the plane to suit Beachey, and Tickell and the others set to work, changing things around, until Hillery was satisfied with the control system. On the afternoon of May 22, Beachey made a very successful flight of 18 minutes from a large field near the Waterloo Manufacturing Company's plant. Although he went no higher than 100 feet, it was a very creditable flight. Another short flight was made on the following morning, but the tip of one of the propeller blades was damaged in landing, bringing to a conclusion the first airplane flights ever seen at Portage la Prairie.

The two machines were then shipped back to Winnipeg, and on May 24 Hillery Beachey flew the Hall-Scott powered biplane before a crowd of over ten thousand people. A strong and gusty wind was blowing, making things most uncomfortable for Hillery, and the machine pitched about in an alarming manner. He was forced to

land after only 10 minutes in the air, but it was not the elements that got the best of him, nor lack of nerve. Spectators had begun to swarm out of the grandstand on to the airfield, and if the airman had waited longer he would have had to search elsewhere for a landing. With wind conditions as they were, he had no desire to run into unnecessary danger.

That final short flight brought to an end the flying of the Kenworthy-McGoey Aviation Company, at least in Canada. Under the circumstances, though their flights were anything but epochal, it was an achievement merely to have preserved their necks.

TURPIN AND PARMALEE: VANCOUVER

While Hillery Beachey was doing his stuff at Winnipeg on that 1912 Victoria Day and Billy

TYPICAL NEWSPAPER ADVERTISEMENT
Displays the grandiose statements of some early promoters

THE TURPIN-PARMALEE MACHINE
At Hastings Park, Vancouver, May 24, 1912
Courtesy Vancouver City Archives

Stark was thrilling the crowds at Victoria, two American airmen, Clifford Turpin and Phil Parmalee, were giving a most creditable performance on the outskirts of Vancouver. These young fellows possessed Aero Club licences of very early date, Turpin's being No. 22 and Parmalee's No. 25, both issued in 1910. They had been flying as exhibition pilots for the Wright brothers for most of their careers.

At Vancouver, their flight headquarters was Hastings Park, later described by the young airmen as the most dangerous air field they had ever known. The area was small in extent, surrounded by a high and heavy forest, and was crossed at one end by an electric power line. Yet the flyers met the hazards with courage and skill, making in all seven take-offs and landings.

The machine they used (a second one was left at the border, since Customs duties were too high to admit both) was a well-made, two-seater biplane, with the engine and propeller in front, known then, as now, as a tractor airplane.

The contract for two days' flying included a feature unique in Canadian air history—a parachute jump from a plane in flight. Only four such jumps had been made anywhere in the world and none at all in Canada, so this exploit would constitute a record worth witnessing.

The citizens of Vancouver seemed doomed to disappointment when it was learned that Morton, the American parachutist, was ill and could not come, but by rare fortune a substitute was found in Charles Saunders, who had been parachuting from a balloon a few days earlier at New Westminster, B.C. This was Saunders' first time up in an airplane and it was Parmalee's first time to co-operate with him in a jump—circumstances which added further hazard, and therefore greater interest, to the undertaking.

Compared with today's methods, their preparations were makeshift indeed. A container to hold the folded and bulky cotton parachute was made from a small, round, metal drum. With one end cleanly removed, the drum was then firmly attached to one of the skids of the machine's landing-gear, the 'chute being packed snugly inside.

Attired in bright red tights, and wearing a large leather helmet, Saunders took his place in the machine on a seat ahead of the pilot. Initial disaster was narrowly averted on the take-off, as a very long run was required before the machine became airborne, and the daring couple just managed to clear the high tops of the trees as the machine climbed steeply out of the grounds.

They circled out over the waters of Burrard Inlet to a height of 1,000 feet. Then, as the crowd below watched in silence, Saunders climbed out of his seat, down to the undercarriage. In those days no harness attached the parachute to the jumper's body; the courageous parachutist simply grasped the trapeze bar which was attached to the parachute by suitable cords, and the straining muscles of his own strong hands and arms were all he could rely upon to forestall a sudden trip to eternity when the jolt of the opening 'chute all but wrenched him from his hold.

Straight as a plummet the red-clothed figure was seen to plunge earthward, sheer into space. An instant later the white plume of the parachute emerged from its container. After a few preliminary flutters, during a drop of about 100 feet, the 'chute opened perfectly, and a great sigh of relief went up from the throats of the thousands watching.

Saunders landed on the drying mud flats on the north shore of Burrard Inlet, about 100 feet from the water's edge, and hundreds of picnickers soon surrounded him. They plied him with innumerable questions as he rolled up his parachute while awaiting the coming of a power boat, which he had previously hired to rescue him from the water, if necessary. Half an hour later he appeared at Hastings Park, followed by a great crowd of people, and a number of proud youngsters carrying the rolled-up parachute. The spectators gave him a tremendous ovation.

The programme was repeated the next day. Following exhibition flights by Parmalee and Turpin, Saunders again went aloft with Parmalee as pilot. This time he made a bull's-eye landing right in the centre of the infield, to the great delight of the huge crowd.

Later, Parmalee and Turpin ran into difficulties when trying to return to the United States; the Canadian railway refused to ship their plane as baggage, while to send it by freight would have meant a late arrival for their next contract, which was in Bellingham, Wash. Finally they secured the services of a taxicab owner-driver. They loaded the engine on to the back seat of his big Winton touring car, and on top they packed the fuselage, with blankets and pillows for padding, and the wings protruding fore and aft. Then the three men and the machine, after making a slow, tedious, but safe journey, reached their destination with time to spare.

The dangers which had encompassed the two friends since they had learned to fly were gradually working to a climax, and suddenly luck turned against them. Only five days after his visit to Vancouver, Turpin had the misfortune to crash into the grandstand at the Meadows race track in Seattle. He was not badly injured, but the smash-up resulted in one fatality, and injuries to twenty-one spectators. Then, only two days after Turpin's accident, Phil Parmalee crashed to his death in Moxee Valley, North Yakima, Wash., when his machine went out of control.

These tragic events terminated the flying career of both young men. Turpin, having flown about 400 hours during the most dangerous period of aviation, decided the time had come to marry and settle down on earth. With his many memories of hectic flights, he has since been very entertaining to those who could catch him in the mood to talk about the days when both he and flying were young.

WARD AND MESTACH: WINNIPEG

Frank Coffyn's flying at Winnipeg in 1911 had whetted the public appetite for aviation, and people came in droves to see Jimmie Ward and George Mestach. The programme got off to a slow start, but built up to a flying finish.

On the first day of the exhibition, July 11, Mestach essayed the initial flight, in spite of a violent wind. He was flying an American-built Morane monoplane, a light type of aircraft, fitted with a 50 h.p. Gnome engine. Although well built, the craft proved no match for the stiff prairie wind. Mestach was no sooner off the ground than a particularly hefty gust slammed the machine down on the fence. The

JIMMIE WARD (*dark suit*) AND
MRS. WARD

With his Curtiss biplane *Shooting Star*

pilot was only slightly shaken up, but the machine suffered badly. Wings and landing-gear were severely damaged, three cylinders of the engine were cracked, and the propeller shaft was bent. That was the end of Mestach's flying at Winnipeg. The machine had to be shipped back to the Borel Company's shops at Chicago to be rebuilt.

Jimmie Ward, whose A.C.A. licence was No. 52, then took over, flying several times daily except on occasional days when bad winds kept him down; altogether he was up ten times between July 12 and 19. In his Curtiss biplane, the *Shooting Star*, he made several Canadian records. On his first time up, which was the first flight ever seen over the city of Winnipeg, he made a new height record for Canada of 4,000 feet. Towards the end of his visit he exceeded this mark by ascending to 5,000 feet on the 18th, and to 6,000 the next day, the latter the first time anyone in Canada had flown more than a mile high.

Other flights, made sometimes in spite of very rough wind, were enlivened by Jimmie with

dips, spirals, and glides, and on one occasion he performed with the happy knowledge that his parents were watching. They had come up from their farm in North Dakota to see their son fly.

BROOKINS: PORT STANLEY

During this year, 1912, an electric railway connection was completed between London, Ont., and the Lake Erie resort of Port Stanley. To celebrate the official opening, the transportation company engaged the services of the well-known Walter Brookins—who carried an Aero Club licence No. 19—to fly over a four-day period. The sensational feature which brought people to Port Stanley in throngs was the first appearance in Canada of a seaplane, or hydroplane, as it was first called. This machine was a Burgess-Wright biplane, similar to the one which Brookins had flown at Montreal and Toronto two years previously, with, of course, pontoons instead of wheels.

The first flight, which constitutes a Canadian record, was made on July 17. The flyer delighted

GEORGE MESTACH'S CRASH AT
WINNIPEG, JULY 11, 1912

**WALTER BROOKINS AND
DORA LABATT**
Returning to the beach at Port Stanley,
July 20, 1912

the crowd as he circled about for 9 minutes before alighting on the surface.

On the 18th, a wind made the water very choppy, but Brookins made his three promised flights and even threw in some dips and dives. On his final flight he took up a young woman, Dora Labatt, as passenger—a favoured protégée of the transportation company; on a previous flight he had taken up Lorne Bradley, a newspaper reporter. They were the first Canadian passengers to fly in a seaplane.

Brookins returned to the United States, flying to further fame, but made no further engagements in Canada.

WARD: REGINA

At the end of July, Ward appeared at the Regina Exhibition, where it was hoped that he might try to better his height record of 6,000 feet established at Winnipeg. Unfortunately, weather conditions prevented any such attempt, although the airman flew twice each day as scheduled. His flying left nothing to be desired, and on each of the three days, July 30 to August 1, he circled over the city. When he had completed his flying at Regina, he went to Duluth to fly and Canada did not see him again.

Ward continued to be very active in aviation in the United States for many years after his visit to Canada. During World War I, he became a civilian flying instructor for the United States government.

GLENN MARTIN: SASKATOON

Wherever in the world people speak of aircraft today, the name of Glenn L. Martin is known, but in 1912 he was just beginning to carve out his career. Canada welcomed him when he same to Saskatoon to fly during the annual Exhibition, August 5 to 9.

Ever since St. Henry's short flights the previous year, the citizens of that northern Saskatchewan city had patiently awaited the opportunity to see an airplane in action again. Martin brought with him his original pusher biplane, the machine he had designed and built during 1909 in an old church building at Santa Ana, Calif. This was the machine in which he had taught himself to fly. He passed his flying test at Santa Ana on August 9, 1911, and secured A.C.A. licence No. 56.

On each day of the Exhibition, Martin thrilled the spectators with consistently fine flights, but on the final day of the fair he surpassed everything he had previously done. During a flight which lasted 40 minutes, the American reached a height of 6,400 feet, bettering Jimmie Ward's record of the month before by 400 feet. This altitude record of Glenn Martin's remained unchallenged throughout Canada for almost five years.

To complete the event, Martin cut the engine upon reaching his greatest height, circled down in wide spirals, and made a steep fast glide back to the grounds. Fair officials and spectators alike were unstinted in their praise, and went home happy, but few realized they had watched a record-making achievement.

JAMES MARTIN: VANCOUVER

While Glenn L. Martin was flying at Saskatoon on August 9, James V. Martin, an Englishman, was at Vancouver, assembling his machine for flights scheduled to take place on the 10th. James Martin was one of the pilots who had taken part in the air meet at Hamilton in July 1911, and the bad luck which dogged him and his Farman biplane at that time seemed to follow him on his second visit to Canada.

On this occasion he brought along a well-built machine of his own design and construction, a two-seater biplane with the propeller

GLENN MARTIN
With his pusher biplane at Saskatoon,
August 9, 1912

and engine in front, a machine which he had already flown very successfully at various points in the United States and in which he established a number of records.

Half-page advertisements stated he would fly on August 10, and that Millie Irving, "America's most daring aviatress," would also be on hand. The young woman, however, was not in evidence when the airman was ready for his first flight, during a lull in the horse races being held at Minoru Park. A strong wind was blowing from the west, and as the airman attempted a take-off, engine trouble developed. His sudden forced landing carried the machine into the fence and it suffered considerable damage.

Martin hoped to repair the damage, and it was announced that further flights would be made the following week, but he was unable to effect repairs in Canada and was forced to ship his biplane to Seattle for the job, so his flying efforts in Canada came to a dismal ending. This was regrettable, because both Martin and his machine were capable flyers. With this same machine he established in 1911 a United States speed record of 75 m.p.h., and in 1913 made the first flight in Alaskan territory.

MESTACH: QUEBEC CITY

Meanwhile air fans in Quebec City had been so impressed by Mestach's flights in 1911 that he was engaged for a second appearance in 1912. Mestach's misadventure in Winnipeg had not cooled his enthusiasm, and on August 22 he and his thoroughly repaired machine arrived at Quebec by train.

With the weather consistently in his favour, Mestach flew each day of the 1912 Exhibition from August 26 to 31. His Morane monoplane

proved to be a much faster craft than the Blériot he had used the previous year. Its sensational speed and the daring turns the airman made at low heights brought the crowds to their feet time and again.

Monoplanes were rare in Canada in the early days; in fact, the only monoplanes successfully flown anywhere in Canada in the pioneer flying era were the two Blériots used by Count de Lesseps at Montreal and Toronto in 1910, and the Blériot and Morane flown by Mestach at Quebec in 1911 and 1912.

From Quebec the French airman went to Chicago, Ill., to take part in two big air meets. During an air race, Mestach's Morane tangled with a Wright biplane piloted by Howard Gill. In the resultant crash of the machines, Mestach came out uninjured; Gill, however, was killed.

After that time Mestach seems to have dropped out of flying in America.

PEOLI: SAINT JOHN

Two years had passed in the Maritimes since the pioneer flights of McCurdy and Baldwin, at Baddeck, N.S., when the first flight in New Brunswick was made by a well-known young American pilot Cecil Peoli. Peoli had already acquired a flying reputation in the United States when he arrived at Saint John with the big pusher biplane, *Red Devil*, owned by the famous aeronaut Captain Tom Baldwin (unrelated to, and not to be confused with, "Casey" Baldwin of Baddeck). The *Red Devil* was a massive affair, weighing some 1,050 pounds without the pilot, but it was well powered with an 80 h.p. Hall-Scott engine.

As the exhibition grounds were too small for taking off or landing, flights were made from the Courtenay Bay flats, on the beach near Rockwood Park. They could only be made when the tide was out, consequently it was not always possible for the airman to fly on schedule, but he went up twice a day as his contract required.

Great crowds were present every day. The first flight was made just before 11:00 A.M. on September 2, and as it had to be made three-quarters of an hour earlier than advertised in order to avoid the incoming tide, hundreds of people were disappointed when they arrived too late. They had no cause for complaint, however, as the days passed, for the young airman demonstrated flying at its best, and on every trip would circle low over the fair-grounds to give everyone a chance to view the craft in the air.

September 5 was a highlight of Peoli's visit to Saint John. On the morning he took up as a passenger the secretary of the Exhibition Association, Horace Porter. This privilege called for some courage on the part of Mr. Porter since he had to occupy the usual passenger seat on all single-seater machines—a board fastened to the wing. The flight lasted 11 minutes, and as Peoli took the craft up to 1,300 feet, his passenger had a good view of the city. After they had landed, Peoli asked Porter how he had liked the trip. The latter remarked that he thought the airman had dived rather too steeply before landing. Peoli then confided to him that a front balancing elevator had been removed to help lessen resistance, and when he had nosed down, the extra weight of Porter in front of the centre of gravity had made the biplane almost unmanageable.

On the afternoon of the same day, a reporter on the Saint John *Telegraph*, J. J. Marshall, was also taken on a short flight, without, however, the thrill of an uncontrolled dive.

September 6 brought the fair to an end, and young Peoli went on to carve for himself an illustrious flying career until he met his death in a crash at Washington, D.C., in 1915.

WALSH: HALIFAX

Apparently the Maritime Provinces had quickly become air-conscious, for a few days after Peoli's flights at Saint John, N.B., the city of Halifax, N.S., watched its first aerial barnstormer. Charles F. Walsh flew for the Nova Scotia Exhibition on September 11 to 19, 1912.

The American airman, who was one of the Glenn Curtiss Company's exhibition pilots, had previously flown in Canada at Victoria, B.C., in May 1911.

On September 11 everything seemed to promise a fine first flight, but Walsh had no sooner taken off than something went wrong. A vicious down-draft was thought to have been the cause, but this was never substantiated. When the machine was over the railings at the far end of the infield, it suddenly dived to the ground, struck a fence, and collapsed in a shattered heap. The airman was thrown clear, and landed atop the roof of a near-by cattle barn. Wonderful to relate, he suffered nothing more serious than shock and bruises.

His manager immediately wired to New York for another machine. It came along promptly

THE DAILY PHOENIX (SASKATOON)
The issue of Saturday, August 10, 1912, gave this account of Martin's flight

CECIL PEOLI IN BALDWIN'S *RED DEVIL*
Just before his first flight at Saint John, N.B., September 2, 1912

by train, and was assembled and in readiness to fly by the 16th—and Walsh was ready to fly it! Although his bodily aches were still severe, and a large patch covered a damaged nose, he succeeded in making a fine getaway from the grounds. After flying as far as Bedford Basin to gain altitude, he returned and flew over the city of Halifax. Just before landing he executed one of his daring "death spirals," which was as near a thing to a spin any airman had accomplished in those early days and lived to tell about. When he passed over the city at 2,000 feet he became the first airman to fly over Halifax.

The next day he made two flights, during the second of which the wind was so strong at times that his ground speed was cut down to only 10 miles per hour.

On the 19th he had a nasty experience, and was lucky to come out of it without injury. During the take-off on his morning flight, a down-draft struck his machine as he was clearing the roofs of the cattle barns. Only quick thinking and clever manipulation of the controls carried him under the telegraph wires by the railway track, and the undergear of the biplane was within a foot of a stationary locomotive on the siding as he swept past, to land safely in a near-by field. Strong winds continued to blow, but in spite of them he put on a good show in the afternoon, spending most of the 20 minutes he was aloft in circling over the city.

The 20th was the final day of the Exhibition,

which perhaps was just as well for the airman. When he endeavoured to get clear of the grounds, a down-wind eddy caused the machine to dive earthward at the exact spot where the same thing had occurred the previous day. Walsh managed to land in a small area but the aircraft smacked into a solid board fence at high speed. At the last moment, the airman threw himself from the seat and escaped injury, but when the dust subsided the Curtiss biplane was only a forlorn pile of wreckage by the fence.

It is possible that Walsh's experiences at Halifax had an unnerving effect, because only thirteen days later he crashed to his death at Trenton, N.J.

EDWARDS: NELSON

Situated in the mountainous Kootenay district in the heart of British Columbia, the progressive town of Nelson decided to feature an airplane at its annual Exhibition in September 1912, and a committee arranged for Billy Stark of Vancouver, B.C., to do the job. Stark was unable to fulfil his contract after signing it because he became involved in a crack-up at Portland, Ore., so he passed the opportunity on to his friend Walter Edwards, an American pilot.

Edwards and his manager, F. A. Bennett, arrived by train at Nelson on September 22, bringing along a Curtiss-type biplane, equipped with an 80 h.p. Curtiss engine.

The confined surroundings of the area worried

Edwards, for he realized that he would be beset with treacherous air currents caused by the rugged country, and he was quite frank in saying that he certainly would not have contracted to fly there if he had known what a rough prospect was in store. But he was both a brave man and a good sport, and on September 24 he made his first flight right on schedule. For 9 minutes he stayed in the air. During that time he flew out over Kootenay Lake, preferring the calmer air above the water to the crosswinds near the mountains.

The high fences and telephone and electric wires surrounding the fair-grounds made get-aways and returns a hazardous affair, and Edwards' daring handling of the biplane in and out of the small area of the infield imparted thrills to both airman and spectators.

After the first day, Edwards was very outspoken in his objections to the risk of the further flights. Even Manager Bennett made gloomy forebodings. However, they didn't back out, and on the 25th, Edwards made a 12-minute jaunt in which he again kept well out over the lake.

The following day, after a very fine flight, the airman had to make three attempts before he was successful in returning to the infield, and as a cold and boisterous wind was blowing, he was almost numb when he was at last able to set the machine down. Even then trouble developed. When he saw that the plane was going to strike a fence, he jumped from his seat and allowed himself to be dragged along the ground while grasping the machine, in order to act as a human brake. Thanks to his effort, the only damage was a slightly buckled front elevator which was easily repaired.

On the 27th, Edwards' final day at Nelson, the weather at last decided to be a little more considerate. There was only a slight breeze and the sun shone brightly. During this flight Edwards made an unscheduled landing on the flats some distance out of town. Considerable anxiety was felt, but before a search could be instigated the roar of the engine was heard, and soon the watchers saw the machine winging low towards the grounds, where the pilot settled down safely—a perfect ending to a brief nightmare for Edwards and his manager.

Although recognized as a very good pilot, Walter Edwards never applied to the American Aero Club for a licence, and his name early dropped out of the flying news.

With the last of the fall fairs in September the Canadian flying season came to an end. The year 1912 saw 122 deaths in world aviation. Thirteen barnstormers flew in Canada that year, and seven passengers went aloft. Five pilots and one passenger narrowly escaped death. Five days after his last Canadian engagement Turpin crashed into a grandstand at Seattle, killing one person and injuring many others. A week after his flying at Vancouver, Parmalee crashed to his death in Washington State. In the same year Mestach quit flying after a collision in which the other pilot was killed. Walsh died in a smash-up thirteen days after his Halifax exhibition. Peoli had three more flying years before his fatal crash. Let the present-day passenger remember, as he rides in a luxurious airliner above the clouds with greater safety than he could enjoy in an earth-borne automobile, that his comfort and security are the heritage of those early days when pilots risked their lives in unpredictable aircraft.

WALTER EDWARDS

Makes a good start from his hazardous field at Nelson, B.C., September 24, 1912

13. The friendly invasion wanes, 1913-1914

THE flying seasons of 1913 and 1914 brought to a close the era of American barnstorming pilots in Canada. Nine made exhibition flights. It is significant of the increasing safety of the machines and of the frequency with which two persons were being carried aloft that it is quite impossible today to secure the exact number of passengers who flew during that period. The fact is that in the larger Canadian centres a passenger flight *per se* was no longer news; and the tendency in press reports of those years was to drop specific names of passengers unless they were otherwise newsworthy. So, too, surviving pilots and witnesses of exhibition flights have forgotten the particulars of individual passenger flights, because the number of people taken up in 1913 and 1914 was so large.

THE BRYANTS: VANCOUVER

As in 1912, the season in 1913 began in Vancouver, with the first man-and-wife flying team to visit Canada. They were John and Alys Bryant, advertised as members of the Bennett Aero Company—the same Bennett who had been Edwards' manager at Nelson, B.C., in 1912.

They began their programme at Minoru Park race track on July 31. Johnny went up first, in the Curtiss-type biplane which he and his brothers had built in California. He delighted the crowds with his expert handling of the machine and with a number of breathtaking, steep dives.

The next flight made history: Alys Bryant took the machine up, flying solo to make the first flight in Canada by a woman pilot. During

the 16 minutes she was in the air she reached a height of 700 feet, and her landing in the infield was faultless.

On the third and final flight of the day, Johnny went up to try for an altitude record. He did not surpass the one made by Glenn Martin at Saskatoon the previous year, but he did reach an altitude of 5,100 feet, which was a record for the northwest to that date. A brilliant finish to this flight was achieved when the airman shut off his engine at a height of 2,500 feet and dived almost vertically until he was 100 feet from the ground. He made a perfect landing without resorting to the use of his engine again, which was clever although risky flying for that early period.

On the next day, Mrs. Bryant took the machine up to a height of 2,200 feet. Then two flights made by her husband under bad weather conditions concluded their flying at Vancouver.

The machine was shipped across the Strait of Georgia to the city of Victoria, where a water carnival was to be held the first week of August.

GLENN MARTIN: SASKATOON

Meanwhile, the people of Saskatoon were eagerly awaiting August 4, when Glenn Martin was scheduled to fly again, the experience of the year before having made a satisfying impression on both pilot and public. With him he brought a new machine he had recently designed and constructed—a fuselage-type tractor biplane, fitted with a 90 h.p. Curtiss motor, capable of carrying two passengers as well as the pilot. This meant that local passengers were

ALYS BRYANT
Just before her flight at Vancouver, July 31, 1913

a feature of the visit. One woman and two men went up at different times, Beatrice Mulhullen, George Martin, and W. R. Drennan.

Before Glenn Martin's visit, the First Saskatchewan Aviation Co. Ltd. had been formed. Its promoter was Harold E. Hartney and it was probably the first registered flying company in Canada. The company owned a shed (which was used by Martin during his stay) and they were hoping to get an English airman to come out with a Farman biplane and start a flying school. But the plans did not materialize, and the next year a windstorm destroyed the shed.

Glenn Martin did not break his former Canadian altitude record of 6,400 feet, but later in the year when back in his own country, he climbed to 9,800 feet—a world record—flying the tractor biplane which he had used at Saskatoon.

Glenn Martin was still in Saskatoon when the telegraphs across the country were humming a story of a kind now all too familiar but until

then mercifully absent from Canadian aviation. The story came from Victoria where, on August 5, Alys and Johnny Bryant were scheduled to put in an appearance.

THE BRYANTS: VICTORIA

Alys was the first to take the biplane up. She took off from the Willows race track on the morning of August 5, and flew in the direction of Uplands, one of the city's suburbs. But the notorious ocean winds of Victoria were almost too much for her. Gust after gust struck the machine and the airwoman had great difficulty in keeping the craft on an even keel. Instead of flying over the city, as she had planned, she had to swing the biplane around and head back for the race track in the teeth of the wind. After a hard-fought battle, her expert handling brought the machine back to the grounds, ending an experience which she later said was the worst she had ever had in the air.

It was left for Johnny Bryant to make the first flight over any British Columbia city, when he flew over Victoria at 1,000 feet on August 6. For this event he had converted the craft into a seaplane, substituting a single float for the wheels.

He took off from Cadboro Bay, where the change had been made, and after a brief test, landing a short distance off shore, he again climbed into the sky and set a course directly for Victoria. As his machine appeared over the city, crowds surged about in the streets and cheering broke out from thousands of people. After crossing the city, he alighted just outside the inner harbour and taxied in to a wharf. Every point of vantage was thronged with spectators, and the airman was given a rousing welcome.

At 5:50 P.M. that same afternoon, Johnny Bryant took off from the water near the wharf

MRS. BRYANT IN FLIGHT

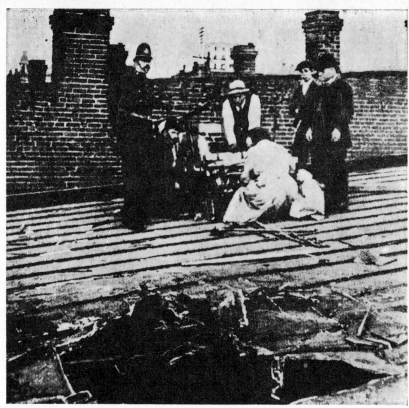

CANADA'S FIRST FLYING
FATALITY
Above: Johnny M. Bryant

Left: Inspecting the engine of Bryant's plane after his crash at Victoria on August 6, 1913

GLENN L. MARTIN
With his tractor biplane at Saskatoon, August 1913

of the Department of Marine and Fisheries, and after circling out over the harbour to gain height, came back over the business section of the city. He had only been in the air 5 minutes, and had reached about 800 feet. The watchers below saw the machine dive sharply when it was directly above the City Hall, the dive growing steeper and faster until, at some 200 feet from the ground, the craft began to disintegrate. The right wing collapsed and broke away, and momentum carried the machine for some distance till it ended up on the flat roof of the Lee Dye building on Theatre Alley. Although motor constables arrived almost immediately, they were too late—Bryant had been killed instantly by the impact. This was the first flying fatality in Canada.

Alys Bryant gave up flying after her husband's death, but she did some splendid work in aviation plants during both world wars, and in 1939 was an honoured guest of the United States government during flying ceremonies at Fort Myer, Va.

Robinson: Montreal

At this period, many countries were making real progress in flying. This was especially true in Europe, where airmen were striving to gain fame and prize-moneys by establishing new records for height, speed, distance, or duration.

The world record for long distance was then almost 650 miles; the United States record was 217.5 miles. In Canada no official record for distance had been attempted. It seems curious, in view of the achievements of later years, to learn that the first one to merit special honour was merely for the distance of 116 miles and that it was by no means a non-stop flight.

Although there were two Canadian pilots, John McCurdy of Baddeck and Billy Stark of Vancouver, who might have been given the opportunity of taking on the job, both were overlooked, and by a policy all too often followed, before and since, an outsider was engaged.

William C. Robinson had been a licensed pilot since August 22, 1912, and was holder of American Aero Club certificate No. 162. He arrived at Montreal bringing, in two huge packing cases, his tractor aircraft, which was a Lillie biplane fitted with a rotary engine of his own design and make.

The flight was to be from Montreal to Ottawa, and was expected to take only a few hours. The pilot left Snowden Junction at 9:50 A.M. on October 8 as scheduled. Mayor Lavallée of Montreal had handed Robinson a number of bundles of the *Daily Mail* to be delivered at points *en route*. In the bundle for Ottawa, four papers were addressed, one each to the Premier, Sir Robert Borden, to Sir Wilfrid Laurier, to Sir Charles Fitzpatrick, and to Mayor Ellis.

Robinson had only been 10 minutes in the air out of Montreal when a broken fuel line forced him to make a landing at Lachine. With the help of a local mechanic he was able to remedy the trouble, and took off again at 1:05 P.M. He had three scheduled stops to make, to deliver papers and to take on gasoline. At these points, an advance man had marked out suitable landing-areas by putting out large crosses, made of muslin, pegged down firmly in the fields. Robinson arrived at Choisy at 1:30 P.M. and left at 2:30 P.M.

His second scheduled stop was Caledonia Springs at 2:45 P.M., and the third, Leonard, where he landed at 4:05 P.M. At each place he stayed half an hour.

The weather was hazy, and the airman had only a rough, blueprint map of the C.P.R. railway line as his guide. Several times he had difficulty in ascertaining his direction, owing to the many different rail lines in that vicinity.

Arriving at Ottawa, Robinson found that Lansdowne Park football field, where he was to have landed, was swarming with people who had come to welcome him, so he flew around for a time, finally alighting on Slattery's field. It was 5:00 P.M.—he was 5 hours behind schedule—but Mayor Ellis and a large crowd hurried over to where he had landed and gave him a warm welcome.

Very few Canadians know anything about this flight, and it does not rank among Robinson's best efforts. A year later he made 390 miles in a non-stop flight in the United States, taking less time than on the Canadian flight. In 1916, at Grinnell, Iowa, he tried to exceed the United States altitude record of 17,000 feet. But the effort cost him his life. His machine rose till it was out of sight, then suddenly reappeared spinning crazily down to earth. It is probable that lack of oxygen caused a fatal black-out to the pilot, in which he lost control of the machine.

MACAULAY: TORONTO

Early in 1914, W. A. Dean of Toronto purchased a flying-boat from the American Curtiss Company, and obtained the services of an expert airman, Theodore C. Macaulay, to pilot the craft. Macaulay had been an instructor with the Curtiss Company in the United States, and was well known as a flyer. His Aero Club of America licence was No. 228.

Mr. Dean's *Sunfish* was the first flying-boat to be owned or flown in Canada.* Macaulay flew it a great number of times from the waters of Lake Ontario, in the vicinity of Toronto and neighbouring cities. His most publicized achievement was a flight from Toronto to Hamilton and back, on May 15, 1914. He carried as passenger the well-known sports writer, Lou Marsh, who was then with the Toronto *World.*

Macaulay and Marsh left Toronto from Lake Ontario at the foot of Bathurst Street at 8:27 A.M., and landed near the Burlington piers at Hamilton 32 minutes later. During the stop-over at Hamilton, the pilot took up a number of

*A flying-boat differs from a seaplane (hydroplane) in having a fuselage like the hull of a boat. A seaplane has floats (pontoons) attached to the regular fuselage.

THEODORE C. MACAULAY IN THE *SUNFISH*
Flying over Lake Ontario, near Hamilton, May 15, 1914

passengers on short hops, first Mrs. L. Zimmerman, and then three others, H. H. Biggert, A. C. Lindgren, and M. M. Robinson, all of Hamilton. On the return flight, the flyers stopped for a short time at Oakville, where other passengers were taken up, before the journey was successfully completed to Toronto.

One other passenger flight which took place at Toronto some time later is deserving of special mention. Norman Pearce was taken up in the *Sunfish*, and as Macaulay flew above Toronto at 1,200 feet, Pearce took eight pictures with a Graflex news camera. These, the first aerial photographs ever taken in Canada from an airplane, appeared a few days later in the picture section of the Toronto *Sunday World.*

After severing connections with Dean, Macaulay became an instructor with the Curtiss Aviation School at Toronto during 1915, as described in a later chapter.

HORTON: VICTORIA BEACH, AND MINCHIN: WINNIPEG

When the American airman, Jimmie Ward, was flying at Winnipeg in 1912, his mechanic, Clair G. Horton, formed a friendship with William J. Robertson of Winnipeg. In 1914 the friendship became a partnership. Purchasing two used pusher biplanes from the Curtiss Company at Hammondsport, N.Y., they went into the flying business in Manitoba, operating one machine as a land plane, from Winnipeg Exhibition Grounds, and the other as a hydroplane from Victoria Beach, a summer resort on Lake Winnipeg, north of the city.

Horton had just become a pilot, and he undertook to fly the seaplane. Another pilot, F. F. Minchin, just out from England, was hired to fly the land plane.

The object of their operations was passenger flying, and by the end of June both machines were ready for use. During the weeks which followed, although numerous people availed themselves of the opportunity to fly, business was not really brisk.

On July 13, misfortune overtook Horton during a passenger hop. In a letter to the author dated July 13, 1944—exactly thirty years later—Clair Horton explained what happened:

On taking off with Doctor Atkinson of Selkirk, Manitoba, as passenger, I pulled the controls back for normal take-off, and when I pushed them forward again to regain normal level flight position, I found they had become locked. I glanced around and saw that a metal snap hook used to hold the controls in neutral when the machine was at rest on the beach had caught in a turnbuckle of the control wire, and I had to pull the control column back still further to get the snap free. By the time I had done this the plane had made such an abrupt climb, it had reached stalling speed, and immediately the nose went down and we dived into Lake Winnipeg, about 500 yards from shore.

That's just how simple it was to get into difficulties, no matter how careful a pilot might be.

The plane was damaged quite badly, and Dr. Atkinson received a severe cut on his right arm. Horton was uninjured, but both men suffered from their sudden immersion and from shock.

With one aircraft out of commission and the other not much in demand, the firm of Robertson and Horton decided to move out of Canada, and the shadow of war. At the end of August, they and their two machines left for Detroit Lakes, Minn. Here they flew for the rest of the season.

Cooke: Nelson and Kelowna

During the summer of 1914, Billy Stark was again seen in the air in British Columbia, and the Ellis-Blakely machine was being flown at Calgary. Two Americans also made flights in Canada that summer.

A carnival was being held in July at Nelson, B.C., and the committee had invited a promising flyer, Weldon Cooke, of Oakland, Calif., to perform. He had passed his Aero Club of America test in January 1912, obtaining licence No. 95. He owned a Curtiss-type seaplane, powered with a six-cylinder Roberts motor. He stated he would operate it from the waters of near-by Kootenay Lake, which offered safer take-off and landing possibilities than the fair-grounds from which Edwards had so precariously flown in 1912.

Cooke undertook to make daily flights during the carnival dates, July 14 to 18, and on the first day, under ideal conditions, he made two very successful ascents. On the 15th a strong wind hampered his take-offs by stirring the lake into an angry mood, and only after three attempts was the airman able to make one short flight.

Rain prevented any flying during the next two days but Cooke delighted the few spectators who turned up with an exhibition of speeding along the surface of the lake. On the

CLAIR G. HORTON (*right*)
With his two-seater Curtiss hydroplane at Victoria Beach, Manitoba, July 1914

final day, the airman satisfied the throng with two flights, the second of which was admittedly the best he made during his visit to Nelson.

A month later, Cooke and his seaplane were back in British Columbia for a one-day stand, to fly from Okanagan Lake during the Kelowna Regatta, on August 13, 1914. During the afternoon he made a fine flight far out over the lake, coming back to circle over the Exhibition grounds, but engine trouble developed when he attempted to make a second hop later in the day and forced him to cancel further efforts.

Although Cooke was a clever pilot, his machine, which he had constructed himself at Sandusky, Ohio, apparently could not by any stretch of imagination be described as an expert piece of workmanship. Reports indicate that the airman had little regard for his own safety.

Only a month after his visit to Kelowna, Cooke lost his life at Pueblo, Colo., when coming in for a landing after a short test flight in the same machine he had flown in British Columbia, but with wheel landing-gear fitted to it in lieu of a float. A strut was seen to break in the air and fall to the ground where it was later picked up, and the machine went out of control, diving into the ground and killing Cooke almost instantly.

There is no doubt his machine had had a hard life. By the time he had arrived at Pueblo, one of the wooden engine bearers was cracked its entire length, right through the three holes into which the engine bolts fitted, and a cable attached to the rear elevator from the control column, found to be too short, had been lengthened simply by the addition of a piece of ordinary bale or hay wire. The left aileron had a broken rib, and to cap it all, a leak in the gasoline tank had been plugged by a bit of wood about the size of a lead pencil! Cooke had stated he had "flown it that way up to now and had gotten away with it," but facts prove that he tempted fate just once too often.

BEACHEY: WINNIPEG

The pre-war barnstorming era closes appropriately with one of the earliest, most enterprising, and undoubtedly the most sensational of the pioneer airmen on this continent—Lincoln Beachey.

"Link" had made a name for himself in Canada in 1906 and 1907 when he had flown a powered airship of his own construction (chapter 7). Since then he had graduated from airships to airplanes, and had become a member of the Aero Club of America in May 1911, with certificate No. 27. A born airman, his thrilling exhibitions had soon earned him recognition as the foremost pilot on the continent.

He was the first flyer in America to perform loop-the-loops and inverted flying, which he demonstrated first at Los Angeles in November 1913. He used a Curtiss biplane built especially to withstand the terrific strains put upon it. Equipped with a seven-cylinder Gnome rotary engine, it had a wing span of only 21 feet, and the entire ship, with engine, weighed only 517 pounds.

This was the flying marvel that came to highlight the annual Exhibition at Winnipeg, July 11 to 17, 1914. Every day of the fair he astounded the gaping thousands with his rolls, loops, spins, dives, and upside-down flying. In addition to these stunts, performed at from 3,000 to 5,000 feet, Link also joined in the motor racing. He flew around the track above the roaring cars at a height of less than 30 feet. On one circuit of the mile track, he made the distance in 49 seconds, which was really clipping off the air miles for that period!

When he left Winnipeg he was, according to press and people, "king of the air," admired enviously by all the men, fearfully by all the women, and worshipped by every boy.

Beachey owed his preservation not only to skill and luck but also to judgment and sense. He took every precaution and verified every preparation. But it was some of these necessary precautions which led to his death.

He was flying at San Francisco on March 14, 1915, in a specially built Eaton monoplane, made expressly for stunting. A number of straps and belts held him firmly in his seat—even his shoes were tightly fastened to their rests. He was over San Francisco Bay when the wings of the machine suddenly gave way, and the craft plummeted straight down into the sea. When his body was recovered, it was found still strapped firmly to the seat, and uninjured. It was proved that the airman had died by drowning; he had been unable to extricate himself in time.

LINCOLN BEACHEY

THE EARLY BIRDS

With the outbreak of World War I on August 4, 1914, exhibition flying in Canada practically ceased. Barnstorming, however, continued in the United States for another two years, so a change was made in the date of admission to the Early Birds, an organization to which I referred at the beginning of this book. It was founded at Chicago on December 3, 1928, to keep a record of living pioneer aviators, who "piloted a glider or airplane, gas balloon or airship prior to December 17, 1916, upon evidence deemed sufficient by the Membership Committee, and approved by the Board of Governors, except that: nationals of countries other than the United States engaged in World War I must have met the foregoing conditions prior to August 4, 1914."

This fellowship was described in an address by Albert B. Lambert, holder of F.A.I. certificate No. 61, as follows:

THE BADGE OF THE EARLY BIRDS

"I know of no group scattered all over the country who have so instinctively held together in sentiment, mutual regard and loyalty. Comparatively few in numbers and only one or two in each location, the Early Birds constitute a rather romantic and distinguished organization and one of treasured membership."

A membership list of the Early Birds to the end of 1960 showed 222 still living, including only two Canadian air pioneers—Frank H. Ellis, and William Wallace Gibson; 241 have been called by death since the organization was formed; 28 remain unaccounted for.

PART 3

_____

THE WAR YEARS, 1914-1918

14. The war birds learn to fly

WHEN the shot was fired at Sarajevo in 1914 that sparked World War I, the Royal Flying Corps of Great Britain was already an established organization, and a few weeks after Britain's declaration of war, British airmen were employed in reconnaissance operations on the Western Front.

Like hundreds of other young Canadians across the country Tom Blakely and I in Alberta were all for getting into the thick of the fight. But it was easier said than done. Throughout the winter of 1915–1916 we kept making applications and hoping for favourable replies, but it soon became clear that there were only two ways then for a Canadian to become a war pilot. One was to enlist in the regular military forces in Canada, and hope for a transfer to either the Royal Naval Air Service or the Royal Flying Corps on arrival in England. The other was to pay our own fare to England or France, or work our passage over, then present ourselves at Farnborough or Rouen, and hope for the best. It was not until the Royal Flying Corps was fully established in Canada in 1917 that the majority of those who wished to serve as pilots were afforded the opportunity.

A limited number were luckier than the rest, however. These were the pupils of the flying schools which operated at Toronto and Vancouver during 1915.

THE CURTISS AVIATION SCHOOL, TORONTO, 1915

The Curtiss Aviation School which opened at Long Branch, near Toronto, in May 1915, was an offshoot of the Curtiss Company of Ham-mondsport, N.Y. The Company began production of airplanes on a commercial scale in Canada in that same year, establishing its plant, Curtiss Aeroplanes & Motors, Limited, on Strachan Avenue, Toronto, under the management of John A. D. McCurdy. This factory turned out a number of two-seater wheel-equipped JN Curtiss training planes, some of which were the first machines to be used by the Curtiss Aviation School at Long Branch. Three fine hangars were erected, giving the flying field an imposing appearance, and the sight and sound of planes rising, flying, and alighting attracted spectators from miles around. A flying-boat base was established near Hanlan's Point, on the island which lies across Toronto harbour. Two hangars were set up and several two-seater Curtiss flying-boats shipped from Buffalo. During the entire period of operations of the Curtiss Aviation School, popularly known as the Curtiss Flying School, there were neither fatalities nor serious accidents, although 54 students were trained.

The Curtiss Aviation School was the first flying school in Canada; its base at Hanlan's Point was Canada's first seaplane base; and its airdrome at Long Branch was Canada's first flying field.

One reason for the dearth of historical data on the pioneer period of Canadian aviation is that until 1915 there were no flying fields in the country where organizations could be formed whose records would furnish historical data. Originally called airdromes, such flying centres had been established in the United States and Europe before the second decade of the century.

FIRST TEN GRADUATES OF THE CURTISS AVIATION SCHOOL, TORONTO

Left to right, standing: Douglas Hay, Eric McLachlin, Homer Smith, Jimmy Day (mechanic), Claire MacLaurin, Innes Van Nostrand, and Douglas Joy
Seated: Grant Gooderham, Strachan Ince, Victor Carlstrom (Pilot Instructor), Charles Geale, and Warner Peberdy

England had Brooklands and Hendon; France had Pau and Buc. In Germany, the flying field at Johannisthal came into being at an early stage, and Russia had established an airdrome on the outskirts of St. Petersburg long before World War I. In the United States, Mineola Field and Nassau Beach, on Long Island, N.Y., Hammondsport, N.Y., and Kinloch Field, Mo., were in operation. At other American points, the Curtiss and Wright flying camps were hives of activity. But until the establishment of the Curtiss Aviation School, Canada had no flying field worthy of the name.

While the actual training of students was conducted by the school, it was the Aero Club of Canada that provided the examining board. Before the outbreak of World War I, a number of Torontonians who had an amateur interest in aviation had come together and formed the Aero Club of Canada, with a list of 26 members. Rules and by-laws were the same as those of the very active Royal Aero Club of the United Kingdom, with which the Toronto club was affiliated. Adam Penton, publisher of a motorboat magazine, was the first president, and the first secretary was Norman C. Pearce then a reporter with a Toronto newspaper, whose experience during a flight over Toronto with T. C. Macaulay is described in an earlier chapter.

When the Curtiss Aviation School began, the government paid the cost of training selected pupils, on the understanding that graduates would proceed overseas at government expense to become members of either the R.N.A.S. or the R.F.C. As neither the British nor the Canadian government had officials in Toronto who were qualified to examine the students, they asked representatives of the Aero Club of Canada to act on their behalf, that is, to pass or reject students as pilots. The first thirteen students who graduated were examined in this manner.

The president and the secretary used to journey out to Long Branch or take the ferry to Hanlan's Point to witness various tests, patterned on Royal Aero Club rules. To graduate, a student had to take off alone, fly, make a reasonably accurate figure-eight while in the air, and then return to the field and make a good landing. Perhaps this sounds easy nowadays, but it was not so elementary in 1915, when flying still offered plenty of unforeseen hazards.

In July 1915, the Aero Club of Canada was reorganized to admit a number of young pilots who were eager to join. At a banquet held July 28, 1915, at the Walker House, Toronto, Bert Wemp was elected president, A. T. Cowley vice-president, C. H. J. Snider secretary-treasurer, and G. K. Williams overseas secretary. As the years passed the Club grew into an organization which did yeoman service in Canada in the development of aviation. By the time the Royal Flying Corps finally came into being in Canada in 1917, the Aero Club of Canada had become widely known.

By July 11, 1915, the first two pilots had completed their course of instruction at the Curtiss school and passed their tests; they were A. Strachan Ince and F. Homer Smith, both of Toronto. These graduates served overseas throughout World War I. Smith served again in World War II, reaching the rank of Wing Commander. During much of the time between 1939 and 1945, he served under Air Marshal

W. A. Bishop, V.C., on recruiting missions for the R.C.A.F.

On the day following Ince's and Smith's graduation, three more finished the course: Douglas A. Hay of Owen Sound, Ont., Grant A. Gooderham, Toronto, and C. Innes Van Nostrand, also of Toronto. Hay was killed on active service overseas. Gooderham served in the air with distinction, later returning to Canada to re-enter civilian life. He is now deceased. Van Nostrand also became a war bird and lived to tell the tale.

July 20 produced five more airmen: Douglas Joy, Toronto; Claire MacLaurin, Ottawa; Charles Norman Geale, Peterborough, Ont.; Eric H. McLachlin, Ottawa; and Warner H. Peberdy, Rugby, England.

Before the end of the month, these ten original pupils of the Curtiss Aviation School proceeded to England as a group, where they were posted to R.F.C. or R.N.A.S. stations for additional training before going on active service.

Douglas Joy had been an officer with the Ninth Mississauga Horse at a prisoner-of-war camp at Kapuskasing, Ont., when he requested his Commanding Officer, General Sir William Otter, to allow him to go overseas to join the Royal Flying Corps. His wish granted, Joy proceeded to Ottawa where he was interviewed by the Chief of Staff, accepted, and given his transportation abroad. As it was expected that he would be the first airman to be sent overseas to be trained with the R.F.C., Joy was interviewed by the Duke of Connaught, then Governor

FIRST TRANSPORT PILOT'S
CERTIFICATE

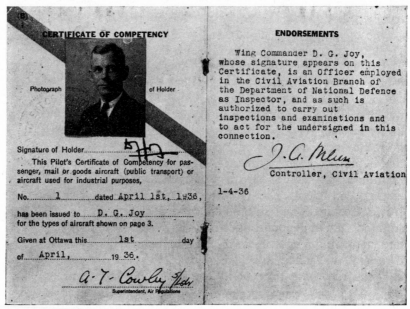

111

CURTISS JN (JENNY)
Used at the Flying School at Long Branch

General of Canada. During the ensuing conversation, it came out that Joy had never seen an airplane! So they took away his transportation vouchers and sent him to the newly formed Curtiss Aviation School instead. He quickly adapted himself to the air, went overseas, and served with distinction for the duration. He continued his career in flying; in 1936 he was granted the first transport pilot's certificate in Canada.

Of the others,* all their names are known, with the dates they graduated and the branch of the flying services they were absorbed into overseas. Thirty-five out of a total of 129 died for their country.

Several outstanding instructors were responsible for the fine record of training at the Curtiss Aviation School. Victor Carlstrom came to the Toronto school with a fine flying record already established. He was born in Sweden but had become an American citizen. He first began flying as a Curtiss pupil at San Diego, Calif., and received his A.C.A. licence No. 144 in 1912.

One of his greatest flights was on November 15, 1915. He took off from Toronto to fly to New York, in an endeavour to establish a cross-country record of some 485 air miles in one of the school's wheel-equipped machines. It was the first flight of any importance to be made between Canada and the United States. His route the first day followed a course via Hamilton, Ontario, then across the international boundary into New York state where he made landings at Bath, Elmira, Binghamton, Port Jervis, Cornwall, and Ridgefield Park, N.J. Fog stopped him only once on the hop, when he landed at Binghamton to verify his whereabouts. The spot he landed in at Ridgefield Park was so small that he could not fly the craft out. It

*See list of names in Appendix E.

had to be dismantled, moved to a larger area near by, and reassembled before he could continue on the 16th. His total flying time for the entire flight from Toronto to New York was 6 hours and 40 minutes. Carlstrom made many American aviation records from that time onward, chiefly altitude flights. He met his death in a crash near Newport News, Va., on May 10, 1917.

T. C. Macaulay, a flying-boat pilot, also came to the school well recommended. His brief career in Toronto as the pilot of the flying-boat, *Sunfish*, has already been described.

John Guy ("Goggles") Gilpatric was both a flying-boat and a land-plane pilot. He passed his tests on a Deperdussin monoplane at Garden City, Long Island, N.Y., September 25, 1912, and received A.C.A. licence No. 171. He became well known as an airman and instructor with the Curtiss Company in the United States. After leaving the Toronto school, Gilpatric entered the American air service as an instructor. Later, he became one of America's well-known writers, the creator of "Mr. Glencannon" of the *Saturday Evening Post* stories. He is now deceased.

Victor Vernon came to the school chiefly to instruct on flying-boats. As holder of A.C.A. Seaplane Pilot's certificate No. 27, he was well fitted for the job. Of the surviving instructors of the school, he is the only one actively engaged in aviation today. In 1948 he held the position of personnel director with American Airlines, Inc.

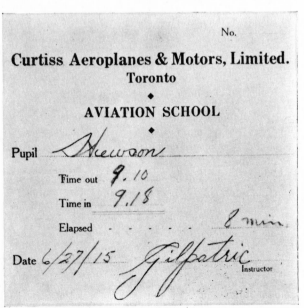

CURTISS AVIATION SCHOOL FLIGHT VOUCHER

Stephen MacGordon, one of the part-time instructors, came as well recommended as the others. As pilot and owner of a flying-boat, he possessed A.C.A. Seaplane licence No. 10. He became an instructor in the aviation section of the United States Signal Corps in 1917. MacGordon lost his life at the same place at which Carlstrom was killed, when a machine he was flying over Newport News caught fire, crashed, and burned.

Antony Jannus earned A.C.A. licence No. 80, flying a Benoist flying-boat at St. Louis, Mo., on December 27, 1911. He did not give much instruction to students while with the Toronto school, but as chief engineer and test pilot for all the aircraft turned out at the Toronto plant, his position was important. In 1916 he went to Russia with a number of Buffalo-built Curtiss flying-boats. While he was flying one of them from the waters of the Black Sea, near Sevastapol, on October 12, 1916, the machine crashed and burned, taking him to his death.

One of the foremost mechanics of the school was Bertram Acosta, who had taught himself to fly in California several years before. When with the Curtiss Company at Toronto, he was given some flying instruction, and allowed to go solo, although he was not a student of the school. In after years he became world-famous for spectacular flying achievements, and as test pilot for numerous firms. He was one of the pilots with Admiral Richard Byrd during the non-stop flight across the Atlantic in 1927. Acosta also flew in the Spanish War in 1937.

John A. D. McCurdy, as manager of the Canadian branch of Curtiss Aeroplanes & Motors, Limited, and the Curtiss Aviation School in Toronto, should be remembered as having a leading part in helping ambitious Canadians to represent their country with distinction in the grim days of 1914–1918.

To many who survived the ordeal, it was the first step in careers which subsequently were closely associated with aviation in Canada.

THE AERO CLUB OF BRITISH COLUMBIA, 1915–1916

The Vancouver flying school was very different from the one near Toronto. It began as the work of one airman in co-operation with a group of Vancouver business men. It proved to be a praiseworthy but not very practical undertaking on the part of patriotic civilians.

CURTISS FLYING-BOAT AT TORONTO ISLAND
Left: Grant Gooderham, and (*sitting*) Strachan Ince

The key man was William M. (Billy) Stark, Vancouver's pioneer airman, whose story is told in an earlier chapter. Stark had learned to fly in 1912 at the Curtiss Flying School in San Diego, Calif., and had brought back a Curtiss biplane, which by 1915 was still in good condition.

In the summer of 1915, Stark and a number of Vancouver business men met to discuss the possibility of forming a flying school to train pilots. Among those who attended were J. W. Pattison, who became the secretary, J. H. Senkler, K.C., W. E. Burns, Judge Schultz, and H. O. Bell-Irving, Sr. The outcome of the meeting was an application to the Fédération Aéronautique Internationale in France for official recognition and a charter, which were promptly granted. The Aero Club of British Columbia thus officially came into being.

With funds raised by public subscription, the club purchased Stark's aircraft at a cost of $2,500 and retained the airman himself as instructor at a fee of $25 a week. Letters poured in from hopeful applicants, and from them a number were selected who became the first pupils. Each pupil was expected to pay a fee of $200 to help defray expenses. The club had the assurance of the Canadian government that all students who graduated as pilots would be given full assistance to go overseas, for entry into either the R.N.A.S. or the R.F.C.

Training began near what was then the Minoru Park race track on Lulu Island. Through the kindness of Imperial Oil Limited, 200 gallons

EARLY FLYING INSTRUCTORS AT TORONTO

Left: Victor Carlstrom, Senior Pilot Instructor   *Right*: Victor Vernon, Flying-Boat Instructor

of gasoline and 20 gallons of lubricating oil were donated for the use of the club.

As the aircraft was only a single-seater, instruction was very slow. The method adopted showed ingenuity and boldness. First the student was allowed to taxi slowly up and down the field. During the preliminary runs, Stark would sit on the leading edge of the lower wing, close to the student who was at the controls, and yell instructions as the machine bumped and bounced across the field. Each pupil was provided with a wooden wedge with his name thereon, which was fixed firmly under the foot throttle control while its owner was operating the machine. As a student became more proficient, Stark would shave down the wedge, thereby making it narrower and allowing slightly increased throttle and more engine power. This method prevented a pupil from giving the motor too much energy, either from accident or intention. Instructor Stark took charge of all wedges when not in use.

It soon became apparent that the field they were using was too small for the lengthy taxiing required for training, so another site was located about a mile away. The ground was harrowed until it was very smooth, and a hangar, 30 by 40 feet, was erected. The new field was christened Terra Nova, which was the name of a near-by salmon-canning factory on the banks of the Fraser River.

Training began in September and continued until late in November, when winter conditions put a stop to activities at that location. Before that time, two of the pupils had graduated. They were Murton A. Seymour and Phil H. Smith, and they were passed by Lieutenant Colonel C. J. Burke of the Royal Flying Corps, who had come out from England to interview prospective Canadian applicants. At Terra Nova Field, the Colonel was surprised when shown the 1912 vintage training plane which the students were using; nevertheless, Seymour and Smith did their stuff in the air very creditably and the Colonel passed them as applicants for the Royal Flying Corps.

During the winter months, the aircraft was moved to the motorboat works of the Hoffar Brothers, on Burrard Inlet. Here, with the generous help of the Hoffars, the craft was converted into a hydroplane. Bad weather, however, prevented much use of the seaplane for training purposes and the climax came when the main float of the craft struck a drifting log while taxiing at considerable speed, with Stark and a young trainee aboard. Men and machine received a complete ducking, but the men were quickly rescued by the crew of the West Vancouver ferry, *Sonrisa*. The aircraft was salvaged, but it was a sorry mess, and no further training of any kind took place until the following year.

By the time the accident happened some of

the embryo pilots were running out of funds. To tide them over until spring, arrangements were made with the Commanding Officer of the 58th Battalion, then recruiting in Vancouver, to place them on the strength of his unit. This entitled them to draw pay and allowances, while still remaining actively connected with the Aero Club of British Columbia. Arrangements were also made for them to study wireless and nautical astronomy under the tuition of Captain Ward of the North Vancouver 6th Field Engineers.

As the old Curtiss was in bad shape after its dip in the ocean, and funds were practically non-existent, the Aero Club was completely reorganized. It emerged in December 1915 as the British Columbia Aviation School, Limited, and stock in a patriotic and non-profit venture was offered for sale. Additions to the original group of directors were Sir Charles Tupper, Hon. James Duff, H. H. Stevens, M.P., C. H. Cowan, K.C., G. C. Crux, T. F. Hamilton, G. E. McDonald, and William B. Burnside. J. W. Pattison continued as secretary, giving unstintingly of his time.

Expectation was high and plans were very ambitious. It was hoped that $15,000 could be raised and an approach was made to the Hamilton Company in Seattle, Wash., to provide four aircraft of different types. But funds did not pour in as had been expected, and the planes were never bought.

One machine was, however, constructed in Vancouver—a two-seater military-type Curtiss tractor. A tag-day was then arranged in the city, during which the new airplane, hauled by manpower along down-town streets, led a military parade. This effort brought in about $1,500.

The new machine was finally taken to a wide area of farm land at Pitt Meadows near Coquitlam, about 25 miles east of Vancouver. Although the pupils received a great deal of ground instruction in 1916, no additional graduates passed officially, but the experience that they received stood them in good stead when they eventually became members of the R.N.A.S. or the R.F.C. Additional pupils in 1916 were Norman Drysdale, Knox Martin, "Happy" Ray, and Jack Wright.

Stark had severed his connection with the Aero Club in the winter of 1915, and two other pilots were engaged, both from the United States. Although they both came well recommended, neither Clarence Hilborn nor N. B. Robbins was at that time a licensed pilot.

The Curtiss tractor biplane was finally cracked up at Pitt Meadows, while under, or more probably while not under, the control of Charles Raynor. Since funds were at a low ebb, the accident brought to an end the active career of the school.

Although the work done by the organization could not be considered highly successful, it was a beginning, and the effort was generously commended. It was evident, however, that work of such magnitude, importance, and expense called for greater financial resources. The careers of the 15 charter pupils of 1915 justified the effort. All of them later served in some branch of His Majesty's armed forces. Their brief biographies follow.

Arthur H. Allardyce was gazetted a Flight Lieutenant in England in June 1916. He served with the R.N.A.S., saw considerable service on submarine patrols over the North Sea, and later was selected as one of the first instructors at the Manston Flying School. He was also one of the original group who were sent by the British Admiralty to report on Colonel Smith-Barry's special system of training at Gosport. In the later years of the war he carried out experimental work on deck landings and development of rockets for defensive work against observation balloons. He also chalked up time for night flying in BE2C's. After being demobilized, Allardyce entered the newspaper world and became well known as advertising director of the Winnipeg *Tribune*. He continued to keep in touch with various aviation organizations in Canada.

"Chad" Chadwick did not graduate from the

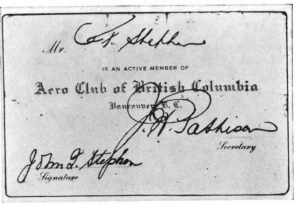

MEMBERSHIP CARD OF THE AERO CLUB OF BRITISH COLUMBIA 1915

club, but later enlisted in a militia unit. He lost his life during the Mesopotamia campaign.

J. B. Crawford was an American national, residing in Vancouver at the time of his affiliation with the club. His age prevented his entry into the air services, so he enlisted with the Canadian armed forces and went overseas.

W. E. Damer was with the club from its inception, but he did not graduate as a pilot. He enlisted later in the Royal Canadian Engineers, and served in dispatch boats in Mesopotamia. He returned to Vancouver in 1919.

Gerry Hodgson did not graduate as a pilot in Canada, but was sent overseas, and became a Flight Lieutenant with the R.N.A.S. He was fortunate in realizing the ambition of all R.N.A.S. personnel—sinking a German submarine—for which exploit he received the D.S.O. Later he was forced to make a crash-landing in neutral Holland, where he was of course, interned. He lost a leg in this accident, and on his recovery was paroled to England, where, since he could not fly again, he was able to continue his law studies. In later years he became one of Vancouver's best-known lawyers.

Robert Main eventually entered the Royal Flying Corps in Canada in 1917. After training, he was commissioned a Second Lieutenant, and proceeded overseas. He was soon fighting in France, and on November 23, 1917, he crashed behind the German lines. He suffered very severe injuries and the loss of a leg, and was a

prisoner of war until hostilities ended. After his return to Vancouver, he became one of the members of McCleery and Watson, Ltd. of that city.

William Gordon McRae was born at Agassiz, B.C. Billy also had to await the coming of the R.F.C. to Canada before his ambition to become a fighter pilot was fulfilled. He went overseas as a Second Lieutenant and for a time was posted with the 39th Reserve Squadron at Montrose, Scotland. Afterwards he was sent to France as a pilot with a Spad fighter squadron. Three short weeks later he was listed as missing. It was believed that McRae's machine collided with one of the enemy's and that both crashed, killing the pilots.

Sidney Mowat left the Aero Club before graduating, enlisted with the Canadian Army Service Corps in 1916, and finally got into the Royal Flying Corps at Cairo, Egypt, where he became a Second Lieutenant in May 1917. For thirteen months he served with distinction in the eastern forces, and was then posted back to Canada to become a flying instructor. Between World War I and World War II, Mowat was in civilian occupations, but on May 21, 1941, he entered the R.C.A.F. and served as an instructor in Canada for the next four years.

Charles ("Cy") Perkins became a cadet with the R.F.C. at Toronto in 1917. After being commissioned, Lieutenant Perkins served as an instructor all through the war, chiefly at Camp

CURTISS PUSHER BIPLANE OF THE B.C. AERO CLUB, 1915
Left to right: William McRae, Cyril Scott, Instructor Billy Stark, Philip Smith, Murton Seymour, and Vick Phillips

VOLUNTEERS IN TRAINING
With the second aircraft, built by the
Aero Club of British Columbia

Borden. In 1918 he suffered severe injuries in a crash but after months in hospital he resumed flying as a stunting instructor. He was demobbed in 1919 and returned to British Columbia, to his home at Port Kells.

J. V. W. ("Vick") Phillips of New Westminster, B.C., was a very active member with the original Aero Club until 1916, when he went overseas, joined the R.F.C., and became a Second Lieutenant. Of the entire group of original Aero Club members, he was the first to see action and the first to be wounded. He was shot down over Allied territory on September 2, 1916, and suffered the loss of a leg.

Charles Raynor was an active member of the club until his enlistment in the Royal Canadian Engineers late in 1916. He lost his life in a motorcycle accident while on active service in England.

Cyril Scott left the Aero Club in 1916, got himself a job on a freighter, and sailed for England. On arrival, he immediately enlisted in the artillery. In less than three weeks he was in France. He was attached to the 2nd Division Supply Park, loading and unloading ammunition—certainly a far cry from flying! During the first battle of the Somme, he and his fellows worked night and day handling the stuff, sometimes going for 15 to 22 hours at a stretch, to make sure the vital ammunition was kept on the move. After six months of this, Cyril was invalided out of the army as unfit for further service. He then worked in England for a large film supply house, returning to Vancouver at the end of the war. When World War II broke out, Cyril again came forward and served in the C.M.S.C. for duration.

Murton A. Seymour was the first student to graduate from the school; he received his certificate in November 1915. He was sent overseas for further training, and later received a commission with the R.F.C. After considerable work as an instructor, he was posted back in 1917 to assist in organizing the R.F.C. throughout Canada. He remained on the Toronto headquarters staff for the duration, reaching the rank of Captain. Murton has been actively connected with civilian aviation in Canada throughout the peace years. He was awarded the McKee Trophy for 1939 "for the leadership he gave the Flying Clubs of Canada" in that year.

Philip H. Smith graduated from the Aero Club on the same day as Murton Seymour. He went overseas with Seymour and was commissioned as a Second Lieutenant in the R.F.C. He survived the war, but died in the crash of a machine in which he was stunting during the Armistice celebrations.

John F. Stephen was with the club from its inception but, like many of the others, he had not reached the solo stage of instruction before things folded up with the dunking of the training plane in Burrard Inlet. John thereupon signed with the Army Supply Corps, Motor Transport (Imperial), and after proceeding overseas, served with them for three years, mostly in France. In the final year of the war, Stephen obtained a transfer to the R.A.F. shortly after being posted to Egypt, but before he could earn his coveted wings, the war ended. He returned

117

to Canada, eventually settling at Oyama, on Long Lake in the Okanagan Valley of British Columbia, where he became a fruit grower.

Canada's first two aviation schools, though of short duration, were a means of concentrating in two Canadian cities the flying skills and varied experiments of a significant number of North America's most competent airmen. From that point on, the self-taught pilot, who learned things—if he lived—the hard way, was an anachronism. The gap in time between the government's awareness of a need for trained aviators, and its provision of full facilities to satisfy that need, had been most competently bridged. With the arrival from England of the spearhead of the Royal Flying Corps staff to set up a Canadian military air training centre a new chapter in Canadian aviation began.

AN EARLY "WAR BIRD"

Six weeks after World War I began, Canada's Minister of Defence, Colonel Sam Hughes, cabled the British government asking if the services of aviators were desired, and received the reply that six expert pilots could be used. As Canada had neither military aircraft nor military pilots the suggestion was somewhat premature. However, the Minister at once formed the "Canadian Aviation Corps," of which Mr. E. L. Janney was appointed provincial commander, and commissioned a Captain. As a plane was a necessity to complete the one-man air force and none of a suitable nature was available in Canada, Janney was authorized to proceed to the United States to procure one. The government allocated $5,000 to buy a plane and accessories, and after visiting several centres in the eastern states Janney was able to obtain a second-hand, Burgess-Dunne, tail-less float-plane from the company factory at Marblehead, Mass. Because immediate delivery was imperative, the firm sold him the only machine they had.

As Canada's new air commander was not a pilot, company pilot H. J. Weber was assigned to fly the craft to Canada. It was first shipped by train to Isle la Motte, at the southern end of Lake Champlain, Vermont, where company mechanics assembled it for flight.

With Canada's Valcartier Military Camp, Quebec, as their objective, the two men set off on September 21, 1914, following the Richelieu River. An 80-mile flight took them to the hamlet of Sorel, where they stayed the night and the next day they crossed the international border into Quebec. History was made: it was the first delivery of a military airplane from the United States to Canada. Over the St. Lawrence River, they turned west, only to experience engine trouble and to be forced down to the river for a landing. After drifting for thirty minutes, they were towed by motor boat to a wharf at Deschaillons, where three days were spent fixing up the motor and completely changing the fuel in the tanks which had been taken on at Sorel and was of inferior quality. The final 80 miles of their journey was completed on September 26 with a good landing at the city of Quebec.

There was just time to have the seaplane dismantled and put aboard the liner *Athenia* before the ship sailed for England carrying the troops of Canada's First Contingent. The Burgess-Dunne company pilot was replaced by a Canadian, Lieutenant W. F. Sharpe, who, together with the Provincial Commander, proceeded overseas with the military forces.

The sequel is disappointing and tragic. On arrival in England, the airplane was at once placed in storage by the authorities, as not suited to military purposes, and its war service was over even before it had begun. Without an Air Corps to command, Captain Janney was ordered to return to Canada, and Lieutenant Sharpe was attached to the British Royal Flying Corps for flying training. On February 4, 1915, he lost his life in an air accident, at Shoreham Aerodrome, to become Canada's first military airman casualty in World War I.

THE FIRST CANADIAN WAR PLANE

The Burgess-Dunne seaplane which accompanied the First Contingent, Canadian Expeditionary Force, September 30, 1914

15. The Royal Flying Corps in Canada, 1917-1918

In the spring of 1914 Jean Marie Landry of the Province of Quebec was so determined to fly that he packed up his belongings, and accompanied by Mme Landry, journeyed across the Atlantic to France. He at once enrolled in the Blériot Flying School at Buc, and was taught to fly in a Blériot monoplane under the supervision of the famous Louis Blériot himself. He passed his tests in a few weeks and received a Fédération Aéronautique Internationale Licence, No. 1659, issued by the French Aero Club, and dated June 20, 1914.

The outbreak of World War I caught M. Landry in France, and he enlisted under the French flag. He was one of the earliest pilots to go into battle flying Blériots and Morains and was probably the first Canadian to fly in World War I.

As explained earlier, there was only one alternative to the course taken by Landry for Canadians who wished to become war birds early in the war: to enlist in a regular military unit and apply for transfer to the flying services on reaching England. This was a risky procedure for those who wanted, above everything else, to fly. A striking example of how difficult it was for airmen to get accepted for flying early in the war is the experience of William Fray, who, like Landry, flew early enough to qualify as an Early Bird.

Bill Fray had been a fireman with the Canadian Northern Railway working out of Mirror, Alta., for a number of years. A native of Missouri, he had never got around to taking out naturalization papers. By the spring of 1914 the urge to fly overcame even the fascination of flunkeying to a powerful but earth-bound locomotive; and May of that year saw him enroll at the Glenn Martin School of Aviation in Los Angeles, on leave of absence from the railway. He quickly became adept, and made his first solo in June. On July 12, 1914, he passed his flying tests for his Aero Club of America certificate, receiving No. 306, dated September 16, 1914.

Meanwhile war had been declared. Immediately upon his return to Canada, where he again took up his duties with the railroad, he offered his services as a pilot to the Canadian government. Americans were not being recruited at that time, and he was rejected. When he tried again a year later, he received the same reply, with the added information that, at thirty-three, his age precluded any possibility of his being accepted! So Bill stayed on his job, working up to engineer, later taking out full Canadian naturalization papers.

In spite of all obstacles a surprising number of Canadians actually succeeded in transferring early into the Royal Flying Corps in England, where they were speedily trained and posted to active duty in France and other war zones.

There is little doubt that the early prominence of Canadians who trained in the R.F.C. overseas, making a name for themselves in the air, turned official eyes to Canada as a source of additional excellent flying material. It will be recalled that in November 1915 an R.F.C. recruiting officer, Lieutenant Colonel C. J. Burke, had crossed Canada looking for eligible recruits,

JEAN MARIE LANDRY'S LICENCE

and had examined the first two graduates of the Vancouver school. Blakely and I had received word from Ottawa that Burke would be stopping off in Calgary and would interview us there when he arrived. We never did find out whether he stopped off or not; in any case we missed him and so lost our earliest and best chance of becoming airmen overseas.

R.F.C. ESTABLISHED IN CANADA

Towards the end of 1916, a rumour began to circulate that the R.F.C. was going to recruit and train men in Canada. The rumour became fact when the advance group arrived in Toronto on January 22, 1917, followed, at the beginning of March, by the first of the many administrative and instructional personnel that were brought over. All of these men were highly trained in their duties, which included everything from administration and clerical work to mechanical transport, aerial gunnery, photography, and innumerable other jobs. Headquarters was established in Toronto, and recruiting officers were dispatched to large cities throughout Canada, and to several major points in the United States such as New York, Chicago, San Francisco, and Los Angeles.

Long before recruiting got under way, plans had been approved by the Canadian Parliament for the construction of airdromes and equipment, and things began to hum. Beginning in January 1917 the work of preparing the ground

and putting up buildings was started at several sites chosen as flying fields—Camp Borden, Long Branch (where squadrons X and Y were at first stationed), Leaside, Armour Heights, Rathbun, Mohawk, and on the School of Aerial Fighting (called at first the School of Aerial Gunnery) at Beamsville, Ont.

During the winter of 1916, the Canadian government had sponsored the sending of a number of university graduates to a flying school at Pensacola, Fla. Some of these men returned to Canada the following spring to join the R.F.C., and concluded their training in Canada with Squadrons X and Y at Long Branch. Under the command of Captain Doré, they formed the vanguard of military trained pilots of the Corps in Canada.

The area at Long Branch soon proved to be too confined and too muddy for mass flying training, and the school was moved to Armour Heights, north of Toronto. Captain Doré flew the first machine to land at Armour Heights airdrome, but because the ground was sodden, his machine tipped up on its nose and was badly smashed. Thus did that busy flying-field-to-be make its début.

At that time the airdrome at Leaside was not yet finished and no one who saw it then could have believed its later flying activities possible. The enlisted men moved into the top storey of the munitions factory which was to be their barracks while shell cases were still being made

WILLIAM FRAY'S LICENCE

FÉDÉRATION AÉRONAUTIQUE
INTERNATIONALE

AERO CLUB OF AMERICA

No. 306

The above-named Club, recognized
by the Fédération Aéronautique
Internationale, as the governing
authority for the United States of
America, certifies that
William Fray
born 27th day of January 1882
has fulfilled all the conditions required
by the Fédération Aéronautique
Internationale, for an aviator pilot,
and is brevetted as such.

Dated Sept. 16th 1914
Alan R Hawley
President
Howard Huntington
Secretary.

Signature of pilot:
William Fray

below. Hordes of civilian workmen operating dozens of graders, ditch-diggers, and steam-rollers, quickly put the finishing touches to the airdrome, as scores of carpenters completed the hangars and buildings. Leaside blossomed into a fine airdrome, with nine hangars capable of housing three squadrons—a squadron, apart from personnel, comprised three flights, A, B, and C, each with its own hangar and six aircraft, making a full complement of 18 machines to a squadron. Other large structures sprang up, including the wing headquarters building, officers' and cadets' mess and quarters, a garage capable of housing 20 large motor vehicles; an airplane and engine repair shop, and numerous smaller buildings; all of which rose from the ground in the short space of 7 weeks! The same feat of incredibly rapid construction was repeated at all the airdromes in order that training should get under way as quickly as possible.

Recruits streamed in from all parts of the continent including Mexico, and some came from as far away as the West Indies and Hawaii. Those who were accepted as cadets in 1917 were housed in the beginning in buildings at the University of Toronto, where they also received their primary ground training. Enlisted men first went to the old Givins Street School in Toronto, which had been fitted out as a barracks, and there the rookie cast aside his civvies to don a stiff new uniform, with its distinctive double-breasted tunic, "cheese cutter" cap, and the cane swagger stick which all enlisted ranks were obliged to carry when wearing their walking-out uniform. Many a young fellow around five foot four in height received underwear or other apparel marked clearly "for men 5 ft. 10 ins., to 6 ft. 4 ins.," but that was a minor detail! In any event the underwear was of pure wool, heavy enough to stand up by itself. It was very handy to sleep in, for the waistband came up to the armpits, making an almost perfect sleeping bag!

BADGE OF THE ROYAL FLYING CORPS

All operational recruits, irrespective of their trade categories, were classed as air mechanics, known better as A.M.'s, and unless they later earned the rank of corporal or higher, air mechanics they remained in name, whether they were clerks, cooks, policemen, truck drivers—or actual mechanics!

All those joining a mechanical branch were allowed to choose the work they considered themselves best fitted for, but budding "ack emmas" (A.M.'s) had to undergo preliminary tests at the old Toronto Armouries under British R.F.C. experts, whose keen appraisals quickly determined if the applicant was really suited to the branch of work he had selected.

The test over, recruits were bundled off to the training camp at Leaside. There, for three weeks or so, they slept under canvas, suffered daily drill, and more drill, with lots of "physical jerks" and plenty of kitchen and sanitary fatigues. As rookies were licked into shape, they were sent off to various points to fill the hundreds of jobs that awaited them, and the Royal Flying Corps swelled towards its final training strength of well over 10,000 men. When the Corps was fully organized, the demand for air mechanics ceased, but the call for cadets for

flying training went on unabated up to the last day of the war.

CADET TRAINING

By this time, Long Branch had been converted into the initial housing and training camp for cadets, where they were outfitted with uniforms, then formally introduced to the frequently choleric British non-commissioned officers, who sought to stiffen them up by running them off their feet.

Cadets were formed into classes and sent off to the University to take up residence and receive their ground schooling for the flying to come. Life took on more interest as they were initiated into the mysteries of aerodynamics, elementary gunnery, engine and airplane construction, and so forth. About six weeks was the average period of tuition before cadets were considered to have absorbed sufficient knowledge to be sent on for flying instruction. Class examinations were rigid, and the student who flunked had "had it"!

After a short leave, cadets were posted to various squadrons, where several were placed under the care of each flying instructor, and the long-anticipated flying at last began. Student

LEASIDE AIRDROME
An aerial view showing a plane just taking off (*centre*)

HAZARDS OF THE SERVICE

Left: The cow has fared worst in this forced landing in 1918
Right: Although the airplane is a total wreck the pilot has emerged unscathed

flyers were rushed through their training efficiently, but with what might now seem like suicidal haste. Many a lad took off nervously on his first solo after less than six hours' dual instruction. But the demand for pilots from overseas was cruelly urgent.

A cadet first received dual flying instruction. As soon as he could handle the machine in the air, and make landings without bouncing too much, he was dispatched on his first solo. Here it may be pointed out that the most critical event in a learner's life was not so much the first take-off and flight, but the first solo landing. It was indeed an achievement to bring your roaring Juggernaut back to Mother Earth, without splitting it and her wide open.

Intermingled with solo flying was more dual flying, with stunting of every description. Later still came aerial photography, cross-country flying, and finally, a stiff course in aerial gunnery. By the time the young airman had passed his final tests he knew how to fly.

The greatest day of all was when the white cap-band of cadet days was reverently removed from the old cap and the commission as a Second Lieutenant became an actuality, with all the trimmings that went with it, shoulder "pips," "wings," and Sam Browne belt.

After a couple of weeks' leave, the majority of these young men went overseas. A few, much against their will, were retained in Canada, as additional flying instructors were urgently required. The instructors never did receive their just due—training rookies to become pilots in double quick time was a risky business at best and could be dangerous in the extreme.

SOME CAMP CLOSE-UPS

The efficiency and speed with which training was effected by the R.F.C. was not achieved overnight, however. A great many applicants all over Canada, unless they were handy to the newly set-up recruiting centres, cooled their heels through the earlier months of 1917. The contrast between Tom Blakely's experience and my own is an instance in point.

Early in 1916 Tom parted ways with his employers, and started up a small box factory. When this didn't succeed he returned to Winnipeg. As soon as the R.F.C. recruiting station opened there he was camping on the doorstep, and early in 1917 he got me travel vouchers and instructions to report at Winnipeg. By the time I reached the 'Peg Tom had gone. We met briefly in Toronto, but by November he was overseas. Even then he never qualified as a fighter pilot, and spent most of the last year of the war on submarine patrol out of Newcastle over the North Sea. Curiously enough, the narrowest escape he had was when, on returning from a patrol, he had to make a forced landing at a bathing resort. The crowd on the beach made a ground landing impossible, so he smacked into the shallow water, demolishing his plane in the process. At that he was lucky. He had aboard his machine two live bombs, the detonating charge of one of which was, freakishly but neatly, sheared off in the smash-up.

In my own case, though I reached Toronto only a few weeks after Tom, I was not to fly for many months. Apparently—I never did learn exactly what happened—my medical was unsatisfactory, and I found myself designated

ONE-WAY COMMUNICATION BY VOICE PIPE

This device permitted the instructor to talk to his pupil while in the air. Author Ellis at the mouthpiece speaks to Air Mechanic Bill Thorburn

"Motor Cyclist," and had a wonderful time speeding about the Ontario landscape as a sort of glorified messenger boy, but getting exactly nowhere as an airman. It was not until the summer of 1918—perhaps because of a relaxation of rigid standards in the face of urgent need —that I was finally admitted for flying training.

I well remember my own first impressions of the R.F.C. organization. Spring was in the air when the train rolled through the Ontario countryside approaching Toronto, a perfect setting of peace and rural contentment which I have never forgotten. The war seemed far away. As a matter of fact, as the year passed and grew into 1918, many men who had voluntarily enlisted in the ranks of the Corps, with the expectation of quickly going overseas, soon became disappointed, and to them also the war seemed more a myth than anything else.

My first impression of Leaside Camp, a mile north of the airdrome, was not very favourable, and as its discipline descended upon me I was none too happy. However, when a young man is thrown together with a number of fellows his own age, all in the same boat, so to speak, it doesn't take long to get into the spirit of things.

Soon I was up at the crack of dawn with the best of them, hurrying through ablutions right out in the open, and then off for a half hour of physical jerks before breakfast was "served." Served, did I say? Well, call it what you wish, but when you finally received your share after a long stand in line it wasn't always too satisfying. Fortunately, I had camped out quite a bit and roughing it came easy. In any case, if you didn't like the grub, you didn't have to eat it, and the solution was to fill up with cake and milk later on at the canteen near-by.

But for a young man whose boyhood had been spent dreaming of flight in heavier-than-air machines, it was a real thrill to watch the cadets try out their wings in the trim Curtiss biplanes flying round and round the distant airdrome. Even though I was not flying I felt a bit repaid for joining, and resigned myself to the delay, little realizing how close to the end of the war it would be before I was sitting in the cockpit of a Jenny for my first solo.

Generally speaking, the enlisted men of the R.F.C. of those days and the R.C.A.F. today were cut from the same cloth. But conditions certainly have changed. The food undoubtedly has improved—as well it might—though I don't suppose the grumbling is less. But in the matter of discipline...!

Let me illustrate. Sooner or later it fell to everyone's lot to be placed on guard duty, and the one and only time that happened to me, the result was farcical. I had never held an army rifle in my hands until the morning of the day I was assigned to guard duty, and five other lads dedicated to the same job were no better qualified. The British sergeant who did his best to whip us into shape for the ordeal certainly earned his pay that day. When he turned us loose, we knew about as much of rifle drill and correct guard procedure as an equatorial native might know of a snowball.

Nevertheless, on guard we went. I, for one, kept very much alert, not because I was zealous but chiefly to keep clear of any high ranking officer who might come along, when I should have had to salute him with a presentation of arms. At the merest inkling of the arrival of any V.I.P., I adjusted my steps so as to be well around a near-by corner of my beat when he passed by.

My first taste of real discipline came in 1917 when least expected. I had been ordered to do some minor job or other by a fellow who ranked just one class higher than I. He was a 1st Class Air Mechanic, I a 2nd Class. Not liking his attitude, I told him where to go, and things seemed settled. The next morning I found out differently. It appeared that I had almost disrupted the war by disobeying the command of a senior soldier. I was hauled off to headquarters office, between two stalwart military policemen, and paraded before the Officer Commanding the wing. As I marched into his office, stiffly erect, some unseen hand whipped off my cap; it is a heinous crime for a prisoner to wear one when going before the "great man." The outcome was sudden and shocking: my crime was read, and without further ado the O.C. stated brusquely, "Seven days." I must confess that I enjoyed my spell in durance vile, as there was absolutely nothing to do in the "clink," with one exception. It was the winter of 1917 at the time, an old-fashioned Ontario winter, and everything was frozen solid. Near the guardhouse an unfortunate dog had been killed by a passing truck; I was delegated to bury it. I tried, but as the pick and shovel were only made of steel and I was not made of iron, I found myself unequal to the task. Finally, the guard and I buried the dead animal in a snowdrift. I have no doubt that when it emerged from its grave with the advent of spring my superiors were not slow to get wind of its presence, and some unhappy later inmate of the Leaside jail had to give the animal a second and more durable interment.

When the United States entered the war on April 6, 1917, its air service was of minor proportions compared with those of the then warring nations. In July 1917, a detachment of 1,500 men from the Aviation Section of the U.S. Signal Corps arrived at Leaside Camp. They were soon distributed throughout the various units of the R.F.C. to receive initial training as enginefitters, riggers, and so forth. United States cadet members of the Aviation Section also came up for training under the R.F.C. After four months of instruction in Canada the unit was re-formed as a detachment, becoming the 22nd Aero Squadron, U.S.S.C. Then it left for duties under its own flag at airdromes which had been partly prepared for it in Texas.

The American boys were a good crowd, but many got into minor scrapes because they balked at the comparatively strict discipline of the R.F.C.

Once, a cadet named Freund, who had had only an hour's solo flying, took it into his head to pay a visit to his parents back home in Pennsylvania. While up on a solo hop one day, he turned his plane southward and headed out over Lake Ontario. He managed to reach Buffalo safely but then he had to come down to refuel. He had never before alighted anywhere but on an airdrome, and when he selected a swampy field to land in, his machine promptly nosed over on its back. Freund was soon missed at Toronto, but it was assumed he had met with an accident or had had to make a forced landing. It was not until he telephoned from Buffalo that the astounded camp officials learned where he was. As punishment, he was kept from going overseas for three months—a severe blow to a young airman impatient for action.

On another occasion, shortly after the Ameri-

WINTER FLYING
Early tests showed that our airmen could still carry on despite deep snow and extreme cold

can boys arrived at Leaside, several of them ran foul of regulations and were consequently confined to barracks. In revenge, they kidnapped the huge St. Bernard belonging to Lord Innes-Kerr, the camp commander. They shaved all the hair off one side of the big dog (it was hot weather, and he probably enjoyed it), painted his name, "Bruno," on the bare skin in large blue letters, and turned him loose. Next day the O.C. stormed about the camp, the "evidence" trotting serenely behind him, but the culprits were never found.

R.F.C.–R.A.F., 1918

By the end of the summer of 1917 the R.F.C. in Canada was functioning smoothly at full strength. Planes and engines had been arriving at the airdromes in increasing numbers. The manufacturers shipped them by rail and truck, and riggers and engine-fitters of the R.F.C. assembled them and put them into flying trim.

With the approach of winter, the authorities realized that it would be impossible to carry on flying training at Borden and Deseronto, which in winter were more or less isolated. So the airdromes were closed and their personnel sent down to Texas for continued operations—a big undertaking. From November 1917 to April 1918, the 42nd and 43rd wings of the Royal Flying Corps sojourned in the warmer, though windier, climate of the deep south, where flying went on unabated at Hicks, Everman, and Benbrook fields.

Although the 44th Wing stayed in Canada at Leaside and Armour Heights, it was not definitely known whether flying training could continue successfully during the severe Canadian winter. The answer was not given until the first real cold snap and a heavy fall of snow arrived in November. Captain Acland of 89th C.T.S. (Canadian Training Squadron), took off in a wheel-equipped machine from the Leaside airdrome, which was blanketed with an eighteen-inch fall of snow. The success of this flight clinched the government's decision to have the 44th Wing remain in Canada. Later, skis were designed and fitted to many machines; they proved very serviceable in deep snow. It was tough going for all ranks of the 44th that winter, particularly for those who had to fly, but regardless of frozen fingers and toes and frostbitten ears and noses, training went forward without interruption, much of it at zero temperatures, and all of it in open cockpits!

When the two wings of the R.F.C. returned from Texas to Canada in the spring of 1918, they left behind them 29 members who had lost their lives in accidents. Eleven of them were interred in the Greenwood Cemetery at Fort Worth, where a memorial to their memory now stands.

In 1918, with the full brigade strength back in Canada, a reshuffling took place; the 44th Wing was moved to Camp Borden, the 43rd took its place at Leaside and Armour Heights, and the 42nd Wing went to the Deseronto airdromes at Rathbun and Mohawk.

Cadet recruits were still pouring in, and the Jesse Ketchum School in Toronto had been turned into a receiving depot for them. There they were given uniforms and put through elementary drill; later they went to Long Branch for additional drill and physical training. Other sections which had been established and were in full swing in Toronto were the engine and airplane repair parks; a large mechanical transport garage; the stores depot; No. 4 school of aeronautics; and the armament school at Hamilton. The last two were for preliminary cadet instruction in aerodynamics and gunnery.

On April 1, 1918, the Royal Flying Corps and the Royal Naval Air Service in England were amalgamated and became the Royal Air Force, operating under one command. The R.F.C. in Canada was a unit of the Imperial Army (although all personnel received Canadian rates of pay), so the change in name became at once effective in Canada also. The reorganization brought some new regulations. From the R.N.A.S. came the custom of saluting the flag as it was raised and lowered, morning and evening; also the ringing of bells, shipboard style, to announce the hours; and the use of the nautical terms starboard and port instead of right and left. The continental method of marking the day from 1 to 24 o'clock, eliminating A.M. and P.M., was also adopted.

FURTHER CAMP CLOSE-UPS

Life in the flying camps was, of course, not all spent in the air. Sundays were entirely free from flying and all major duties, and could be spent away from camp. Evenings were enlivened with movies, concerts, sing-songs, basketball, and other entertainment, held in the spacious hall which was part of every airdrome. During 1918 large cement swimming pools were constructed at the different camps, the work of excavation being done entirely by cadets with shovels and wheelbarrows. Although this exercise was all to the good in strengthening flabby muscles, it was a subject of grousing with some boys, which at times was expressed in lusty singing of the following lines to the well-known tune of *Mother*:

> S is for the soup they often feed us,
> H is for the hash that we all know,
> O is for the 'orsemeat they put in it, and
> V's the voice of sergeants, sweet and low,
> E is for the end of our enlistment,
> L is for the last day of the war,
> Put them all together and they spell Sho-v-vel,
> The emblem of the Royal Flying Corps.

At one time, great rivalry existed between the office staffs of the different squadrons, each claiming it had the most beautiful garden patch in front of its office building. When the competition was at fever height, the adjutant of the 86th Squadron bethought himself of a near-by and apparently deserted apple orchard. He requisitioned a large truck, all the available shovels, and every cadet he could round up. They picked out a large tree, in full leaf, and laden with fruit. With superhuman efforts, the cadets uprooted the tree, lifted it from the ground, loaded it on the truck, hauled it to camp, and planted it in front of the squadron office. Next morning, however, before the adjutant had time to even lean out of his office window and pick himself an apple, an irate farmer appeared in camp, roaring for the return of his property. The boys had to dig the tree up again and restore it to its orchard home, and the interest in gardening fell off remarkably thereafter.

Another composition sung—or yelled—to the tune of the then popular song, *Smiles*, was also highly enjoyed:

> There are stunts that make you happy,
> 　There are stunts that make you blue,
> There are stunts that steal away your breakfast,
> 　As the waves of the ocean sometimes do,
> There are stunts pulled off for exhibition,
> 　For the eyes of all around to see,
> But the stunts that keep the Hun from off my tail,
> 　Are the stunts that appeal to me-e-e.

"Pulling" unauthorized stunts for exhibition, especially if lady visitors were in the vicinity,

THIRD TIME UNLUCKY

The wreckage of the plane crashed by Cadet Milton at Collingwood

was a temptation to which many youthful cadets yielded, and such flying, often as not, ended by getting them into hot water.

Young pilots too full of vim for their own good would indulge in stunts strictly off the record, and rarely were caught. One practice as far back as the summer of 1917 was for a pilot to wait in the air at the time the Toronto express for Barrie was due, and then to come in from behind at a fast clip, low down above the tracks. As he and his plane swept by over the length of the train, the crazy man at the plane's controls would deliberately bump his wheels on the roofs of several of the coaches before reaching the engine, when he would speed over the top and actually dip down in front of the fast moving train, flying directly in front of it along the tracks until far enough in front to swing away and yet not have his identifying numbers on the rudder seen by passengers or train crew. It was supposed to be great fun along a particular section of the C.P.R. track between Baxter and a small town named Ivy, where the telegraph wires and poles were amply spaced to allow for such mad exploits.

One lad, while supposedly on a high, cross-country flight, flew down over the town of Collingwood, on Georgian Bay. Spotting a large, bridge-type derrick in a shipbuilding yard, he decided to give the workmen a thrill by flying *under* it. Swooping low, he made a run for it, and put his plane through as neatly as could be. A second try was equally successful. A third

BRIGADIER GENERAL C. G. HOARE

time he tempted fate, but with dire results. A wing tip smacked one of the great latticed iron uprights, and down went the aerial gladiator, the machine striking the corrugated iron roof of an adjoining building with a resounding bang. The cadet suffered only slight injuries, but a good aircraft was totally wrecked. The culprit earned for himself a lengthy spell in the camp "clink."

Unintentional damage to aircraft was of daily occurrence, and minor injuries to pilots came often enough, but sometimes weeks would pass without a fatality; then there might be a number of them in quick succession. In the early summer at Camp Borden in 1918, to quote but one instance, 19 airmen lost their lives in flying accidents in 21 consecutive days.

Whenever there were planes in the air, a high-powered Packard ambulance was kept standing by, ready to speed away the moment a crash was reported. The ambulances were nicknamed "Hungry Lizzies," and there was work enough for them at all the airdromes.

During 1917 and 1918 accidents were, of course, numerous, and planes were demolished almost daily. Unfortunately but inevitably, there were numerous fatalities, intermingled with the minor crack-ups. Death from aerial misadventure claimed the lives of many pilot-instructors and student flyers, but considering the large total of hours the entire corps flew daily, the list of casualties was held to a low average.

Parachutes had not yet become part of a pilot's equipment at that period and it is well to bear in mind that in World War I not one Canadian-trained pilot even saw a parachute, let alone wore one while in the air. When a pilot went aloft in those days, he stayed with his machine no matter what happened; there was no choice.

The number of planes destroyed in accidents during training in Canada has been roughly estimated at 2,000. When the cost of these is added to the total cost of training 2,539 cadets (the number trained in Canada and Texas during 1917 and 1918 who went overseas), and an average is struck, the fact emerges that to transform a rookie into a first-class pilot cost approximately $9,800, "f.o.b. Halifax."

The official figure for the total hours flown by all Canadian training squadrons during their period of operation was 243,566. At the cruising speed of 75 miles per hour credited to the Jennies, that is an impressive number of air miles to chalk up!

When the war ended, the Royal Air Force in Canada was of creditable strength, with its three training wings; its cadet wing; its schools for aerial fighting, special flying, armament, and aeronautics; its repair parks, garages, and depots of various sorts. The General Officer Commanding the Force during its entire operation was Brigadier General C. G. Hoare, C.M.G.

On September 5, 1918, the Royal Canadian Naval Air Service was formed, completely separate from the R.F.C. or R.A.F. in Canada, and bases were established at Dartmouth and Sydney in Nova Scotia. Some of the flying-boats for R.C.N.A.S. use were built at Toronto, and accompanied the first patrol of a ship convoy out of Halifax on August 25, 1918, along with two United States flying-boats. Air convoying of ships was getting well established by the time the war concluded, and at the Armistice there were 14 flying-boats in use, operating out of Halifax.

In the fall of 1919, when the long work of demobilization was at last ended, and the men who had come out from England to organize it had gone back, the Royal Air Force in Canada came to an end.

By the end of 1919 the Canadian Air Force (the prefix Royal was not added until 1923) was put on a permanent basis. A year later three stations were added to the original Camp Borden centre: Vancouver, B.C., Morley, Alta., and Roberval, Que. By the end of 1920 the personnel of the Force at Camp Borden totalled approximately 70 officers and 300 men of all other ranks. This was the nucleus of the huge organization known today as the Royal Canadian Air Force.

The newly organized force was well established when a temporary gloom was cast over its members by the death of two of its most popular pilots. Stunting at low levels on April 1, 1921, cost the life of Flying Officer J. A. Le Royer, and ten days later a similar crash killed Squadron Leader Keith Tailyour. They were the first two fatalities suffered by Canada's newborn Air Force.

In the ranks of Canadian airmen overseas, casualties were high, but training at home in Canada was also a hazardous business, claiming its toll up to the very eve of Armistice Day.

On the afternoon of November 10, 1918, Cadet Wilson of Camp Borden was flying solo when through an unknown cause he lost control of his machine. After a short vertical spin, it turned on its back and came down in a long glide, to crash on the outskirts of the airdrome.

Per ardua ad astra, the stirring motto of the original Royal Flying Corps, still lives as the motto of the Royal Air Force and the Royal Canadian Air Force today. Freely translated, "Through difficulties to the stars," it is surely a fitting phrase with which to recall the glory and the anguish of those gallant young Canadians who gave everything that was in them for the King they honoured and the country they loved.

16. Other aviation highlights during World War I

Two stories remain to be told of the war years in Canada, both of which had their origins in the pre-war years and are woven into the texture of the post-war period. In spite of the gigantic struggle that went on overseas and the heavy drain on air-minded personnel, there was still some private and exhibition flying. The fact that so much of the latter was done by airwomen was symptomatic of the change being wrought in the status of women by the war-created dearth of men in all walks of civilian life. That the friendly invasion of American barnstormers should have persisted at all is indication enough that the pre-war air-mindedness of the civilian population was far from dead.

The other story is of a major contribution to the flying war effort by civilian industry and workers co-operating with the government and the air force. This effort, in the two-year span of its operation during 1917 and 1918, built 2,900 Curtiss two-seater training aircraft, at a cost to the Canadian and British governments of over $14,000,000.

The only Canadian aircraft plant of World War I developed out of the Curtiss Company of Hammondsport, N.Y., which has already been mentioned as having established a branch factory at Strachan Avenue, Toronto, in 1915. In this factory were built the training planes used by the 1915–1916 Curtiss Aviation School at Long Branch. In addition, the plant received an order from the Spanish government for twenty Curtiss JN type machines, each equipped with a single centre float, so that they could operate as seaplanes. These formed the first consignment of Canadian-made aircraft ever to be dispatched from Canada. The engines of all airplanes built in Toronto, both for the School's use and for export to Spain, were made in the company plant at Buffalo, N.Y.

Early in 1917 the Curtiss plant at Toronto was taken over by the Imperial Munitions Board, and the manufacture of Curtiss JN4 aircraft for the R.F.C. began. The reorganized company was called Canadian Aeroplanes, Limited.

Many of the parts were made in a two-storey building on Strachan Avenue, Toronto, which later became an engine repair depot. During the early stages of the industry, the office and assembly workshop were in another small building a few blocks east, where there was barely sufficient space to erect a complete machine.

By the end of April 1917, after only three months of intensive work, a large plant covering some six acres had been constructed at Dufferin and Lappin streets in Toronto. The factory was located in the centre of a district of workmen's homes, and there was no lack of applicants for jobs when full production got under way. There were only two shifts, 7:00 A.M. to 7:00 P.M. and 7:00 P.M. to 7:00 A.M., with a half hour off at noon and midnight for lunch. Only work of an emergency nature was ever done on a Saturday afternoon, and none at all on Sundays.

The rates at first were 25 cents per hour for helpers, 50 cents for machinists, and $1 for toolmakers, but production lagged under this scale, and a piece-work system was put into effect. The output of planes increased and wages swelled; it is said that one man quit work at noon one Friday because he had already made over $50 that week, and he "thought it wouldn't look good to earn any more!" Few women were employed in the aviation industry in Canada during World War I, although thousands worked in munition-factories.

The president of Canadian Aeroplanes Limited, F. W. Baillie, was a prominent executive who had previously handled a large contract for making shells in his own factory. When the Canadian government placed orders for the manufacture of shells, they were obliged to be generous in the amounts they paid, as none of the concerns had previously done such work and as a rule much new equipment had to be obtained. Firms which did not have to purchase much new machinery were few, but those few made very handsome profits, thanks to the additional subsidy. Mr. Baillie's factory was one of them, but he very patriotically turned back his excess profits to the government. Thus the government had every confidence in putting him in charge of the aircraft construction plant. A further proof of the general esteem in which Mr. Baillie was held was the bestowal of a knighthood upon him about a year later.

The plant manager was E. T. Musson, formerly of the Canadian Russell Motor Car Company, and the master mechanic of the organization was W. E. Tregenza from the Packard Motor Car Company of Detroit.

The machines turned out embodied a great many improvements in design over the original type of Curtiss tractor biplane. Credit for this must go to F. G. Ericson, chief engineer, who for his work was later appointed to membership on the International Standards Board, the British Committee on Inventions, the British Committee on Aeroplane Design, and the Canadian Air Board. The engines for the machines— 90 h.p. Curtiss OX5's—came from Detroit. The rest of the fabrication was Canadian—spruce from British Columbia for the wooden parts, and fabric from the Wabasso Plant at Three Rivers, Que., for wing coverings and other surfaces.

By the end of 1917 sufficient JN4's had been turned out to fill Canada's training requirements, a fine achievement in view of the fact that at the beginnnig of the year there was no factory, and only a handful of experienced employees.

With the entry of the United States into the war, production began in American plants on

KATHERINE STINSON

machines for training purposes. Although plans were pretentious, production fell far short of the mark, and at the end of 1917 deliveries were much behind. In desperation the American government got in touch with Canadian Aeroplanes Limited, with a request for 1,000 machines in a hurry, so the employees of the Canadian plant rolled up their sleeves and really went to it. A dozen airplanes a day were turned out regularly. Once or twice the word went around the plant that 15 had been finished in one day. In three months the thousand training machines were built and delivered.

In January 1918, at the request of the United States Air Board, Chief Engineer Frithiof Ericson was sent to England to study and obtain plans of the large twin-engined Felixstowe flying-boat. Upon his return, Canadian Aeroplanes Limited were asked by the American government to build 30 of the huge craft for delivery as quickly as possible. During the preliminary stages of their construction they were known as "Canadian Bombers," but their official description boiled down simply to F3's and F5's.

They were huge machines even in comparison with such World War II monsters as Lancasters and Cansos. Their wing span from wing tip to wing tip was 102 feet, equal to that of the Lancaster and just under the Canso's (PBY's) 104 feet. In performance, they were, of course, no match for present-day craft, for their two Liberty engines gave a top speed of only 87 m.p.h., and they required ten minutes' flying to gain an altitude of about 850 feet with a full load of four crewmen and four 250-pound bombs. With a full crew, and sufficient fuel to cruise for 11 hours, only two such bombs could be carried.

KATHERINE STINSON
At Camp Hughes near Brandon, Manitoba, July 22, 1916

The Canadian factory turned out the first of these large flying-boats three months from the date the contract was signed. The entire thirty were delivered in seven months. Three flat-cars were required to carry the three cases containing parts of one boat. All had to be shipped from the factory in that manner as there were no facilities near by where they could be test-flown.

The building of Avro training planes powered with 130 h.p. Clerget rotary engines also got under way before the Armistice, but although parts for one hundred were in readiness, only two were fully completed when the war ended. The factory also built one experimental De Havilland fighter, but that type never went into production.

The whole effort, from start to finish, reflected great credit on Canada and on the executives and staff who contributed to it.

With the signing of the Armistice the whole organization was dissolved and the factory dismantled. At the same time as the R.A.F. disposed of its vast stock of equipment, thousands of planes of varying condition were also sold for a song. Ericsons Limited, of New York and Toronto, bought up the entire stock of air equipment, reselling airplanes at top prices of $3,000. Engines were dirt cheap; spare parts were a drug on the market.

Gone were the old obstacles to flight of the pioneer days; the time-consuming chore of building the machine by hand, the pocket-draining expense of procuring suitable materials and hand-tooled parts, the inadequate power and general unsuitability of pre-war gasoline engines, and the lack of trained pilots to train other flyers.

In a few months the 2,500 or more Canadians estimated to be overseas in the R.A.F. when the war ended began flooding back. Planes and engines which they knew how to fly, service, and repair were waiting for them by the thousands. It is small wonder that the post-war period saw a tremendous new barnstorming boom across Canada.

But before that story is recounted, a few further facts are needed to record the few flights that kept the lighter side of flying active through the grim years of war, and the first attempts to adapt the airplane to the needs of a nation at peace.

It was chiefly in western Canada and mostly by Americans that the most successful exhibition flying was done during the war. In Ontario, of three events planned in 1915, the first failed to come off and the second was disastrous; the third, however, was noteworthy.

The first was to have featured an authorized mail flight from Windsor, Ont. The Patterson aviators from across the border had made arrangements with the mayor and the postal authorities, and considerable publicity had been given to what was hoped would be a record-making event. Souvenir postcards were printed for the affair and Postmaster Wigle of Windsor was given authority from Ottawa to swear the airmen in as "Special Canadian Aerial Mail Carriers." But the engine of the plane proved to be faulty, and when the second of two trial flights resulted in injury to the pilot and damage to the plane, the affair had to be cancelled.

The next event was at Sturgeon Falls, Ont., on September 6. An American pilot named William S. Luckey was engaged to give a demonstration of flying at the exhibition grounds. This area proved to be utterly inadequate as a flying field, and the airman proved to be more plucky than lucky when he attempted to use it. His first flight was successful, but on his second, he flew into the telegraph wires that infested the place, crashed badly, and was picked up with a broken back. He was taken by train to Montreal, where he was placed in hospital. His injuries proved to be severe and he died three months later.

Lake Ontario was the scene of the third private flying event, the performer being Theodore Macaulay, previously mentioned as an instructor at the Curtiss Aviation School near Toronto. When the school closed for the winter, Macaulay obtained a position as test pilot with the Curtiss Company at Buffalo, N.Y., and before the year closed he had set a number of unofficial records, all made over Canadian territory.

His most outstanding flight was made that year, with a twin-engined Curtiss flying-boat of the so-called "trans-Atlantic" type. At 3:22 P.M. on November 10, 1915, Macaulay took off from Lake Ontario near the Toronto Exhibition grounds, flying west past Long Branch. Turning southward at Bronte, he reached a point midway between Hamilton Beach and Grimsby, then returned, landing at Toronto after 40½ minutes in the air. The flight was notable by reason of the number of passengers and the weight carried. Besides eight bags of sand weighing 640 pounds, four men were aboard as passengers, bringing the total load up to 2,100 pounds.

This was the first successful flight of a twin-engined flying-boat or twin-engined aircraft of any type, in Canada.

KATHERINE STINSON: THE PRAIRIES, 1916

The barnstorming spotlight in 1916 centred on a single person, but a very remarkable one—the renowned American airwoman, Katherine Stinson.

Just turned twenty, Miss Stinson was a most unassuming young woman, yet her ability in the air was phenomenal. Everywhere she performed, crowds went wild with enthusiasm. She came to Canada with four years' flying experience, having earned her A.C.A. in 1912.

Her first flying in Canada took place at Calgary, Alta. On June 30, and on the four days following, she thrilled the crowds with three flights daily which included steep turns, spirals, dives, and loops. She so impressed the military that she was entertained as guest of honour at a dinner at the Sarcee military camp, several miles south of the city.

Miss Stinson gave equally generous and sensational flights during the next few weeks at Edmonton, Alta., Brandon, Man., Regina, Sask., and Winnipeg, Man. At Brandon, she went the whole social gamut. The day after she was made a princess of the Sioux Indian tribe, by Chief Waukessa in an exhibition grandstand feature, she was arrested as a spy, when she paid an unannounced visit to Camp Hughes!

Katherine seemed to specialize in nosedives. On her last flight at Regina she took such a steep dive from a great height that people who had witnessed it from a distance telephoned to the city in great numbers to inquire if an accident had happened.

Her flying at Winnipeg was a civic money-raising attraction in aid of the Manitoba Patriotic Fund. The advance notices said she would fly between 8:30 P.M. and 10:30 P.M. on August 3, but repairs to her plane, which had been damaged in a storm when being assembled, caused a day's delay, and it was not until 10:30 P.M. of August 4 that all was in readiness. When at last she sped into the air, the sky was pitch black. Owners of automobiles parked around the infield were asked to switch on their headlights to assist in illuminating the grounds, and soldiers were stationed at intervals to light flares at a given signal. Huge bonfires were set alight at both ends of the oval, to help guide the flyer safely down.

The machine flew away from the grounds and vanished into the night; only the sound of the engine could be heard as Katherine circled above. After gaining an altitude of 1,000 feet, she pressed a switch which ignited flares attached to the rear edges of the wings, and the full outline of the airplane was clearly illuminated. Cheers rang out in the night air as her

THE HOFFAR SEAPLANE

machine circled again and again, sometimes cutting through the cloudy smoke trails of the burning flares. When the lights on the machine finally burned out, Miss Stinson cut the engine to a low hum and came drifting down on the wind, to end with a faultless landing in the centre of the infield.

To compensate for the cancelled flights of the first days, Katherine gave additional flights on the 5th. On this occasion she "bombed" a dummy fort which had been erected on the grounds while soldiers, scattered about the area, opened fire with blank cartridges, producing quite a warlike atmosphere.

Miss Stinson's popularity was due not only to her courage and skill as a pilot, but also to her conscientious determination to fulfil all her contracts.

Such spectacular flying required, in addition to an expert pilot, an absolutely reliable machine. It is of interest to learn that the engine owned by Miss Stinson had been used by Lincoln Beachey when he flew in Winnipeg two years earlier, and also when he had crashed to death in 1915. It was a seven-cylinder, rotary Gnome engine. The aircraft itself was a single-seater biplane designed and built by an American named Mattie Laird. The craft was one of the few to survive its flying years, and eventually was placed on exhibit in the Ford Museum at Dearborn, near Detroit, Mich.

The Boat-Building Hoffar Brothers: Vancouver, 1917

By 1917, training for war purposes had absorbed every ambitious airman in Canada and the United States. Furthermore, military regulations prohibited private flying without special permit. Several exhibition committees tried to obtain civilian pilots to perform at their fairs,

but without success. Then the Calgary Exhibition Board wrote to Ottawa to ask if a military airman might be permitted to fly at their fair. They were told in reply that "the suggestion was most improper and could not be considered" —an official way of saying "Don't you know there's a war on!"

However, one private enterprise in aviation which is well worthy of description was carried on in 1917.

The boat-building brothers, Jimmie and Henry Hoffar of Vancouver, who had assisted the British Columbia Aero Club in 1915, had become more and more air-minded until the urge blossomed into action in the fall of 1916. They undertook to construct a two-seater seaplane, fitted with a 100 h.p. water-cooled Roberts engine. When completed by the early summer of 1917, it was a beautiful piece of workmanship. A large single pontoon took the main weight of the craft while on the water, with a small float at either wing tip and a similar small float at the tail.

Then Jimmie Hoffar spent several weeks teaching himself to fly. The most outstanding flight he made was on July 16, 1917, when he took up Bruce J. McKelvie, a reporter on the Vancouver *Daily Province*. Ascending to 2000 feet, they made a wide circular flight directly over the city of Vancouver, the first pilot and passenger to fly over the down-town area of that port. Some weeks later, as the seaplane taxied swiftly along the water, its pontoon struck a submerged log, and the craft was totally immersed. It became so badly water-soaked that it was never flown again.

Jimmie Hoffar has since confided to me that when he and his brother built the machine, their ignorance of aircraft was abysmal. For instance, their plans called for interior cross-wire bracing in the construction of the wings; but they decided not to bother with it as it seemed unnecessary. Certainly, the wings they built must have been particularly well constructed to stand the rigours of strain in flying without such bracing. But Jimmie has shivered since, realizing that the wings might easily have folded backwards at any moment during one of his numerous high flights.

At any rate the Hoffar brothers happily escaped personal injury, and by the next year they had another aircraft ready, a flying-boat, to which further reference will be made.

KATHERINE STINSON: THE PRAIRIES, 1918

The civilian flying performed in Canada during the last year of World War I, though not extensive, was remarkable. It consisted chiefly of exhibition flying by two American airwomen.

Katherine Stinson returned under a special permit from the Canadian government to fascinate her friends in the West. She first entertained the Exhibition crowds at Calgary for a week, then, on July 9, she made a memorable flight to Edmonton. She flew the 175 air miles in 2 hours and 5 minutes, and established a western Canadian record for endurance.

From July 9 to 13 Miss Stinson amazed the crowds at Edmonton not only by her flying but by her ability as a racing motorist! Invited to drive the racer owned by Leon Duray, Miss Stinson, with Leon as passenger, whipped twice around the half-mile track, thereby creating a new record for women drivers in Canada, of one minute, fifteen and one-fifth seconds for the mile.

Miss Stinson then went on to give exhibitions at Saskatoon, Sask., and Red Deer and Camrose, Alta. Although she once had engine trouble, and difficulties and damage from windstorms, at no time did she run into serious danger, and when she said good-bye to Canada on August 3, she left nothing but happy memories.

She afterwards became Mrs. Michael Otero of Santa Fe, N.M., and has never since taken to the air as a pilot.

RUTH LAW: TORONTO, OTTAWA, MONTREAL

While Katherine Stinson was performing in the Prairie Provinces, another equally remarkable American airwoman was thrilling the crowds in central Canada.

Ruth Bancroft Law appeared first at Toronto on June 29, in conjunction with motor races at the Exhibition grounds. She raced the speeding cars in her plane, flying low over the course. In one race against Gaston Chevrolet her handicap was six laps to his five, but since Ruth had to fly wide of the track on the turns the auto won by one-third of a lap.

This daredevil programme was afterwards carried out at Ottawa and at Montreal, under the auspices of the War Veterans' Association. During a demonstration race, Miss Law raced the driver of a high-powered Marquette-Buick around the track for two miles, keeping only a few feet above his head all the way.

At these three centres, the airwoman also performed at high levels, putting her machine through loops, vertical turns, spins, and other acrobatics. R.A.F. pilots who watched her were unstinting in their praise.

Miss Law had earned her A.C.A. licence in a Wright biplane in 1912. The machine she used in Canada was a Curtiss biplane fitted with Wright controls. The reason for this combination is interesting. When Ruth tried to purchase a machine from the Curtiss Company, Glenn Curtiss made excuses and politely told her that he could not supply one. His real reason for refusing, as he confessed years later, was his concern about the young girl's life. He could not believe that a person of such slim build, weighing only a little over one hundred pounds, could handle an airplane with safety.

Later, Ruth procured a 1913 Curtiss biplane, which she converted to Wright controls. It proved to be thoroughly satisfactory. In June 1918, she used the aircraft to establish a non-stop record flight from Chicago, Ill., to Hornell, N.Y., a distance of 512.1 miles—an achievement which won her a special award of $2,500 from the Aero Club of America.

Ruth Law was as fortunate as Katherine Stinson in surviving her flying experiences safely. She later married and as Mrs. Oliver took up residence near San Francisco.

THE HOFFAR FLYING-BOAT FOR FOREST PATROL

While the American women were thrilling crowds with their flying in the interior of Canada, the Hoffar brothers at Vancouver had been busy constructing another aircraft. This one was a flying-boat equipped with a Green

RUTH LAW

At the controls of her 1913 Curtiss biplane fitted with Wright controls

WRECKAGE OF THE HOFFAR
FLYING-BOAT
On the roof of Dr. J. C. Farish's home
on September 4, 1918

engine, and it was intended for use by the
Forestry Department of British Columbia for
forest patrol.

When the boat was completed in August 1918,
the government appointed an R.A.F. pilot home
on leave to take it up on its maiden flight over
Burrard Inlet. The pilot made a successful
flight, but recommended certain control adjustments which the Hoffars forthwith made. At
this point the provincial government took over
the machine on a year's lease, and appointed
Lieutenant Victor A. Bishop, another R.A.F.
man home on leave, to make a second test
flight. Bishop was a very proficient airman, and
experienced in many types of craft, so the outcome of the flight cannot be put down to any
incompetence on his part.

On September 4 the pilot took the machine
up, and was flying above the west end of the
city at 1,500 feet when the engine suddenly
went dead. The airman immediately put the
craft into a glide in an endeavour to reach a
landing on the sea, but since it was without
motive power of any kind the flying-boat became nose-heavy and hence unmanageable.
Almost at once it went into a spin, and crashed
with terrific force into the sloping roof of a
house situated at 755 Bute Street. The nose
smashed completely through the roof, but the
wings and tail were left protruding. Thanks to
the strong construction of the hull and the
"give" afforded by the roof, the pilot was only
slightly hurt, and when it was all over he found
himself in the bathroom of the residence, which
was owned by Doctor Farish. Except for the

engine—which had failed—the flying-boat was a
total loss.

The most serious effect of the crash was the
setback it gave to forestry flying in British
Columbia. Instead of being the first, the province was one of the last to use airplanes in forest
patrols.

The accident was only a temporary setback
to the Hoffars, because the government had
already assumed ownership of the flying-boat,
and therefore was responsible not only for the
cost of the machine, but for all damages. The
Hoffar brothers continued with aircraft construction and in time became affiliated with
Boeing's Limited of Seattle, in connection with
a branch works in Vancouver. It was located on
the site of the original Hoffar workshop on
Burrard Inlet, and during World War II it
became a very busy place.

With the spectacular crash of this flying-boat,
civilian flying in Canada during the years of
World War I came to an end.

CANADA's FIRST AIR MAIL, 1918

In 1918, R.A.F. training was in full swing,
and Captain Brian A. Peck of the Leaside airdrome in Toronto wanted very much to pay a
week-end visit to his home in Montreal. He
"wangled" it by suggesting to Brigadier General
Hoare that an impetus might be given to recruiting in Montreal by putting on a bit of
exhibition flying over the down-town area. The
General "bought" the idea.

So, early on Friday morning, June 20, 1918,
Captain Peck, with Corporal Mathers as passen-

ger, set off from Leaside. The aircraft used was No. 230, one of the regular JN4 Curtiss training planes.

In spite of low, scudding clouds and a strong south wind which forced the pilot to fly crabwise most of the way, the flight to Montreal was made safely, with a stop at Deseronto for fuel. They landed at Bois Franc Polo Grounds at noon, where the Officer Commanding Military District No. 4 had the aircraft put under guard.

Continuous rain made the recruiting flight over Montreal impossible and prevented the flyers from returning to Toronto on Sunday, as they had intended. Meanwhile, the idea of sending mail back, under postal authorization, had occurred to two air-minded persons—George Lighthall, president, and Edmund Greenwood, treasurer, of the Montreal branch of the Aerial League of the British Empire. Peck was more than willing, and contact with post office officials did the rest. Mr. R. F. Coulter, Deputy Postmaster General at Ottawa, authorized Mr. Greenwood to act as a local postmaster, and it was he who did the rubber-stamp cancelling of the 120 letters, taken at random from the regular mail for Toronto.

When Peck took off on Monday, he was forced by zero visibility to return to the polo field. This is the reason that all the air-mail envelopes are dated June 23, 1918, although the flight took place the day following, starting at 10:30 A.M.

A feature of this flight which, for obvious

CAPTAIN PECK ARRIVES BACK AT LEASIDE
Carrying Canada's first air mail, June 24, 1918

reasons, was not made public then, may now be revealed. Quebec at that date was "wet," while Ontario was very "dry." A lieutenant in charge of stores at Leaside was to be married, and had asked Peck to bring back something suitable for celebration. So when the plane finally took off on Tuesday morning, Corporal Mathers was sitting on a case of "Old Mull" in the passenger seat and clutching the bag of mail. Loaded also with full fuel tanks, the plane was too heavy to be lifted quickly into the air. Captain Peck had to duck under the telegraph wires along the railway tracks which bordered the south end of the polo field. Seeing more wires ahead, the resourceful pilot banked sharply and followed the railway right of way just a few feet above the rails. With a bridge now coming at him fast, and wires all over the place, he banked out over the river spanned by the bridge, and was at last in the clear, but the machine almost settled down into the water. The airman then followed along the river to the Lake of Two Mountains, going about five miles before he reached a forty-foot altitude, which then enabled him to turn overland and head for Toronto.

A strong west wind was blowing, accompanied by heavy rain squalls, and once a plugged fuel line caused some anxiety, but the trouble suddenly cleared up and the two men were able to continue to Kingston, where they

CAPTAIN BRIAN A. PECK

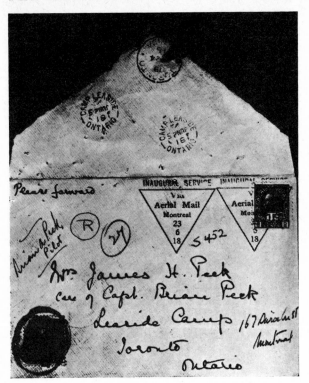

ONE OF THE AIR-MAIL LETTERS

were then made for her to take along a small sack of mail, and the 259 letters which she flew from Calgary to Edmonton were carried under postal authorization. Each piece of mail bore a *cachet* stamped in violet ink, bearing the words, "Aeroplane Mail Service. July 9th, 1918. Calgary —Alberta." Many stamp catalogues report this flight incorrectly but most of them agree that individual letters carried on the flight are now worth approximately $200 each.

Miss Stinson received the mail bag from the hands of Postmaster King of Calgary, and then took off on her momentous flight to Edmonton, starting at 1:30 P.M. Because of a forced landing *en route*, caused by engine trouble, she did not reach her destination until 8:00 P.M., but her actual flying time for the 175 air miles was 2 hours and 5 minutes. She was met by Postmaster Armstrong of Edmonton upon her arrival, and a great crowd of people was also on hand to extend a noisy welcome.

It was big news in the West, even overshadowing war topics on the front pages of the newspapers of that date.

landed almost out of fuel. There they could procure only ordinary automobile gasoline, but they took on half a tank-full, then flew on to Deseronto, where the "hit and miss" stuff was drained out, and a supply of good aviation gas was poured in. From there the flight was uneventful.

Arriving at Leaside at 4:55 P.M. Captain Peck requisitioned a car, and delivered the mail sack to Postmaster W. E. Lemon, in person, at the General Post Office in Toronto. Peck was presented with the bag as a souvenir. Presumably, the Old Mull reached its destination, too.

Thirty years later, single specimens of the cancelled envelopes carried on this flight were listed in stamp catalogues at $200 each, and those bearing the pilot's signature at $225. One envelope was included in the collection of King George V.

The second official air-mail flight in Canada was made by a woman. During the summer of 1918 the pioneer American airwoman, Katherine Stinson, was making numerous exhibition flights at fairs in western Canada, flying a single-seater Curtiss biplane, fitted with an eight-cylinder Curtiss motor. Upon completion of her flying contract at Calgary in July, she arranged to fly to Edmonton, her next exhibition point. Plans

In August and September 1918, three more air-mail flights were made, all by R.A.F. pilots and planes. These flights were all planned in advance, not impromptu as were the flights of Captain Peck and Miss Stinson.

The flights were all between Toronto (more precisely, Leaside) and Ottawa. On August 15, Lieutenant T. Longman flew to the capital in a Curtiss aircraft, No. C280, returning two days later, both times with mail. Lieutenant A. Dunstan made the same trips August 26 and 27,

KATHERINE STINSON

Handing the bag of mail to Postmaster Armstrong at Edmonton, July 9, 1918

WILLIAM LYON MACKENZIE KING (*wearing straw hat at right*)
Watching Lieutenant Dunstan prepare to take off for Toronto on August 27, 1918

carrying mail in both directions, the flights taking 3 hours and 40 minutes each way. On September 4, Lieutenant E. C. Burton flew from Leaside to Ottawa and returned on the same day. A total of approximately 3,000 letters was flown on the three flights just mentioned. The Aero Club of Canada sponsored the flights, and on each letter was affixed a sticker picturing the destruction of a German Zeppelin by a British plane. Around the margin were the words "Aero Club of Canada—First Aerial Mail—Service per Royal Air Force—August 1918—25¢." Stamp catalogues today list these envelopes at values ranging from $75 to $125 each.

If civilian flying accomplished nothing else during the war years it demonstrated in the spectacular flying of pilots such as Katherine Stinson the tremendous advance in manœuvrability of the flying machine and in skill on the part of the pilot since the beginning of the war. And the last air-mail flights of 1918 made it clear that the unreliability of pre-war engines was becoming a thing of the past.

The stage was set for the remarkable achievements of the next ten years.

Not many historical aviation events have as yet been suitably acknowledged in Canada by the erection of plaques or cairns. Interest in such measures fortunately is on the increase, and additional outstanding aerial accomplishments worthy of being permanently remembered may yet receive their just due.

The Ontario Archaeological and Historic Sites Board deserves praise for being instru-

mental in the placing of a marker to commemorate the fortieth anniversary of the first official air-mail flight in Canada. Although it cannot be classed as permanent, as it is only a painted sign on a sheet-metal background, nevertheless, it is a reminder until the time when a plaque of a more solid nature can be set up in its place.

The marker can be seen on the site of the old Leaside aerodrome northeast of Toronto on the north side of Eglinton Avenue East slightly to the east of Brentcliffe Road. It was unveiled by the Honourable William Hamilton, Postmaster-General of Canada, on September 5, 1958. In gold lettering on a dark blue background is inscribed the following:

CANADA'S FIRST AIRMAIL
1918

In June, 1918 the Montreal branch of the Aerial League of the British Empire persuaded postal authorities to sanction an air mail delivery to Toronto. A JN4 Curtiss aircraft from the Royal Air Force detachment at Leaside aerodrome was selected for the attempt. Piloted by Captain B. A. Peck with Corporal C. W. Mathers as passenger, it took off at 10:30 A.M. June 24th from Montreal's Bois Franc Polo Grounds. After refueling at Kingston, Peck landed here with his cargo of 120 letters at 4:55 P.M., thus completing Canada's first air mail flight.

Erected by the Ontario Archaeological and Historic Sites Board.

It would have added greatly to the occasion if Captain Brian Peck could have been there himself, but unfortunately the placing of the marker came a little too late for Canada's first air-mail pilot to enjoy the honour in person; he had died in Montreal only six months before.

PART 4

THE DOLLAR-A-MINUTE DAYS, 1919-1920

17. Dollar-a-minute days in the East

NEVER again can there be flying times like the two years following the end of World War I.

The boom in aviation was world-wide and Canada had her fair share. Canadians everywhere had read and heard about the doings of their fighting airmen but surprisingly few civilians had seen an airplane close up. Even thousands of soldiers who had watched military aircraft in full flight overhead had never been any closer to an airplane than that.

On May 17, 1919, the first shipload of airmen returning from overseas disembarked at Montreal where they were royally welcomed by the Aerial League of the British Empire. There were possibly 500 in the group, to be followed in July and September by groups of 850 and nearly 1000 respectively. At a reception for this last arrival one of the League officials regretted "that no effort has been made to retain their [the returned men's] interest in aeronautics by our Canadian authorities. . . ." Hundreds of Canada's most experienced airmen (for only trained pilots had been accepted from Canada for service overseas) thus went back into civilian occupations, and their flying knowledge and experience were lost forever.

In fairness to the government I should point out that the permanent Canadian Air Force was organized later in the year. But even then, though hundreds resumed their pre-war ground activities, or discovered others, hundreds were left over, many of them with the restlessness of the war still in them, and the itch to fly unsatisfied. However, there were any number of cheap planes to be had. Many former war pilots who could rake up enough cash immediately purchased one or more aircraft, while a great many others persuaded someone else to invest money in new-born flying companies. Airplanes in increasing numbers thus made their appearance throughout Canada, and few were the towns and villages of any size which did not receive a flying visitor during 1919 and 1920. Aviation for everybody was suddenly made available to a very receptive public. On top of that there were no government restrictions on flying in Canada in 1919. Anyone with the urge, and the money, could buy a plane and fly away in it if he could, or try to if he couldn't.

At Camp Borden the news of the signing of the Armistice was barely off the wires before the boys were in the air celebrating with every stunt at their command. At Camp Mohawk several exuberant young pilots winged over the neighbouring towns to put on hair-raising displays. One of these—the Officer Commanding a squadron no less—outdid his fellow madmen by flying down the main street of Oshawa, and clipping off a number of flagpoles with the undercarriage of his machine. In the two years that followed, some pilots forgot that they were not still flying on war service, and their recklessness dotted the countryside with wrecked aircraft.

Many of the veteran flyers, however, kept their heads; and began to figure ways and means of making a serious livelihood out of flying. At Toronto, for instance, two of Canada's most famous airmen were combining to form a flying company. They were Lieutenant Colonel W. A. Bishop, V.C., and Lieutenant Colonel W. G. Barker, V.C., who between them had destroyed 122 enemy aircraft in World War I. The Bishop-Barker Flying Company was formed with the object of operating a passenger service between Toronto and the Muskoka resort area, but the public was not ready for so radical an innovation and the Company collapsed. This was only one of a number of flying firms that sprang up almost overnight, and sometimes vanished as quickly.

During World War I, some Canadian cities had donated aircraft for training purposes to the R.A.F., and when the war ended, replacement machines were returned to them. The three machines Toronto received bore the names of *City of Toronto*, *Mercer*, and *Lundy's Lane*. The last of these was purchased by a Lieutenant Holmes of Toronto for his newly formed flying company and flown to a field at Mimico, west of Toronto. With it he did some spectacular flying, above and along the Niagara gorge. Photographs taken on the flight were published in hundreds of North American newspapers.

THE FIRST PRIVATE AIR PILOT'S CERTIFICATE

Allied Aeroplanes of Brantford purchased the two other machines from the city of Toronto. This company was backed by Brantford business interests, all the shares in it being owned by members of the Dowling family. Lieutenant Don Russell, a nephew of the president of the firm, was pilot, Arthur Mason was engaged as rigger, and I was taken on as a mechanic.

My own demobilization will serve here, perhaps, to represent the stories of hundreds of other airmen whom Armistice caught unawares without plans or prospects for the future.

Several weeks after the war was over, a call went out for volunteers to go to the Records Office situated in Toronto, where a large staff was set in motion to facilitate the quick discharge of men of the force. I became one of those who took on a job at Records, and was in Toronto for Christmas. My chief work was in connection with the discharge branch dealing with cadets. The work was slowly moving to completion when, in May 1919, as I was walking down one of the corridors, I bumped into Don Russell, whom I had known and flown with at Camp Borden. After preliminary handshakes were over, I asked him, for a joke, if he was trying to join up. His reply was "No," but he was looking for a mechanic to join up with him, to set up two Curtiss Jennies owned by Allied Aeroplanes, Ltd., a company owned chiefly by his uncle. Right then I told him he need look no further. I obtained my discharge within the week, and was off with Don to the Toronto Exhibition grounds to shake the dust off the two dismantled machines stored there, and get them in flying trim to take off for Brantford. In two days we had them ready to fly to Brantford where they lost their Air Force identity under a coat of royal blue paint, with the name "Brantford" inscribed in large letters on both sides of their fuselages. We thought the finished jobs looked very smart indeed.

Lieutenant Cook was engaged to fly the second machine, and on July 1, the two machines with the four of us aboard set off for Crystal Beach, 70 miles to the southwest, a Canadian summer resort on Lake Erie, where we were to spend the summer taking up passengers at $1 a minute.

Cook wrecked his machine on his second landing in the undersized field we originally used, but with the remaining Curtiss, Don Russell

EXHIBITION GROUNDS,
TORONTO

Used to assemble two machines of
Allied Aeroplanes, Limited

carried on throughout the season, and hundreds of passengers were flown without further misadventure. A number of trips were made over Niagara Falls, and several across the lake to Buffalo.

Generally speaking, the barnstorming flyers who roamed the country in 1919 seldom indulged in risky flying if they owned the airplane themselves, being more interested in doing stunts that could be seen—and paid for—by the public, than in flying wild just for hellery. Russell and I both liked our necks too much to tempt Providence, but on some of our flights over Niagara Falls we were none too particular about how low we were when we passed over the roaring waters. While Holmes was taking his photographs of the Niagara gorge, he and his photographer companion had a hair-raising experience. Getting down into the gorge and flying under the International Bridge was easy enough, but once down in the gorge their Curtiss Jenny was flung about in all directions and the down-currents of air were so strong that nothing Holmes could do would make the machine climb up and out of the turbulence. So they were forced to go right on down the gorge, not many feet above the white, raging waters; and when they finally emerged, at the down-river end, though they had some wonderful photographs to show for it, no amount of money would have influenced them ever to try it again.

While at Crystal Beach, I frequently went up with Don during the afternoon, when passengers were a little tardy at coming over to where we were operating, and we would fly along the beach only feet above the bathers' heads in an effort to drum up business, usually ending up with a run straight at the pier, pulling up in a steep climb just at the moment when to those below an impact seemed inevitable. The govern-

ment and the public surely would take a dim view of such exploits now; proof enough that flying and its ways have changed for the better. Actually, that same summer the government was taking action to create some kind of control over flying activities.

Although the first Canadian Air Board was constituted by Order-in-Council on June 23, 1919, effective air regulations were not drafted until December 31, 1919. On April 19, 1920, the new Air Board was constituted by Order-in-Council, with full powers to act.

Private pilot's certificate No. 1, the first to be issued under the new regulations, was made out to James Stanley Scott on January 24, 1920. At the time his No. 1 pilot's certificate was issued, Major Scott was superintendent of the Certificate Branch, at Ottawa.

The Canadian Air Board functioned admirably until 1923, when the National Defence Act of 1922 came into effect. Under this Act the Minister of Defence was responsible for all matters relating to defence and these included all military, naval, and civilian flying activities in Canada. This assumption of full control of civilian operations by the government resulted in the drafting of regulations which did much in the years to follow to promote safety in air transport.

HITTING THE SILK

On July 4, 1919, a noteworthy event took place. The Irvin Parachute Company of Buffalo, N.Y., was in the throes of initial experiments, and Leslie Irvin, inventor of the air 'chute which bears his name, telephoned from Buffalo asking if Allied Aeroplanes would collaborate in testing a new type of parachute. Don Russell at once consented to fly the machine, and early on the 4th Irvin arrived at Crystal Beach accompanied

PARACHUTING 1919

Left to right: Leslie Irvin, maker and designer of parachutes; Chilson, and Don Russell. Crystal Beach, Ontario, July 4, 1919

by a Mr. Chilson, who was to make the test jump. They brought along the precious silk bundle. (Irvin would have made the jump himself, but was prevented by a sprained ankle which he had suffered some days previously, after parachuting from a balloon.)

This parachute, known as the "back-pack" design, was the grandaddy of the modern types. Compared with the compact equipment of the present day, it was a cumbersome contraption. The whole thing was held tightly to the jumper's back by three stout straps, which fastened with hefty buckles. The shroud lines from the 'chute were attached to the harness by canvas straps. It looked and weighed more like a bag of cement than an aerial life preserver.

As the descent was to be made into Lake Erie, Chilson wore only a bathing suit, with an inflated inner tube fastened around his chest to keep him afloat, or, as we suggested, to "mark the spot" where he might vanish.

With the apparatus strapped tightly in position, one belt around his chest and one around each thigh, Chilson climbed laboriously into the rear cockpit of the Allied Aeroplanes machine, where he was forced to kneel on the seat in a most uncomfortable position, since the bulk of

the pack prevented him from sitting down. Don then took off from the Crystal Beach field and set off over the lake, circling steadily to gain altitude. About a quarter of a mile off shore, at a height of 2,000 feet, Chilson battled his way slowly over the side of the cockpit against the violent air stream from the propeller. Then, as Don remarked later, "with a sort of good-bye look on his face," he let go.

In that original pack, the cord used to operate the release panel and allow the parachute to emerge from the container was not controlled by hand, as is a modern rip-cord. Instead, there was a fifteen-foot length of strong, thin rope, one end of which was fastened tightly and carefully to some rigid part of the aircraft. For Chilson's jump, it was attached to a bracer-bar inside the fuselage. All that the parachutist was required to do was to clamber out of the cockpit with the pack on his back and launch himself into space, with a prayer that everything would work according to plans.

Chilson's jump was a complete success; he alighted in the lake as planned and was picked up by a motorboat which happened to be near. His own boat, hired for the job, was almost a mile away!

One important thing learned from that jump, apart from the fact that it fully demonstrated that a back-pack could be used for a jump from an airplane, was that silk thread would have to be used to sew the 'chute together. Linen thread was used in the original, and after immersion it shrank so badly that every seam of the 'chute was puckered beyond belief. All later 'chutes were silk sewn, until nylon came into use.

The heavy water-soaked 'chute was left with us at Crystal Beach, to dry out before being returned to Irvin at Buffalo. But before we returned it, we made good use of it ourselves. On the 5th, although badly wrinkled along the seams, it was quite dry enough to be replaced in its pack, ready for use. That afternoon I put on a bathing-suit and the inner tube used by Chilson, the pack was buckled on, and I climbed into the rear seat of the plane.

I was at an age then when danger did not come into my thoughts: the important thing was to make sure that everything was correctly done to ensure a successful "drop." In spite of precautions, I discovered after I was in the cockpit that due to the largeness of the parachute on my back, I could not fasten my safety belt. It was none too warm in a bathing-suit as the wind whistled past, although it was the 5th of July, and I was exceedingly cooled off by the time we had got "upstairs." At 1,800 feet Don circled to a point about a half mile off shore, where we spotted a pleasure launch lazying along the lake. Getting out from the cockpit to the outside step built into the fuselage was not too easy, as the wind thrust from the propeller ripped at me with every apparent intent to tear me loose from my hold before I was ready. As soon as I had gained the position outside, I reached forward and tapped Don on the shoulder to let him know I was ready to go; he raised his left hand and I gripped it for a second in a handshake.

He then put the machine into a steep climbing turn to the left, as we had arranged he should do, to give me every opportunity to leave the machine without the apparatus' being caught anywhere, and to allow me to get well away before the tail assembly might pass over me. As it happened, such a precaution was unnecessary: as Don said afterwards, I was carried well forward with the momentum of the craft on leaving and was many feet below in a few seconds. Of the actual drop until the 'chute opened, I have no recollection at all, because it happened so quickly that there was insufficient time for any feelings to become impressed on my mind. My first knowledge of anything tangible after letting go my hold on the plane was the yank to my body where the three straps of the harness were attached: several red welts later gave ample proof that the jar had been severe. The downward drift was exciting, or I should say exhilarating, and pleasant too; it was very quiet, and not even the air passing by the 'chute reached my hearing, though I have been told since that many parachutists can nearly gauge their rate of descent from it. I was not at all interested in any such thing, but I did not have very much time to enjoy any sensations because the surface of the lake came up to meet me quite fast, and almost before I had accustomed myself to hanging in space there was a resounding splash. I had gone under. The inner tube that was firmly attached around my chest yanked me back to the surface in a split second, and I was lifted quite a distance out of the water by its buoyancy before finally subsiding on the surface in a normal way. I was quite close to a motorboat with several persons

aboard, as we had judged I should be, and they reached me and helped me aboard after I had unbuckled the three straps which attached the 'chute to my body. I had not noticed whether the air was particularly cold or not as I came down, but I quickly learned that Lake Erie was a liquid refrigerator; no one was happier than I to be hauled out of its chilly embrace and into the motorboat. The 'chute was hauled aboard and wrung out before being stowed; and so came to a happy ending the first of a total of five parachute jumps I have made.

The first jump from a plane by a parachutist over Canadian territory had taken place as long as seven years before in Vancouver; but Chilson's jump was the first to be made anywhere with the back-pack type of 'chute, and my own was the first parachute jump from a plane in Canada by a Canadian.

These descents at Crystal Beach were, of course, planned jumps, but only seventeen days after Chilson's jump, the first emergency jumps by airmen wearing Irvin 'chutes took place. When an aircraft went out of control on July 21,

FRANK ELLIS
The author just before his first jump, July 5, 1919

1919, at Chicago, Ill., the two civilian pilots, Henry Wacker and John Boettner, bailed out in a hurry, and drifted safely down as their plane crashed to the earth below them.

First International Passenger Flight

On July 23, 1919, Allied Aeroplanes Ltd. made international flying history when Don Russell flew to Buffalo and brought back a passenger to Crystal Beach. On that date, the entire staff of the Larkin factory and stores, 7,000 of them, embarked from Buffalo aboard the two ferries, S.S. *Canadiana* and S.S. *Americana*, and Don and I were given the job of flying over to the Curtiss flying field at Buffalo, N.Y., to pick up J. D. Larkin, Jr., president of the company, and fly him to Crystal Beach in time to greet the picnickers on their arrival.

The two of us set off from the Canadian resort just after 8:00 A.M., and after flying across Lake Erie to Buffalo we had considerable difficulty in locating the field where the rendezvous was to be made, as a thick ground mist all but obscured the countryside. We had to cruise around for 45 minutes before we at last spotted the place, went down to a landing, and found Mr. Larkin waiting. He was helped into the seat I vacated, and Don took off, while I returned to Crystal Beach by ferry-boat.

Don was able to reach Crystal Beach with his passenger just as the ships were docking, and the official reception went off as planned. One point the newspapers completely missed was the fact that by flying from Buffalo to the Canadian resort, Mr. Larkin became the first known paying passenger to make an aerial trip across the boundary between the United States and Canada at any point along its entire length.

Flying in the Maritimes, 1919

In 1919 flying became so general that it would be impossible to describe in detail everything that was going on across Canada. But before we follow my own trail west in 1920 for further close-ups of a barnstormer's life, a quick survey should be made of typical air incidents in the Maritimes.

Nova Scotia was the scene of a notable flying incident of 1919. Two former R.A.F. pilots of that province, Lieutenant Laurie Stevens and Lieutenant L. L. Barnhill, organized the Devere Aviation Company, and, purchasing a Curtiss biplane, set forth to carry passengers and do stunt flying wherever opportunity offered. Their most outstanding show was in connection with the Charlottetown Exhibition. Arrangements were made for an official air-mail flight from Truro, N.S., across Northumberland Strait to Charlottetown, P.E.I., and return.

The airmen took off from Truro on September 24, carrying 200 officially stamped letters. On each envelope were the words "Via Aeroplane" or "Via Aerial Post." Cancellations bore the Truro post office stamp, dated September 24, 1919, 1:00 P.M., and all were backstamped at Charlottetown, the same date, at 3:00 P.M.

Stevens and Barnhill made a five-day stopover at the Prince Edward Island capital, during which time they made the first flights ever seen in that province. Then on the afternoon of September 29, 1919, they left Charlottetown and returned to Truro. The 30 letters they brought along were stamped with that day's date, 3:00 P.M. from Charlottetown and 4:00 P.M. at Truro. Today, one of those envelopes first flown in the Maritimes is a rarity, and a single cover of the flight from Truro to Charlottetown is valued in stamp catalogues at $150, while those carried on the return flight are priced at $200 each. That makes a $36,000 value for the entire 230 flown, yet the two pilots did not own one envelope between them!

Flying in Newfoundland, 1919–1921

By 1919, the flying history of Newfoundland began to take shape. (It was then, of course, a separate colony, and is not therefore dealt with here as one of the Maritime Provinces of Canada —as it later became.) In the previous year Major Cotton and Captain Bennett of St. John's had organized the Aerial Survey Company, and ordered two aircraft from England. The first to arrive and be put into shape was a Martinsyde biplane, a cabin-type aircraft with an open cockpit for the pilot. This machine was fitted with skis, and made its first test flight at Quidi Vidi Lake near St. John's, on January 14, 1919. Cotton and the Martinsyde thus have the honour of being the first to fly in Newfoundland. A Westland biplane from England was the second aircraft to operate for the Company. For several years both machines gave excellent service, summer and winter, flying to many points of the colony.

Some outstanding air-mail flights were made by Cotton in the Martinsyde machine in 1921, among them being journeys between St. John's and St. Anthony, on February 26, another over the route between the same two cities *via* Botwood, on March 1, and a flight from St. John's to Fogo, Fogo Island, *via* Botwood, which was completed on March 28. A special *cachet* bearing the words "Airpost Fogo" was used on the latter flight, the mail being dated at St. John's March 10, and at Fogo March 28. About 200 letters were carried.

Cotton also flew the first air mail between Newfoundland and Labrador, but some time was required to complete the event. The airman left St. John's on February 8, 1923. Delayed by an accident to the aircraft at Botwood, he was unable to take to the air again until May 16. He reached Cartwright the same day—the first airman to fly at any point in Labrador. The return flight from Cartwright to St. John's was accomplished in its entirety on May 17, 1923. Mail carried was cancelled at St. John's at 2:30 P.M. and all letters bore an additional *cachet* in heavy type: "Aerial Mail."

Very few of the envelopes carried on those initial Newfoundland air-mail flights have survived and their price, therefore, is probably on a par with that of other mail first flown over Canadian routes in 1918, 1919, and 1920.

It might be pointed out that although numerous air-mail flights of an "official" and "unofficial" nature were attempted in Canada and Newfoundland in the decade following World War I, regular air-mail routes did not follow quickly. But those original flights made spontaneously by pilots in various parts of the country did demonstrate that the carriage of mail by air was feasible, and could be a great time-saver.

INTERNATIONAL AIR RACES, 1919

The biggest flying events of the year—and indeed the biggest of their kind ever staged between Canada and the United States to that date—were the International Air Races between Toronto and New York sponsored by the first post-war Canadian National Exhibition. Two distinct contests were under way at the same time, both beginning on August 25.

The Canadian race was from Toronto to Roosevelt Field, Mineola, N.Y., and back, while the American race was the same course in reverse. Pilots of either nation could fly in either race, and the distance of the round trip was approximately 1,142 air miles.

Many types of aircraft were included in the entries, and many well-known pilots took part, but as the weather was exceptionally bad, with rain and dense low-scudding clouds, it can well be understood that accidents were numerous, although there was no loss of life.

Eighteen had entered for the Canadian event, but only twelve lined up on the 25th. The names of the pilots included some which were already famous or were to become so: Bertram Acosta, Colonel W. G. Barker, V.C., Lieutenant W. Campbell, Captain H. W. Cook, Sergeant C. B. Coombe, Lieutenant Moore, Lieutenant J. E. Palmer, Roland Rohlfs, Lieutenant C. A. ("Duke") Schiller, Lieutenant Schlander, Major R. W. ("Shorty") Schroeder, and Lieutenant W. Young.

His Royal Highness, Edward, Prince of Wales, now the Duke of Windsor, who was on a tour of Canada in 1919, officiated at the opening of the

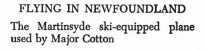
FLYING IN NEWFOUNDLAND
The Martinsyde ski-equipped plane used by Major Cotton

Canadian National Exhibition, and was on hand to wish the Canadian flyers Godspeed when they took off.

The race went to the competitor who flew the route in the fastest time, not necessarily to the first airman to return to the starting point. Major Schroeder, an American, was the winner; he arrived back at Toronto on the 26th, having covered the flight in 9 hours and 35 minutes, flying a Vought VE7, powered with a 150 h.p. Hispano Suiza engine.

Roland Rohlfs, flying a Curtiss Oriole, received credit for the best time made by a civilian pilot. Actually, he accomplished the third best time of the contest, although he had a forced landing on the way back. (A leaky radiator had required attention and he repaired it by plugging the hole with a large wad of chewing gum.)

In the American part of the race, which started from New York, 28 of 47 entries set off on the 25th, flying ten different types of aircraft in which De Havilland 4's predominated. A number of the Americans came to grief or went astray before reaching Toronto, especially at the flying field at Albany, N.Y. This tricky area was surrounded by high wires on every side and city buildings in all directions, and on the day of the race a strong crosswind added to the airmen's troubles. Plane after plane cracked up on the field as they endeavoured to make a landing, and the wreckage scattered about made it difficult for later arrivals to land.

Colonel Clagget had been entrusted with a letter from the President of the United States to H.R.H. the Prince of Wales, and when Clagget cracked up at Albany, the missive was passed on to Lieutenant B. N. Maynard to carry on to Toronto. Maynard became lost on the way, ending up far off his course at Windsor, Ont. As a fellow airman remarked at the time, "Maybe Maynard had heard of the royal castle at Windsor, England, and had simply made a slight miscalculation in the locality!" At that stage the letter went on by post.

Maynard, however, was one of those who made the round trip, being the eighth pilot to touch down at Roosevelt Field. Eleven others also turned in a good performance.

An extraordinary accident took place at Toronto when Major Simmonds came in from New York to land at the Leaside airdrome. A smokescreen sent up to indicate wind direction obscured his vision just as he was about to touch down. As his machine skimmed low over a horse and wagon, the lower wing tip caught on the harness and completely stripped it from the animal, yet neither horse nor machine was injured.

A prize of $10,000 had been announced for division among the winners of the two races, but not one penny in cash was ever awarded. At a dinner held in New York on October 7, 1919, to which all contestants were invited, twelve of the pilots were presented with wristwatches in token of their accomplishments in the race, but the twelve who received the awards were all American pilots.

The Canadian flyers carried air mail during the race. The letters bore special one-dollar stamps depicting an airplane in flight, with the Prince of Wales's feathers some distance above. Around the edges of the stamp was the wording: "The First International Aerial Mail Service, August, 1919—Toronto New York—Aero Club of Canada Commemorative Stamp." Some envelopes also had a rubber-stamp *cachet* "Aerial Mail, Aug. 25, 1919, Toronto, Canada." Just which airman flew the mail is not known, nor the number of letters carried, but there was a considerable volume. Their value has varied since from $3 to $15, according to markings and rarity.

A few more items more or less connected with this race are of interest.

Two of the American flyers, Schroeder and Rohlfs, made history on later occasions. The latter broke the world altitude record in 1919 in the Oriole when he went up to a height of 30,770 feet above Mineola, N.Y. Six months after the race, Schroeder, in an Army LePere biplane, pioneered the way to the stratosphere, when he rose to the great altitude of over 7 miles at Dayton, Ohio, on February 27, 1920. During this flight he lost consciousness from lack of oxygen, and spun down for almost seven miles before he regained control of his senses and the machine. The effects of the heart strain suffered on that terrific adventure inflicted permanent injury, from which he never fully recovered. In recognition of the many flying achievements of Major Schroeder the United States government presented him with the Distinguished Flying Cross on July 14, 1945, at Chicago.

A very indirect result of the race was the invention of one of the most useful aids to aerial navigators—the bank and turn indicator. How this important idea was born on a Canadian exhibition ground as a suggestion from one

American pilot to another, is related in these pages for the first time, as far as I know.

When Major Schroeder left Buffalo on the last leg of his return flight from New York to Toronto, a dense fog over Lake Erie made it impossible to detect any fixed landmark, and he had the greatest difficulty in keeping the VE7 on an even keel. How wonderful it would be, he thought, to have an instrument which would enable a pilot to fly normally under such conditions.

When he landed at Toronto he met his friend Lawrence Sperry, who had had trouble with his Gnome engine and could not get away at the start of the Toronto race. Schroeder had been Katherine Stinson's mechanic in his "Early Bird" days, and was an expert on such engines, so he was able to fix up Sperry's Gnome. It was then that the Major brought up his instrument idea to Sperry, who promised to work on the suggestion.

Not long afterwards, when Schroeder was back at his station at Wright Field, Dayton, a package reached him from the Sperry factory at New York. It contained the first Sperry Bank and Turn Indicator to be manufactured. Schroeder used it on a blind flying test, and proved its worth. The Sperry Company became one of the world's foremost manufacturers of aviation instruments, but Lawrence Sperry, unfortunately, did not live to see it. He lost his life in an air accident in the English Channel on December 14, 1923.

An interesting experience which the race contestants shared with spectators was the sight of the huge British airship, the R-34. She had just crossed the Atlantic, the first lighter-than-air machine ever to do so, and during the days of the International Air Races her impressive bulk floated majestically at her mooring-mast at Mineola.

The world over, 1919 was a momentous year in the air. Across Canada the story was the same: passengers were being flown, not only for pleasure, but on business; and increasingly, experiments were being made in the speeding of mail services by air. By the end of 1920 it was to be demonstrated, beyond any reasonable doubt, that aerial transportation had come to stay.

LABRADOR, 1920

The first aerial forestry survey ever under-

taken in Labrador was conducted in 1920, under the title of the H. V. Green Labrador Expedition of Canada, Limited. It was financed by the Southern Labrador Pulp and Lumber Company, Inc., of Boston, Mass., who then owned a 39-year lease on approximately half a million acres of the finest timber land in southern Labrador. The object of the expedition was to survey and photograph a 2,500-square-mile tract of virgin timber in the areas surrounding Battle Harbour, St. Louis Inlet, and Alexis Bay.

The expedition consisted of thirty men, five of whom were pilots including Mr. Daniel Owens, who was in command of the project. It was an expensive venture, costing $200,000 to finance. Two two-seater Curtiss JN4 biplanes (Jennies) were used for the aerial photography. Five thousand gallons of gasoline and oil went north for the use of the aircraft, together with a large amount of additional supplies.

All personnel and equipment set off from Annapolis Royal, Nova Scotia, the first week in July 1920, aboard the S. S. *Granville*, and a suitable base of operations was eventually selected at Battle Harbour, Labrador, near the Grenfell Mission, where a large cleared area made an excellent flying field.

Three Morse wireless sets went along also, for use in ground to air communications. They were the first sets ever used for such work in Canada. Although their useful range was short, only 125 miles, it was quite sufficient, and in keeping with the flying range of the Curtiss Jennies. An ice cream machine aboard the *Granville* was a part of the equipment, and its happy influence was felt at the Indian Mission established at Battle Harbour where the Indian children had never enjoyed, or even seen ice cream before. The promise of this treat each day performed a near miracle in their behaviour.

The survey was brought to a successful conclusion by the end of August. Not a single accident marred the operations—a rare accomplishment for commercial air undertakings for that era, particularly considering the rugged wilderness over which the planes were flying, which speaks volumes for the abilities of both pilots and maintenance crews. A total of 13,000 aerial photographs were taken, sparking the immense pulp and paper industry now in operation in Labrador, which owes its beginning to two sturdy Curtiss Jennies, and the Canadian personnel which kept them continuously on the go.

IN Winnipeg, Man., two flying companies were established in 1919, one as keen as the other to serve the public and anxious to relieve each eager passenger of $10 for a ten-minute hop, or go bust in the attempt. By the end of the 1920 season one of the two achieved the last alternative, but not before a great deal of interesting flying had been done by both, and additional barnstorming history written. The one outfit, flying from headquarters south of the city, was headed by Lieutenant Mel Dover. The other concern was the British Canadian Aircraft Co. Ltd., with a flying field situated at St. Charles, west of the city. Of the itinerant post-war barnstormers Lieutenant E. A. Alton, owning and flying his Jenny, made his home in Manitoba, foraying into adjacent provinces and states. Others came and went, but Alton and his faithful Curtiss carried on in fair weather or foul, until the first and final crack-up at the end of the season in 1924.

Before we return to the British Canadian Aircraft Company, I can't help digressing on the subject of Alton's mechanic, whom I can only remember as "Slim." Not only was he an expert mechanic, but also a wing-walker *par excellence*. His most daring stunt was to sit atop the centre section of the upper wing, a couple of thousand of feet in the air, hook his long legs around the two front centre struts, at the same time grasping two loops of rope attached to the two rear centre struts. Then Alton would stick the nose of the machine down to earth, and after sufficient speed was attained, over they would go in a tight loop. They would do this several times

on the way down, before they considered they had given the watchers below their full money's worth. Slim would then disentangle himself and scramble back into the rear cockpit, and they would land to receive their well-deserved applause.

In 1920 the British Canadian Aircraft Company was reorganized under the name of Canadian Aircraft Company Ltd., and purchased seven Avro biplanes from England, to enter flying in a big way. Meanwhile, Allied Aeroplanes of Brantford had gracefully folded its wings. I had returned to Stanmore, Alta., to spend the winter, and by the spring of 1920 I had completed arrangements with the Canadian Aircraft Company to become one of their mechanics during the coming flying season. Teaming up with one pilot or another that summer I found myself in the thick of the busiest season of "farmstorming" the West ever saw.

The word "farmstorming" I have coined deliberately, there being no apter way of suggesting the landing-fields. At practically every town where a pilot flew, taking up passengers for a joy-ride at $10 a flip (10 minutes), he usually had to ask permission of some local farmer to use his pasture, stubble field, or "potato patch," as a place from which to operate.

Our usual procedure, after making arrangements for use of a field, was to fly over the town and drop handbills, telling the people that we and our machine were, or would be, at So-and-so's field on a certain date, ready to do business. Also—and this used to bring them on the run —the handbills stated that the first person to hand one of the dodgers to the pilot would receive a free ride. Sometimes the handbills were dropped a week ahead of time, while we were *en route* elsewhere to fill a date. When the great day arrived the whole town was agog, and schools were sometimes closed to give the kids a chance to be on hand. Then as the roar of our engine was heard, and we dropped low, flying at roof-top level along Main Street, the entire populace and all the dogs turned out, racing to the field. In "tin lizzies," buggies, and wagons, on horseback, on bicycles, and on foot, they poured along the highway. Usually the first to arrive was a breathless youngster who triumphantly thrust a ragged but precious handbill into the pilot's hand.

Occasionally we arrived unheralded, and after flying over a few times at a very low level to arouse interest, we selected a suitable landing-spot. A crowd arrived like magic, and soon it was all we could do to keep people from swarming all over the plane, while trying to answer a thousand and one questions at the same time.

At fairs and exhibitions stunting and wing-walking were routine. The latter which was usually the mechanic's role, consisted of climbing out of one's seat when the airplane was in flight, and parading back and forth along the front of the wings, on either side of the fuselage. This of course would hardly do on a monoplane, but practically all the barnstorming aircraft in Canada and the United States were biplanes. It was necessary to remain in front of the wings, as the strong blast of air while flying at approximately 75 m.p.h. would have been rather too much for anyone trying to hang on to the back edges of the wings. By walking in the front, the body of the exhibitionist was kept pressed against the cross-wires between the upper and lower wings and not too much difficulty was experienced in staying put. Sitting astride the fuselage behind the rear seat was another pet exploit of many of us who wished to give the spectators below an added thrill, and it did, as we whipped by over their heads at a very low height, so that they could see clearly what was going on. All this sort of thing was thrown in for good measure by those of us who did such stunts, and I, for one, never received any additional pay for it. Many of the more daring of the wing-walkers, with stiffer nerves and muscles than I possessed, would climb down from the undercarriage and hang there by both hands as the plane sped over the heads of the crowds

A TYPICAL ADVERTISING HANDBILL

below. Others still more daring would get out to the wing tips and hang there by their hands, or even their legs if the urge overtook them. Many, especially in the United States, became specialists at this sort of thing, turned professional, and undoubtedly were well paid for their efforts.

It may be that this kind of behaviour got on our nerves more than we realized. Certainly most airmen of the period were superstitious, and I can't claim to have been an exception. On one occasion Fred Stevenson and I were sitting beneath a wing awaiting the arrival of the crowds, when I spotted a four-leaf clover. As I reached to pick it, I noticed others, and before the patch had been cleaned out, we had gathered 13 of them. Now, undoubtedly we experienced bad luck later that day in having a crack-up; but we survived, which was definitely good luck. I never was able to figure it out.

When a plane stayed at a small town for a day or two, ordinary activities almost ceased, the local school was closed and dances were

CAPTAIN ERNEST HALL (*left*) AND CAPTAIN ALFRED ECKLEY
At Victoria just before their return to Vancouver, May 13, 1919

frequently organized for the occasion. Fred Stevenson and I were invited to one such affair at the delightful farming centre of Treherne in southwestern Manitoba. Unfortunately, we had only the clothes we stood in, which, through much contact with oil and dust, were most unpresentable, Steve's being particularly grimy. With his gracious manner and his good dancing, however, he was extremely popular with the young ladies. As the evening wore on, it was easy to spot the girls he had danced with; their flowing white dresses showed the smudges transferred from his dirty flying-togs. But there were no complaints.

Looking back on those days it is difficult to know what memories to select, so many come thronging. A few will suffice to pass on a glimpse of the times. Every day of the short flying season would have its incidents; some days too many.

Passengers were sometimes good for a laugh. Once, during a routine flight over Winnipeg, the pilot of one of the three-seater Avros, carrying two men as passengers, heard his engine cough, and then go silent right over the heart of the city. He made a masterly landing, past buildings and wires, right into one of the small parks almost in the centre of the city, cutting through low bushes and several flower-beds before the machine came to a halt. He was about to exchange congratulations with his passengers on their close call, when they climbed out of the machine, shook hands in turn, thanked him enthusiastically for the ride, and went on their way, without the slightest show of concern. Apparently for them the landing was normal and usual. Perhaps they were air-minded, but they surely didn't possess any air sense!

On another occasion, near the little town of Hamiota, Man., Pat Cuffe was slugging away one sweltering June afternoon in 1920, hopping off with passengers and landing again at regular intervals, almost like clockwork. While he was up with one of them, making a ten-minute circuit, a buxom woman of about 300 pounds' weight came along, and asked the mechanic to book her for a flight. Nothing daunted, he signed her up, thinking it a good joke on Pat. However, when Pat landed and saw the size of the next passenger, he shrank down in his cockpit so that only the top of his head could be seen, snarling *sotto voce*, "Get her to heck out of here; if she ever tries to get in, she'll wreck the kite." He was adamant, so the joke backfired on the culprit—who happened to be myself. I had to refund her money with apologies. The woman took it in good part, although deeply disappointed.

To relate another instance of how easily trouble could crop up, here is a sample which befell this same pilot, Pat Cuffe, and myself at Portage la Prairie one summer morning in 1920.

We had left Winnipeg in an Avro machine which was fitted with a Le Rhone rotary motor. Now a rotary motor is quite different from the type known as a radial. The former rotates together with the propeller around a rigid crankshaft, so that in motion the whole engine is a rotating blur. In a radial engine the cylinders are arranged radially around the crankshaft, which turns while the engine remains stationary. Our Le Rhone began to splutter and miss as we were winging past north of the prairie town, and finally got so bad we were forced to land in a wide pasture. Pat lit into me at once, blaming me for the trouble, although I had not had anything to do with the servicing of that particular engine prior to the flight. In return I gave him a piece of my mind, and told him that unless he apologized no engine would be fixed by me. It was a fine morning, and we two young hotheads sat more or less back to back in the farmer's field, and there we remained for over two hours until finally Pat cooled off enough to admit he had "blown his top." We apologized to one another, and after shaking hands, I soon had the trouble with the engine remedied. All that was wrong was that several lengths of brass wire that carried the spark from the magneto to various spark plugs had snapped, the wire used having been too brittle. Afterwards, Pat and I became the closest of friends, forming a happy comradeship which continued during many flying trips.

The efforts of the mechanics or, as they later came to be named, engineers, made a chapter of our flying history that has been sadly neglected. In the pioneer barnstorming era, before World War I, engines, being temperamental, were treated accordingly, and mechanics and pilots never trusted them very much. On the ground or in the air, we listened continuously for any sign that might indicate a motor was not functioning properly. It is an awkward habit I have kept to this day, and even while riding comfortably in a four-engined airliner, winging safely along at thousands of feet, the old urge to listen takes over, and I look around sheepishly at the other passengers, as if they could tell what was passing through my mind.

Life was never dull in those distant days. Across the border, the state of Minnesota was in the throes of Prohibition, but the province of Ontario was very "wet." Hector Dougall and I were on a return flight to Winnipeg from Fort Frances, and it was necessary to route our journey *via* a northern portion of the American state. On the way, we were obliged to land for fuel, and came down in a ploughed field near the town of Roseau. We had not long to wait for people to arrive, and down the highway from the town, plumes of dust quickly began to rise, as automobiles came speeding out filled with the townsfolk. From the first car to arrive, a man jumped out hastily, and came running up to us. But did he want to know whether we were going to fly passengers, or to see if we had broken down? Not at all. He held in his hand a roll of money and as soon as he could catch his breath he gasped, "You got any booze? Here's twenty-five bucks for a bottle if you have." When he learned it was the other way round, and that we had landed to buy our engine a "drink," he was utterly disgusted, and so were many others who thought we were bootleggers *de luxe*. While we were there, we were told that a plane frequently landed in those parts, bringing in liquor from some Canadian source, but we were never able to pin the story down.

VANCOUVER AERIAL LEAGUE, 1919–1920

While all this activity was going on in the middle west, much was happening on the Pacific Coast. Spearheading the drive to establish flying as a useful and profitable peacetime occupation was the Aerial League of Canada. Branches had been formed in Montreal and Toronto during the war but other Canadian cities were not far behind, and the branches formed at Victoria and Vancouver in 1919 were particularly active.

On May 13, 1919, two members of the Vancouver branch, Captain Alfred Eckley and Captain Ernest Hall, made a flight to Victoria and back, across the Strait of Georgia: a risky trip, because it was made in a wheel-equipped Curtiss JN4, but quite successful. It is on record as the first flight between the Canadian mainland and Vancouver Island.

Lieutenant "Miny" MacDonald, another Vancouver Aerial Leaguer, had the honour of taking aloft the mayor of Vancouver, R. H. Gale, on July 9, 1919. Macdonald also took up George and Harry, the two young sons of the mayor, for short flights.

One of the outstanding efforts of the Victoria

LIEUTENANT R. RIDEOUT
RECEIVES A LETTER FROM
MAYOR PORTER

While Lieutenant H. Brown stands by,
on May 18, 1919, at Victoria

branch was a flight from Victoria to Seattle and return. The Curtiss machine, *Pathfinder*, was flown by Lieutenant Robert Rideout, accompanied by Lieutenant H. Brown. They left Victoria at 11:00 A.M. on May 18, 1919, and flew by way of Whidbey Island, where they were obliged to land at Coupeville owing to adverse weather, and eventually reached Seattle at 5:50 P.M. They carried an invitation from the mayor of Victoria to the mayor of Seattle, inviting the latter to attend the 24th of May celebrations at the British Columbia capital. On the 19th the return flight was made, non-stop, without incident. The airmen carried back a number of letters, the first actual although not official air mail to travel across Puget Sound.

Two other pilots of the Victoria branch, Captain James Gray and Captain Gordon Cameron, flew another machine, *Pathfinder II*, from Victoria north along the coast of Vancouver Island to Nanaimo on August 16, 1919. They carried official mail. About 90 letters were flown, bearing only regulation postal cancellations. Today, single copies are valued at $75 each. The machine was flown back to Victoria the same afternoon, but through a misunderstanding no mail was carried on the return trip.

Early in 1919, the first official air mail was carried between points in Canada and the United States. This flight, of international importance, was an American achievement. The route was from Vancouver to Seattle, and a machine was brought from Seattle for the purpose.

An American, Eddie Hubbard, was the pilot, the machine a C3 Boeing seaplane. W. E. Boeing, president of the Boeing Company, was a passenger. The worst part of the venture was getting to Vancouver. Twice they were delayed, and once the plane was damaged, by bad weather. But finally, on February 28, 1919, they reached Burrard Inlet, moored their seaplane at Coal Harbour, and were rowed ashore to be greeted by members of the Vancouver War Exhibition Committee, under whose auspices the flight had been arranged.

At 12:30 P.M. on March 3, after receiving a sack of mail containing 60 letters from Postmaster R. G. Macpherson of Vancouver for delivery to Postmaster Battle of Seattle, the two airmen said good-bye to their Canadian friends. They circled the down-town area of the city, then, heading into a stiff west wind, they flew out over the harbour and on past the First Narrows, outward bound. After a short stop at Edmonds, Wash., to refuel, they reached Seattle, three hours after leaving Vancouver, with 125 air miles behind them. Single specimen envelopes carried on the flight are priced at $75 in current stamp catalogues, but it is practically impossible to purchase them as they are already in the hands of collectors, who rarely part with them.

Eddie Hubbard later became the first air-mail pilot to fly a regular international route on the North American continent, when a daily air-mail service went into operation between Seattle, Wash., and Victoria, B.C., beginning

October 15, 1920. On that date, Hubbard flew
5 sacks of mail from Seattle to Victoria, arriving
in time to connect with the Japanese ship, the
Africa Maru, which sailed the same day for the
Orient. Hubbard made the flight of 84 miles in
50 minutes, but his return flight to Seattle, after
dark, against a stiff headwind, took 1 hour and
50 minutes.

On the 16th Hubbard again left Seattle, to
pick up mail from the Canadian–Australasian
mail liner, *Tahiti*. In this manner, official air
mail began over a regular route between Canada
and the United States, and for seven years
Hubbard successfully flew the trip almost daily,
summer and winter, in both the best and the
worst weather the Pacific Coast had to offer. He
had logged the amazing total of 350,000 air
miles before he withdrew as a pilot on that air-
mail route.

Not all the flying in British Columbia, however, was to and fro across the border. Typical
of the hazards to which the coastal flyer was
exposed is the experience of Lieutenant W. H.
Brown of the Vancouver Island Aerial Service,
flying alone in a float-equipped Curtiss from
Alert Bay to Prince Rupert.

About half an hour out of Alert Bay, Brown's
engine suddenly conked out, and the airman
made a forced landing on the sea not far from
Nalau Island. Darkness overtook him adrift on
the open water, and, as if that were not enough,
a gale came up, lashing the sea and carrying the
machine towards the rocky shore of the island.
The aircraft soon was caught on a reef where
Brown abandoned it and with difficulty swam
ashore. By morning the Curtiss had disappeared
beneath the waves. The island was devoid of
life and habitation, and after remaining on it
for twenty-four hours, the airman realized he
would have to act, or die of starvation and
exposure. Brown had earned the Military Cross
in the air over enemy lines, and he still had the
fortitude which had gained that decoration.
Selecting a suitable log from the beach, and a
piece of flat driftwood for a paddle, he straddled
his makeshift canoe and set off to get into the
steamer lanes, where he would stand a chance
of being picked up. After twenty-four hours of
weary paddling, he was spotted by the steamer
Hidden Inlet, bound for Seattle from Alaska. He
was pretty well exhausted when discovered, and
was far out on Queen Charlotte Sound, about
25 miles southwest of Bella Bella, where the
coastline is both rugged and desolate. It was

only clear thinking and a determination to go on
living that enabled the airman to extricate himself from a very tight spot.

ALBERTA FLYING, 1919–1920

It must not be supposed that flying was idle
in either Alberta or Saskatchewan in the 1919–
1920 period. At Edmonton, Alta., Captain
"Wop" May and his brother, Court May, had
established May Airplanes Ltd. in April 1919.
It was a one-plane flying company at the start,
with Lieutenant George Gorman as their first
pilot, since Wop May was under contract to fly
for Captain Fred McCall of Calgary.* The May
brothers procured the *City of Edmonton*, a
Curtiss JN4 biplane which had been given to
the city by the R.A.F. in return for one which
Edmonton had donated during the war. With
Pete Derbyshire as mechanic, the craft ranged

*Fred McCall shared with Captain Claxton the honour
of being Canada's sixth ranking ace. Each man brought
down 37 enemy planes during World War I.

W. E. BOEING (*right*) AND EDDIE HUBBARD
At Seattle after their flight of March 3, 1919

ALL THE FUN OF THE FAIR
Captain Fred McCall's mishap at the Calgary Exhibition, July 5, 1919

the northern Alberta skies, visiting many farming districts. It was the first airplane to be flown in the Peace River area, and the means by which thousands of farmers and their families made their first contact with the age of flight.

In Calgary, Captain Fred McCall had established his flying company, the McCall Aero Corporation Limited. With Captain May and Lieutenant Donnelly as additional pilots, his machines were flown that season to dozens of southern and central Alberta towns and villages.

One of his aircraft came to grief in a spectacular and unusual accident. During the 1919 Calgary Exhibition, McCall was taking off from the infield on July 5, with the two young sons of the exhibition manager as passengers. A few seconds after they were airborne, the engine cut out. McCall had to choose between landing in the middle of a track where an automobile race was in progress, or going down into a crowded midway. He chose the latter. With fine judg-

ment he stalled the machine so that it dropped on the very centre of a merry-go-round, which was going full blast! No injury was suffered by pilot, passengers, or public; the roundabout withstood the impact, but the plane sustained considerable damage.

Two wandering Alberta airmen in a JN4 were Lieutenants Hollbrook and McLeod, of Youngstown and Hanna respectively. Their headquarters were in Hanna, but they were away most of the time, visiting innumerable points during 1919, 1920, and 1921, and they can be credited with having had as much success as any of the early one-plane operators.

SASKATCHEWAN FLIGHTS, 1919–1920

Saskatchewan had its quota of pioneers, too. At Lieutenant Stan McClelland's headquarters in Saskatoon were pilots Lobb, Dick Mayson, Angus Campbell, and Angus MacMillan. The two Curtiss machines they owned visited many towns and villages in the province. Down south at Regina Lieutenant Roland J. Groome and Edward Clarke formed the Aerial Service Company, Ltd. Groome was the pilot on the flight between Regina and Moose Jaw, May 26, 1919, when specially printed copies of the Regina *Leader* were carried between the two cities. Soon after, Groome became the first licensed commercial pilot under federal government regulations; his No. 1 commercial air pilot's certificate bore the date July 31, 1920. One of the Curtiss planes which the company used bore the first registration letters to be used in Canada—G-CAAA. These were allotted on April 20, 1920, and two days later the next letters in sequence, G-CAAB, were allotted to the Curtiss JN4 owned by McClelland and Lobb of Saska-

CANADA'S FIRST REGISTERED
COMMERCIAL AIRCRAFT

The Curtiss of Aerial Service Company, Regina

CANADA AIR REGULATIONS 1920

Left: The first air engineer's certificate   *Right*: The first commercial air pilot's certificate

toon. Groome's mechanic, Robert McCombie, was the first registered air engineer in Canada, receiving licence No. 1, dated April 20, 1920.

After a distinguished aviation career, Roland Groome lost his life in an accident while he was instructor with the Regina Flying Club, when he and a student spun down to their death on September 20, 1935. To that date he had taught 175 young men and women to fly, without a single accident.

AN AIRPLANE SERVES THE PRESS

One last incident remains to round out the story of the growing diversity of uses to which the airplane was being put, as a result of the fortitude and resourcefulness of the farmstormers. Back in Winnipeg the first Canadian newspaper to wake up to the potentialities of the plane as a swift news gatherer was the Winnipeg *Free Press*.

Early in the morning of October 13, the Canadian Aircraft Company city office telephoned their St. Charles airdrome west of Winnipeg to have a machine ready for instant take-off on the arrival of a reporter, who was already speeding to the airdrome by automobile.

The reporter, Cecil Lamont of the Winnipeg *Free Press*, was not long in arriving. As he donned the flying togs we had ready for him, he hurriedly told us that bank robbers had blown the vaults of the Union Bank of Canada at Winkler, a village 74 miles away; that all lines of communication with the town were dead; and as far as was known, the bandits had escaped with all the cash.

Pilot Hector Dougall, Lamont, and I immediately climbed aboard the Avro and took off into a west wind, swinging south in a climbing turn without wasting time in gaining altitude. Our course carried us over some of the rich sections of Manitoba's wheat belt, where threshing operations were in full swing. We could see innumerable dust clouds rising from points where the work was in progress.

Forty-two minutes after leaving the air field the plane landed smoothly on the edge of the village of Winkler, and Lamont hurried to the battered bank. The interior was a mass of débris,

THE PRESS DISCOVERS THE
ADVANTAGES OF AIR TRAVEL

Left to right: Hector Dougall, Cecil Lamont, and Frank
Ellis start for Winkler

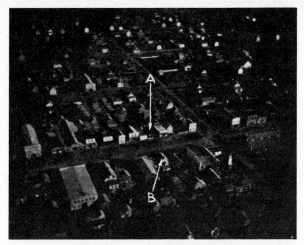

THE SCENE OF THE CRIME

(A) The Union Bank   (B) The blacksmith's shop
Taken from the air on the way back to Winnipeg, October
13, 1920

and the vault had been blasted almost beyond
recognition. The wrecked safe lay in a clutter
of bricks and mortar, its door blown away, and
the inside smoke-begrimed and empty. The contents, $19,000 in cash and bills, had vanished.
Claude Williams, the teller, had been asleep in
his quarters at the rear of the bank. He was
awakened by several armed and masked men
who bound him hand and foot. Learning that
they intended to blow the safe, Williams asked
the bandits to move him some distance away
from the bank, as he had been severely shell-
shocked overseas during World War I. The
leader of the gang said he, too, had served overseas and with comradely consideration had the
teller carried far from the building, and covered
him with a mattress and blankets in order to
lessen the noise of the explosions.

The robbers first dynamited the vault, then
prepared the charge for blowing the inner safe.
They smashed off the combination knob with a
sledge hammer, and poured the nitro-glycerine
"soup" into the hole. Almost an hour after the
blast that blew the vault, the second shot was
touched off. Terrific charges were used in both
instances. At the first explosion the entire popu-

lace had awakened to find the place patrolled
by masked men, armed with sawed-off shotguns.

The village at that time numbered some 600
people, mostly Mennonite farmers. Only one
man ventured forth—the blacksmith, who, under
the impression that a fire had caused the explosion, was rushing to sound an alarm. He was
brought down by gunshot but managed to crawl
back to his house. After that, everybody remained indoors. When the robbers finally
cleared off in high-powered cars *en route* to
Dakota, the villagers had to go eight miles to
the nearest town in order to send word to
Winnipeg.

Lamont rapidly jotted down this story while
Dougall and I took photographs. The Winkler
people were still so excited about the robbery
that very few of them were aware of our machine, which was still a great novelty in small
places.

The Winnipeg *Free Press* came out with the
full story of Manitoba's biggest bank robbery
on that very evening, a scoop which established
a precedent in newsgathering by plane in
Canada.

19. Through the doldrums, 1921-1923

THE flying boom which developed in Canada following World War I had passed its peak by the end of 1920. Joy-riding and exhibition stunting had lost their novelty and financial returns were so small that many of the new companies went out of existence.

Thus the flying doldrums set in. From the end of 1920 until 1924, the only flying of importance was done by the government, a few companies that had been able to carry on, and a handful of ambitious individuals who had sufficient financial backing to do some commercial flying. Although there were a number of notable achievements during the doldrums, the regrettable fact remains that Canada lost several valuable years during the lull, and the development of commercial aviation in Canada was much slower than it should have been.

LAURENTIDE AIR SERVICES

In 1921 and 1922, however, there were one or two ambitious undertakings. Ellwood Wilson, chief forester of the Laurentide Pulp and Paper Company, was responsible for one of them. Thanks to his recommendation and to a small government subsidy, airplanes had been put into use in the St. Maurice Valley in 1919, when two HS2L flying-boats were loaned to the company by the federal government, and were flown from Halifax to Grand'Mère, as will be recounted in detail on a later page. This was the beginning of forestry patrol in Canada, and the base on Lac à la Tortue was the first licensed air harbour in Canada. This, too, was the first time photography was used in forestry work,

one plane being equipped with an Eastman 10-inch K-1 camera. During 1920 and 1921, aerial fire patrols by the company increased in number, and operations were reorganized under a separate branch. In 1922 Laurentide Air Services Limited came into being, and more aircraft were added each year as conditions warranted. The first pilots of the new company were Roy Grandy, Roy Maxwell, and G. A. Thompson.

In 1922 a large contract was obtained from the Ontario government for work in connection with a detailed map showing lakes, waterways, and forest types in Northern Ontario. The work was so well done that next year the company was awarded all the flying required in the area for transportation of fire-fighting personnel, forestry patrol, and mapping. Thousands of square miles of country were surveyed and sketched, as far west as the Lake of the Woods district and as far north as James Bay; twelve aircraft were in use, 550 passengers were carried, and an area of approximately 20,000 square miles was covered by sketching or visual inspection. Over 400 fires were spotted and reported during the season. In the fall of 1923 the main base was moved from Grand'Mère to Three Rivers, which, unlike Lac à la Tortue, was free of ice early in the spring.

OTHER COMPANIES

From 1920 to 1922, Price Brothers of Quebec operated a Martinsyde seaplane, G-CAEA, powered with a 270 h.p. Rolls-Royce Falcon engine and flown by H. S. Quigley. Flying from a main

AERIAL FORESTRY PATROL

The Laurentide Pulp and Paper Company crew, Pilot Roy Maxwell (*right*), Pilot Roy Grandy (*centre*), Engineer Irenée Vachon (*left*)

base at Chicoutimi, Que., Quigley made valuable forest patrols until July 1922, when his aircraft was destroyed in a crash in the Saguenay River. A company engineer, unlicensed as a pilot, had attempted to fly the Martinsyde without the owners' consent. The accident was fatal to himself and to his passenger.

Captain Quigley then formed a new company under the name of Dominion Aerial Exploration Company. The government of Quebec loaned the first two machines used, HS2L flying-boats, and the company began flying from a base at Roberval on Lake St. John, Que., carrying on work such as had previously been conducted on a smaller scale by the Canadian Air Board. C. L. Bath and F. V. Robinson were engaged as additional pilots.

Meanwhile, the Ontario Pulp and Paper Company hired an airplane from the United States to operate from a base on Franquelin Bay, on the lower St. Lawrence. With this plane the Company tested in 1922 the possibilities of aircraft in connection with its huge tract of forest land and timber rights in that area. Results proved so satisfactory that an HS2L flying-boat was placed in commission the next year. Throughout the season, men, supplies, and mail were kept continuously on the move, some 97 flights being made. Entire operations absorbed 144 flying hours, including considerable forest reconnaissance over a range of about 6,000 square miles. The season's work was most successful.

In 1922 the Fairchild Aerial Surveys Company (of Canada) Limited came into being as the result of farsightedness on the part of Ellwood Wilson of the Laurentide Company. It was organized as a branch of the Fairchild Company of New York and its purpose was to supply the increasing demand for aerial photographs in forestry. Two machines were assigned to the service and work began in earnest in 1923. The first pilots were H. M. Pasmore, K. F. Saunders, and A. G. McLerie. The engineer was Bill Kahre, who had been Stuart Graham's mechanic on the Halifax–Grand'Mère flights in 1919 (pp. 197–8).

One of the machines was a Curtiss Seagull flying-boat, the other a wheel-equipped Curtiss Standard. The latter has some claim to additional fame. It was originally registered in Canada by the Curtiss Company, and had been used in British Columbia for shooting aerial sequences in a moving picture entitled *The Unseeing Eye*, starring Lionel Barrymore—the first aerial movie photographs ever made in Canada for use in an actual picture.

The Fairchild aerial camera used by the Company in 1923 was the finest of its kind then made. In 1924 the Fairchild photographic work in Ontario and Quebec covered 1,425 square miles of country, and extremely accurate maps were made from it.

In 1923, only eight civilian flying organizations were registered in Canada. The work of the four largest firms has been outlined. The other four which managed to keep their heads above water during the doldrums were the Commercial Aviation School of Victoria, B.C., J. V.

Elliot of Hamilton, Ont., R. J. Groome of Regina, Sask., and O. H. Clearwater of Saskatoon, Sask. In Newfoundland two companies carried on, the Aerial Survey Company and Aircraft Transport and Travel Ltd.

A few comparative figures (which do not include military operations) indicate how sharply aviation declined during the years of the flying depression. In 1921, 23 firms were engaged in flying, in 1922, 15, in both 1923 and 1924, 8—the lowest ever reached in the post-war years. A corresponding decline took place in the number of certified pilots, from 139 in 1922 to 31 in 1924. The number of registered aircraft showed a great reduction also, from 60 fully licensed machines of all types in 1922 to only 32 in 1924.

But aviation for thrills was really giving way to aviation for profits. The prospect of a great commercial expansion could be glimpsed in the increase of freight and express shipments carried. Although only one-half the number of aircraft in operation in 1922 were flying in 1924, 77,385 pounds were flown as compared with 14,681 pounds in 1922, and flying hours almost doubled, rising from 2,541 in 1922 to 4,389 in 1924. These figures augured well at a time when other prospects seemed anything but favourable.

THE R.C.A.F.

Meanwhile, military flying progressed steadily across Canada. Until January 1924, the activities of the Canadian Air Force had included all types of work in the air, and a civil branch had been created for certain types of flying. Much valuable survey and photography work was also performed.

In 1923, flying operations which could be carried on by private enterprise were discontinued, and the Royal Canadian Air Force (it received its "Royal" prefix on February 15, 1923, but the R.C.A.F. officially came into effect on April 1, 1924) confined its activities to military work and to flying for the federal and provincial governments. This included flying for the Department of the Interior on forestry and water-power development, and patrol work for the Department of Marine and Fisheries for the prevention of illegal fishing. The Department of Agriculture also used Air Force machines as a means of combating forest blight and insect pests by dusting large areas from the air. The Department of Indian Affairs employed military aircraft for transporting agents to many outlying districts, and the Department of Mines made numerous demands for the aerial transportation of men and supplies.

By the end of 1923, five major R.C.A.F. stations were in full operation across Canada, in addition to the large training airdrome at Camp Borden. At Dartmouth, N.S., Canada's first seaplane harbour, Squadron Leader A. B. Shearer was in charge, with two officers and ten men under his command. A total of 92 hours was flown during 1923. One flight of particular interest, made at the request of the Post Office Department, was from the Dartmouth Station to the Magdalen Islands in the Gulf of St. Lawrence, the object being to ascertain the practicability of air transport to the islands, since ice in the Gulf made them almost inaccessible by water during the winter. The outward trip from Dartmouth, *via* Charlottetown, P.E.I., was completed to Grindstone Harbour, Magdalen Islands, on August 29, 1923, and the return trip was made the following day. The flight was the first ever made to the Magdalen Islands.

The Ottawa Air Station, on the Ottawa River at Rockcliffe, Ont., possessed both land and water facilities. It was in command of Flight Lieutenant C. McEwen, M.C., D.F.C., and consisted of three additional officers and fifteen men. The total flying time chalked up by that station during 1923 reached almost 63 hours.

The Manitoba area was served by a main base set up at Winnipeg, with additional sub-bases at Victoria Beach on Lake Winnipeg, Norway House, and The Pas. They were all under the command of Squadron Leader Basil D. Hobbs, who had seven officers and thirty men under him; Winnipeg was, therefore, the largest of the five operational stations. Its season's work was an impressive total of 488 flying hours.

THE MARTINSYDE SEAPLANE

Operated by Price Brothers, with H. S. Quigley at the controls

One major undertaking of the Victoria Beach Station in 1924 was an extensive aerial survey in the Reindeer Lake and Churchill River districts. The single aircraft used was a Vickers Viking amphibian, fitted with a 360 h.p. Eagle 9 Rolls-Royce engine, but it reverted to a flying-boat type when the wheels were removed to cut down weight and allow for additional supplies and personnel. When fully loaded the plane weighed 5,663 pounds. The crew consisted of Squadron Leader Hobbs, Flying Officer Cairns, Corporal Milne, and R. Davidson, district land surveyor. The reports of this photographic survey are quite impressive: between July 18 and August 13 a total of 2,180 air miles were flown in 49 hours, and 17,000 photographs were taken over an area of about 15,000 square miles. The data obtained were of inestimable value to the federal government, filling in innumerable blanks which were apparent on all maps of the area at that date.

The station in Alberta was originally established in 1920, west of Calgary near the village of Morley, but was moved to a new site at High River in 1921. Squadron Leader G. M. Croil, A.F.C., was the first Officer Commanding, but Squadron Leader A. L. Cuffe later took over. Six additional officers and twenty-two men comprised the staff. The flying accomplished during 1923 reached a total of 510 flying hours, the highest recorded at any of the five R.C.A.F. operational stations for that year.

At Vancouver, B.C., the Jericho Beach Air Station on English Bay for waterborne craft was in command of Squadron Leader A. E. Godfrey, M.C., A.F.C., and was staffed with two additional pilots and twenty-five men. Two hundred and seventy flying hours were recorded for 1923. Two incidents will serve as examples of the type of work carried on from this base.

A request for a machine was received from Major Motherwell, Chief Inspector of Fisheries at Victoria, asking for the co-operation of the R.C.A.F. in apprehending three American fishermen who had committed a particularly grave offence. Their boat, the *Slioam* of Seattle, had been discovered fishing illegally in Canadian waters off the west coast of Vancouver Island. When commanded by the Canadian patrol boat *Malaspina* to heave to, the *Slioam* attempted a getaway, and the skipper of the *Malaspina* had to order shots fired across the *Sliom*'s bows to bring her to. One of these shots killed a member of the *Slioam*'s crew. The remaining three men of the fishing-boat then scuttled their craft, but were picked up and taken aboard the Canadian ship. Later, the three offenders stole a row-boat from the patrol vessel and escaped. It was rumoured they had found refuge on another United States boat, the S.S. *Jennie*, bound out of Alberni for Seattle.

The Air Force flying-boat was detailed to search for the fugitives, but after a lengthy reconnaissance on April 26 and 27, which

AIR FORCE STATION NEAR
MORLEY, ALBERTA, 1920

covered the entire coastline south from Barkley Sound to Victoria, no sign of the *Jennie* or her will-o'-the-wisp crew was spotted, and the offenders made good their escape.

Better success was achieved on another occasion when an Air Force machine was requisitioned to search for and apprehend a fast motorboat rum-runner, the *Trucilla*. The motorboat was finally located, and the airmen, alighting alongside the speeding craft, ordered her to stop. A customs man aboard the plane took over, and the *Trucilla* was escorted to port under her own power while the watchful airmen circled above.

At this point in our story we must pause, and return. The lull of the early twenties was to be succeeded later in the decade by a hurricane of flying as vigorous and boisterous as the west wind that ripped the Ellis-Blakely machine from its moorings on the Canadian prairie, and sent it rolling like tumbleweed over the landscape. Much of significance in the early postwar period, however, remains to be told, and we return now to describe the flights which, all through the twenties, were concerned with the experimental side of flying rather than the commercial. These, too, would bear their practical fruit in the years to come.

PART 5

ANNIHILATING TIME AND SPACE, 1919-1926

20. Conquest of the Atlantic, 1919

JUTTING eastward out into the Atlantic, now swept by the off-continent westerlies, now shrouded in fog, are the rugged rock-bound shores of Newfoundland, pock-marked with brief harbours, bristling with reefs and bold headlands. Here for more than four centuries have come the fishermen of the British Isles, of Brittany and Portugal. This was the land that Lief Ericson sighted on his voyage to Vinland: the island discovered by Cabot, searching for a western route to the Indies. And here again, in the twentieth century, a drama of daring was enacted, worthy of that heroic past.

Nothing is more indicative of the change wrought during World War I on man's air-thinking, and of his newly acquired confidence in his powers of flight, than that he should cast a speculative eye across the wild Atlantic.

BY WAY OF THE AZORES

During the final year of World War I the need for large flying-boats increased, and to keep abreast of British design, the United States developed a type known as the Navy-Curtiss (NC), built for long-distance flying. This machine was equipped with four 400 h.p. Liberty engines, and the gasoline tanks held 1,610 gallons. The wing span was 126 feet. Fully loaded, each machine weighed 28,000 pounds or 14 tons. The boat portion of the craft had the appearance of a huge streamlined pontoon. The box-type tail was attached to the wings and boat by large boom-type outriggers. Five or six men made up the crew of each ship.

The original intention was to deliver the flying-boats by air to Great Britain, but the first, the NC-1, was not quite completed when the Armistice was signed. Construction of the other three machines was completed early in 1919, and in turn they were designated the NC-2, NC-3, and NC-4.

In 1919, the United States government sanctioned plans for a flight by the four aircraft from New York to England in a series of relatively short hops, the route being *via* Trepassey Bay, Nfld., to the Azores; thence to Lisbon, Portugal, and on to Plymouth, England. As this would be the first crossing of the Atlantic by any heavier-than-air machine, exhaustive tests were made at Far Rockaway, N.Y., of the radio communication facilities between planes, loading capabilities, and so on.

With preliminary difficulties finally overcome, three of the machines, the NC-1, the NC-3, and the NC-4, took off on May 6 for Trepassey Bay, the scheduled point for the actual beginning of the Atlantic flight attempt. The NC-2 did not have any part in the flight.

Over the entire sea route, spread out between Far Rockaway, Trepassey Bay, the Azores, Lisbon, and Plymouth, 60 United States destroyers were stationed at intervals to give assistance in directing the aircraft and to be on hand in case of need.

As the flight officially began from Trepassey Bay, the whole exploit is related to the flying history of Newfoundland, as were a great many other brilliant and tragic trans-Atlantic endeavours in the years to come.

Shortly after 10:00 A.M., on May 8, the three

THE NC-3 AT PONTA DELGADA
After weathering an Atlantic storm

machines left Far Rockaway on the first leg of their journey, the flagship, NC-3, under Commander Towers, being the first away, then NC-4 in charge of Lieutenant Commander A. C. Reid, and finally the NC-1 under Lieutenant Commander P. L. N. Bellinger. The first section, to Newfoundland, called for hops totalling well over 1,000 air miles. The NC-1 and NC-3 made it in two days in a flying time of about 12 hours at a speed of about 85 miles per hour. The NC-4

developed engine trouble and had to put in at Chatham, Mass., for an exchange of engines. She reached Trepassey Bay on May 15, just in time to see her two sister ships returning to their moorings after attempts to leave for the Azores had been foiled by bad weather.

At 6:06 P.M. on the morning of May 16, all three aircraft took off at minute intervals. The weather was bad, and they were quickly lost to sight.

The NC-4, hitherto the lame duck, checked with destroyer No. 22 at 12:10 P.M., Greenwich time, and received discouraging word of the weather ahead. Some time later, through a hole in the fog, the flyers spotted U.S.S. *Columbia*. Fifteen hours and eighteen minutes out of Trepassey Bay, the plane made a successful landing on the sea off Horta, Azores, after a hop of 1,370 miles.

Meanwhile the NC-3 and NC-1 were having trying experiences. The NC-1 had carried on sturdily throughout the night, but several hours after daylight, U.S.S. *Melville*, stationed at the Island of Ponta Delgada, heard the crew of the plane send out a radio message stating they were

THE MORE FORTUNATE NC-4
Arrives at Lisbon, May 27, 1919

lost in the fog, and intended to come down. They did so at a point about 100 miles west of Flores Island and found themselves in mountainous seas. After hours of terrific battering, they were discovered by the Greek vessel S.S. *Ionia* at 6:00 P.M. on the 17th. The crew were rescued, but an attempt to tow the damaged machine proved fruitless, and she sank.

The troubles of NC-3 were even more protracted. At 9:15 A.M., Greenwich time, her crew radioed to say they were far off their course in dense fog, somewhere between destroyers No. 17 and 18. As they were unable to pick up radio from the ships, the crew had to bring the machine down on the sea to try to locate themselves.

Throughout the 17th and on through the following night, the crew rode it out, buffeted by rain, wind, and waves. After 22 hours of this, the left wing tip float was carried away, and men took turns crawling out on to the right wing to balance the craft, clinging there precariously as each huge comber threatened to wash them off. This heroic fight was all that prevented the NC-3 from turning turtle. Messages came over her radio telling of the rescue of the NC-1, and also the disheartening word that surface ships were searching for the NC-3 west of the Flores, instead of south, where she had drifted. The crew realized that rescue was improbable, and that if they were to live they must save themselves. They had radiator water to drink, and a few chocolate bars and wet sandwiches to eat. On the 18th the rising sun silhouetted the top of the 7,000 foot volcanic cone of the highest peak of the Azores. Observations quickly made showed the plane to be about 45 miles southeast of Fayal Island.

Two hours' supply of fuel still remained in the tanks, and the crew decided not to attempt taxiing towards that island under the severe conditions existing—seas running twenty feet high—but to continue to drift until it might be possible to use their motors to taxi to the island of Ponta Delgada. This proved to be good judgment, for land soon appeared. Shortly after, they were spotted by U.S.S. *Harding*, which came racing up. However, having weathered it to that point, the crew refused assistance, and made the harbour under their own power. Their 52-hour, 205-mile sea journey in a disabled flying-boat under severe conditions in the open Atlantic was a very fine feat of seamanship.

UNITED STATES ARMY AIRSHIP C-5
At Pleasantville, Newfoundland, May 14, 1919

On May 20 the NC-4 made the air hop of 173 miles from Horta to Ponta Delgada. A week later she made the 893-mile flight to Lisbon, Portugal, in 9 hours and 43 minutes, without undue incident. This final hop completed the first indirect crossing of the Atlantic by air.

After a three-day stop at Lisbon the NC-4 left for Plymouth, England, on May 30. A brief stop was made at Figueira, Mondego River, to make minor engine adjustments, but by the time they were under way again it was too late to make England, so they landed at Ferrol, Spain, at 4:47 P.M., with 535 more air miles logged behind them.

The culmination of the flight from America came at 1:30 P.M. on May 31, when the crew of the NC-4 received a tumultuous welcome from the crowds lining the waterfront at Plymouth, as the machine came down to moor off Plymouth Hoe.

The fact that 15 days elapsed from the date the NC-4 left Newfoundland until her arrival in England should not detract from the achievement. The difficulties which defeated the two other planes, destroying one and eliminating the other, are proof enough that the first crossing of the Atlantic by air was hazardous in the extreme.

During the time the three NC flying-boats were at Trepassey Bay, a large twin-engined U.S. Army airship, the C-5, arrived at St. John's, Nfld., from Cape May, and was moored at Pleasantville on the north side of Quidi Vidi Lake.

She was in the command of Captain Coil, who

intended to try to bridge the Atlantic by the lighter-than-air transport. But a gale on May 15 ripped her from her moorings, and blew her out to sea. Two men who were aboard at the time endeavouring to attach more lines jumped the twenty feet to the ground just as the last rope gave way. One of them suffered a broken ankle, but they were fortunate to escape. The dirigible went sweeping out to sea before the wind and was soon lost to sight. One of the American destroyers then in port quickly put out in an endeavour to follow and salvage the blimp, but it was never seen again.

THE FIRST NON-STOP TRANS-ATLANTIC FLIGHT

In England, even before the World War I began, bold men had set their minds on flying the Atlantic Ocean. Plans were quickly revived after the war ended; all the more so when Lord Northcliffe, owner of the London *Daily Mail*,

LIEUTENANT COMMANDER K. MACKENZIE-GRIEVE,
R.N., AND HARRY HAWKER

Seen wearing ill-fitting, borrowed suits after their arrival in Scotland

offered a £10,000 prize to the crew of the first airplane to fly non-stop across the Atlantic.

The first contestants to arrive in Newfoundland from England were Pilot Harry Hawker and his navigator, Lieutenant Commander K. Mackenzie-Grieve, R.N. With them in huge packing cases came their specially built Sopwith biplane, the *Atlantic*, powered with a 375 h.p. Rolls-Royce engine.

They arrived at Placentia Bay, Nfld., on March 28, 1919, on the S.S. *Digby*, and the two large cases containing their machines were transhipped to the *Portia*, a smaller vessel. The harbour at St. John's was still ice-packed, and even at Placentia Bay only the smaller boat could worm its way in.

Once ashore, the cases went first by railroad and later by drays hauled by horses, and so to the field. Time and again they were mired on the road before arriving at their destination, which was about six miles out of St. John's. It had been chosen after a lengthy search by members of the Sopwith Company, headed by Captain Montague Fenn, who had reached Newfoundland some time ahead of the airmen to look things over. The whole country was under snow when the choice had to be made, and when the spring thaws set in, a great amount of work had to be done to put the ground in shape suitable for a take-off. At its best it was a poor spot, but there was nothing better in the vicinity. Its entire length was only 600 yards, and it was L-shaped at that.

While their field was being prepared, Hawker and Grieve tested the waterproof life-saving suits which they later wore on their flight and found very serviceable. The upper rear portion of the *Atlantic*'s fuselage was actually a small boat, equipped with oars, food rations, etc., and this tiny craft was also given satisfactory tests on a small lake near by.

And now some rivals appeared.

On April 19, Pilot F. P. Raynham and Commander Morgan, R.N., arrived by boat from England, bringing with them their Martinsyde biplane, the *Raymor*. They set up shop at Pleasantville, on the north side of Quidi Vidi Lake, north of St. John's. All the personnel of the two machines had quarters at the same hotel in St. John's. Friendly rivalry sprang up and many pranks were played.

A gentlemen's agreement was made between the crews of the two machines that neither

THE *ATLANTIC*

would attempt to take off on the Atlantic flight without notifying the other. Bad weather, however, continued to keep them both grounded until they received the news that the three American NC flying-boats had taken off from Trepassey Bay on an attempt to fly to the Azores. Although the weather was most unfavourable—in fact reports were about as bad as they could be—immediate steps were made to get into the air.

On May 18 Hawker and Grieve climbed aboard the *Atlantic*, and at 3:40 P.M. the wheels of the machine bounced for the last time on the soil of Newfoundland. The Sopwith was so heavily loaded that the utmost skill was required to get her away. There was a crosswind of about 20 miles an hour, and in Hawker's own words, "It was a ticklish job; it meant going diagonally across the L-shaped runway, avoiding a slight hill and a deep drainage ditch." They cleared the ditch by inches, and the trees by what Hawker termed "a respectable margin," but which photos showed to be a matter of feet.

As soon as they were well out to sea, Hawker pulled the release trigger of the detachable undercarriage with which the Sopwith was fitted. As they watched the heavy apparatus go spinning down to the water, they had the satisfaction of noting that the needle of their speed indicator rose another 7 m.p.h.

Unfortunately, their radio went dead almost as soon as they started, and for the rest of the time that they were in the air the waiting world knew nothing of them. They went along fairly comfortably at 105 m.p.h. at 10,000 feet, until 10:00 P.M. Then the weather changed for the worse. Darkness had engulfed them; no glimpse of the sea could be seen, and rain began to fall as they forged ahead into the gloom.

About 11:00 P.M. Hawker noticed that the engine thermometer registered higher than it should, and realizing everything wasn't just

right, he opened the shutters of the radiator. Cloud formations began to loom ahead, massive and formidable, and too high to climb over without wasting valuable fuel, so the pilot was forced to alter his course time and time again in order to go around them.

Grieve had been unable to check their course since darkness settled down, but at last the moon broke through a rift to help him. The engine temperature had climbed to 175, in spite of the wide-open shutters, and they realized something was seriously wrong.

They battled along for hours, Hawker doing everything in his power to keep the temperature from rising further. Eventually he managed to climb to 12,000 feet, at which altitude water boils at a lower temperature than at sea level but the air is much colder. By this time the centre section of the top wing was coated with ice from the water which had steamed out of the hole in the radiator cap. They flew along for many miles with the engine thermometer registering 212 degrees. Occasional glimpses of the stars enabled Grieve to plot their course.

At six in the morning, Hawker and Grieve found themselves approaching an immense mass of clouds which rose to over 15,000 feet, stretching like a huge mounting range to the horizon both north and south. It was impossible to fly over the clouds, and unsafe to fly through them, so in their dilemma the airmen went down to 6,000 feet. There it was blacker than ever, so down and down they went until at 1,000 feet it was possible to fly below the clouds and see with safety.

The water in the engine was boiling constantly now, and they knew the end was near.

The sun was just rising when Grieve's check showed them to be right on the steamer lanes. They had a narrow escape from real disaster at that moment. Hawker had shut the engine off completely on the long glide down, as an aid

THE WRECKAGE OF THE *ATLANTIC*
Salvaged by the *Lake Charlotteville*

to cooling. When he flicked the switch to start it again, nothing happened. He yelled to Grieve to get busy on the hand fuel pump to force some gas into the cylinders, and they were only ten feet off the water when the engine broke into a full-throated roar. With the throttle wide open, their glide flattened out, and once again they began to climb. The wind was behind them at the time, and a very heavy sea was running. Hawker said later that one couldn't have a much closer brush with death than that.

Realizing they would have to go down into the sea eventually, they began a zig-zag course on the lookout for a ship. After two hours of this nerve-racking search, with the engine get-

ting noticeably worse, they glimpsed a ship below. They circled her and fired three red Verey lights to show they were in distress, then flew a couple of miles ahead. Bringing the *Atlantic* round into the teeth of the gale, Hawker set the machine down to a "cushy" landing, in spite of huge seas. The Sopwith rode high in the water, but waves broke clear over her.

With difficulty, they launched their tiny boat and managed to scramble into it. The ship soon reached the scene; she proved to be the Danish steamer *Mary*, commanded by Captain Dunn. It took a lifeboat crew two hours to bring the two airmen to the ship. They were picked up fourteen hours from the time of their take-off from St. John's.

The *Mary* had no radio, and days passed before the world learned that they were safe. As a matter of fact King George V sent messages of condolence on May 24 to the airmen's next of kin! It was with excited delight that the British nation finally heard that their flying heroes were alive.

The *Mary* eventually arrived off the Butt of Lewis, Scotland, and contact with shore communications was made. The British Admiralty despatched the destroyer *Woolston* to pick up the airmen, and they were landed at Scapa Flow. There they received an official welcome from Admiral Freemantle, and a tremendous reception from the men of the Grand Fleet. The

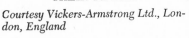

ALCOCK AND BROWN START
THEIR FLIGHT

Courtesy Vickers-Armstrong Ltd., London, England

greeting they received in London the next day surpassed anything of the sort ever seen there before.

Only twenty letters were carried on the flight; the price to send each one was $25. All were salvaged and eventually reached their destinations. Today their value is in the neighbourhood of $500 each.

The battered remains of the *Atlantic* were sighted three or four days after it had been abandoned, and hauled aboard the U.S. steamship, *Lake Charlotteville*. The wings, though battered, were still attached. The plane was later sent on to London, where it was placed on public display.

The cause of the excessive boiling of the *Atlantic*'s engine has been a matter of controversy ever since.

Two hours after Hawker and Grieve had left Newfoundland, Raynham and Morgan attempted to get away.

They managed to get their machine off the ground, but once airborne, it began to drift sideways, and as it was heavily loaded, settled back to its wheels. The strain was too great—the undercarriage was wiped off, and much other damage was done.

Later, the machine was fully repaired, and Lieutenant Biddlecombe replaced Morgan as navigator. On July 17 another attempt was made to take off. The undercarriage again gave way, and the entire enterprise was abandoned.

ALCOCK AND BROWN

During a conversation I had with a young flying officer, who had been on reconnaissance duty on flying-boats out of Newfoundland bases during World War II, I put a casual question. I asked if he ever gave a thought to Alcock and Brown, as he himself flew far out over the grey stretches of the Atlantic. After hesitating a moment, he replied, "Alcock and Brown? I don't know them, where are they based?" So fleeting is fame.

The failure of the first two trans-Atlantic attempts left two other competitive machines still in the running, one a twin-engined Vickers Vimy in charge of Captain John Alcock and Lieutenant Arthur Whitten Brown, the other a huge four-engined Handley-Page bomber.

The Handley-Page was already at Harbour Grace, Nfld., and in process of being rigged

JOHN ALCOCK AND ARTHUR WHITTEN BROWN

when the first two non-stop attempts were made. The Vickers arrived in Newfoundland on May 28, and was first assembled in a field near Quidi Vidi. The field proving unsuited for the take-off of a fully loaded machine, the craft was flown out light to another site on its first test flight on June 9. On the 12th an additional flight test was made, then there were no more flights until the actual start of the Atlantic attempt, which began at 1:58 P.M. on June 14, 1919—a momentous date.

The propeller of their air-driven radio generator sheared off a few hours after they left land behind, so, like Hawker and Grieve, Alcock and Brown were without contact with the human world.

In spite of favourable weather reports received from both sides of the Atlantic, they had not been in the air long before climatic conditions changed completely. The fickle temperament of the skies over the Atlantic is well known, but although the airmen were not taken by surprise, they had a rough and trying trip.

END OF A WORLD-FAMOUS FLIGHT
Alcock and Brown's Vickers Vimy in an Irish bog
Courtesy United Kingdom Information Office, Ottawa

Great masses of black, forbidding clouds rose up to meet them, as they sat in their open cockpit.

One incident was almost their last. During the night, when an airman's sense of balance is least dependable, the craft stalled before they realized what was taking place, and they went spinning down towards the sea. It required the combined strength of both men at the controls to right the Vickers and get her flying again on an even keel. Thus they flew on into the dawn of the 15th, when a waiting world learned that they had crossed the Irish coast at 8:25 A.M.

Low clouds lay over the landscape, and rather than risk flying on to London in such impossible weather over unseen terrain, they decided to land. The spot chosen through the mists proved to be nothing better than a grassy bog. The landing-gear was ripped off the heavy machine as she settled down, and she finished with her stub nose buried in the muck. The Vimy was considerably damaged, but neither pilot was injured. They had come down near Clifden at 8:40 A.M., June 15, 1919, after being 16 hours and 12 minutes in the air, and flying approximately 1,800 air miles.

Their flight not only brought to them the prize of £10,000, but also an audience with King George V at Buckingham Palace, where both airmen received knighthoods in recognition of their brilliant efforts.

Three hundred letters were flown across in the Vickers; in recent years these have come to be valued at $200 each.

Of the first four airmen who left Newfoundland to fly to the British Isles over the direct route, none is living today. Sir John Alcock lived only a few months after his Atlantic triumph. He was killed in an air crash on December 18, 1919, during a forced landing in a fog in France. Hawker also died in a flying accident. Mackenzie-Grieve and Sir Arthur Brown died from natural causes, the former at Victoria, B.C., on September 26, 1942, the latter on October 2, 1948, at his home in Swansea, Wales.

The famous Vimy, first aircraft to fly the Atlantic non-stop, is on permanent exhibition in the Science Museum at South Kensington, London. In St. John's, Nfld., a stone memorial with bronze plaque was erected in 1952 to mark the place where the flight of Alcock and Brown began. A small bronze tablet indicates the exact spot at which the Vimy became airborne.

The fourth contestant being prepared for

THE FOUR-ENGINED
HANDLEY-PAGE

At Harbour Grace, Newfoundland,
July 4, 1919

flight in Newfoundland—the Handley-Page—was in command of Vice Admiral Mark Kerr, with Major Brackley, pilot, Major Trygve Gran, navigator, and Chief Mechanic Wyatt completing the crew. The 4-engined 14-ton monster was one of approximately 200 such bombers built in Britain towards the end of World War I. Its measurements were: length 64 feet, height 32 feet, wing span 166 feet. The crew had no opportunity to try for the trans-Atlantic prize as it had been won before their aircraft was ready to take off.

After relinquishing hope of attempting the ocean flight, the crew made plans to fly to New York. *En route* on July 4, engine trouble forced the plane down at Parrsboro, N.S., and in the resultant rough landing she was so badly damaged that it required nearly four months to put her into flying trim again, and a new fuselage had to be shipped over from England. When the flight was continued to New York on October 9, four passengers—Canadian and American news-men—and four additional crew members brought the number of those aboard to twelve, a large number for those days. A few letters were carried on this trip, but the envelopes bore only regulation postal cancellations. (The few that have survived are valued at $100 each.)

On a later flight from New York to Chicago in the same machine, Major Brackley was forced to make a landing at Cleveland, Ohio, in the gathering darkness. Unable to locate the air field there, he chose a race track, but the space was too narrow for the huge craft and thirteen feet of both wings were sheared off as the landing was made. This brought the Chicago flight attempt to an end. Since winter and other difficulties made repairs impracticable, the crew returned to England by ship. The plane was dismantled and eventually scrapped.

Efforts to use the Handley-Page commercially proved to be premature, for at that period there were no air fields suitable for craft of such huge dimensions.

THE conquest of the Atlantic by Alcock and Brown stirred imagination wherever men flew. In Canada, more than in any other air-minded country, perhaps, the urge to conquer space was confronted with its greatest challenge. Between the populated regions of the East and the prairies lay the rugged tongue of the Canadian Shield, virtually uninhabited. This was obstacle enough. But a second barrier rivalled in ruggedness and treachery the great Atlantic Ocean; this, too, an ocean—not of water, but of rock.

OVER THE ROCKIES

As we recline at ease in the deep soft padded seats of the huge shining metal cabin of a modern Trans-Canada airliner, where a request for a cup of coffee, a pillow, or a warm blanket is quickly filled by an alert and smiling stewardess, we may, perhaps, just take it all for granted. Even so, it is always a tremendous thrill to fly over the Rockies, whether for the first time or the tenth. Winging at a great height over the rugged terrain, as far above the tallest mountain as its own peak is above its deepest valley, the wide vista on a clear sunny summer day, or on a cold moonlit winter night, is magnificent beyond words, although with the proportions so dwarfed by 12,000 to 14,000 feet of altitude, the whole immense and jumbled mass below assumes unreality and is more like a beautifully moulded, contoured miniature map than the real thing.

To the first airman who went over, however, the massive bulk of the mountains and the vicious gusts and eddies were a very real chal-

lenge. Not for him were the great engines of today, with their rhythmic infallible beat, nor the comforts of an enclosed cabin. The Rockies had first to be flown the hard way. In 1919 there were no airports such as now exist at Vancouver's Sea Island, B.C., or at Kenyon Field at Lethbridge, Alta., nor were there any established sites for emergency landings. There was not even an air route, and many people doubted whether such a flight could be made.

The first plans were formulated among members of the British Columbia branch of the Aerial League of Canada, backed by the old Vancouver *Daily World*, the Lethbridge *Herald*, the Calgary *Herald*, and the British Columbia town of Golden. Lots were drawn among those members of the Vancouver branch of the League who had been war pilots, and the first name out of the hat was that of Captain Ernest C. Hoy, D.F.C.

Hoy had had an enviable war record. He first went overseas with the 48th Battalion, and after months of fighting was invalided from France in 1917. When he returned to the scrap, he went as a newly trained pilot in the Royal Flying Corps, and served in that capacity until two months before the cessation of hostilities, when he was forced down behind the enemy lines and became a short-term prisoner of war.

So much for the pilot. As for his plane—it was a puny thing compared with modern air monsters, but it had a stout heart. It was a standard 90 h.p. Curtiss JN4, one of the famous Jennies used to train Canadian and United States pilots for World War I. The only special feature was an extra 12-gallon fuel tank fastened on the seat in the front cockpit, which increased to 40 gallons the total gasoline supply. Under normal cruising conditions of 75 m.p.h. in calm air, this amount was sufficient to keep the machine in flight for approximately four hours.

Hoy attempted a getaway on Monday, August 4, 1919, and flew some eighty miles to the vicinity of the town of Chilliwack, B.C., but there he was forced to turn back, as dense clouds obscured from his view the 7,000-foot Cheam range of mountains which he knew lay right across his path.

Hoy's second departure was at 4:13 A.M. on August 7, from the Minoru Park race track on Lulu Island. At 5:00 A.M. his machine was glimpsed from Chilliwack, going over at a great height. He reached Vernon without undue incident, landing there to refuel at 7:18 A.M. The citizens of Grand Forks welcomed him at 10:34 A.M., and another crowd acclaimed him when he reached Cranbrook at 2:05 P.M. He was doing fine; all these stops were a normal necessity due to the short flying range of his machine.

The airman left Cranbrook, B.C., at 3:35 P.M. and set a course for Lethbridge, Alta. Now came the actual crossing of the main range of the Rockies. He soon found that it was not a case of flying over the mountains but between them, for he could not force his machine high enough to fly freely above them. When going through the Crawford Pass, he scraped by with only 150 feet of clearance between the wheels of his plane and the rocks and treetops below. Try as he would, he was unable to make the Curtiss climb another foot. With her load, the Jenny had a top ceiling of under 7,000 feet, and the airman was forced to steer through passes where the surrounding peaks pierced the sky at much greater altitudes than he himself could gain. In such places, often confined, there were vicious up- and downdrafts, which the pilot was not aware of until he was into them. However, he reached Lethbridge safely at 6:22 P.M., and found almost the entire population waiting for him.

Tired as he was, Hoy lingered less than an hour, for darkness was fast approaching; then he was off again on the last lap to Calgary where he arrived at 8:55 P.M. He made a good landing in the settling dusk, thanks to the aid of gasoline flares and hundreds of automobile headlights switched on by many of the 5,000 Calgarians who were on hand at Bowness Park to witness the great finale.

CAPTAIN ERNEST C. HOY, D.F.C.
Just before his flight across the Rockies, August 7, 1919

The elapsed time of the entire flight from Vancouver, B.C., to Calgary, Alta., was 16 hours and 42 minutes; of this 12 hours and 34 minutes were spent in the air, with only a little over 4 hours' time out for refuelling, though certainly not for resting.

As might be expected, Hoy had carried with him a small pouch containing 45 letters of officially franked air mail, also a batch of newspapers from the Vancouver *Daily World* addressed to the mayors and other officials at Vernon, Cranbrook, Lethbridge, and Calgary. Collectors who are fortunate enough to hold one of the covers of this flight have an item which commands the highest price quoted for any flown Canadian mail to date—$250.

After a stop-over at Calgary until August 11, Captain Hoy set off on what he intended to be a return flight to the coast, planning to fly a more northerly route *via* Banff, Field, Golden, Revelstoke, then southward to Vancouver. He left Calgary at 9:50 A.M. and after circling to 5,000 feet headed west. He reached the mountain town of Golden at 12:30 P.M., the first person to have the privilege of viewing the distant peaks of the mighty Selkirks from the air. As he remarked later, it was awe-inspiring to have so many jagged peaks looking down upon him from far greater heights than he could attain. He was very thankful that the weather was clear, and that there was no danger of encountering a "cloud with a rock in it."

The ground altitude at Golden is 2,583 feet above sea level, which grants scant manœuvrable leeway to a pilot who is forced to make a sudden change of direction during a getaway

CAPTAIN HOY'S CURTISS
After the crash at Golden, B.C., August 11, 1919

from a rough confined area, in a low-powered heavily laden airplane. Hoy had landed in a small hay field, as the show-grounds chosen by the townsfolk were altogether too small. As he was taking off after a two-hour stay, fate played him a scurvy trick. The Curtiss was barely airborne when two Indian boys ran across the field directly in the path of the speeding plane, and Hoy was forced to bank steeply to avoid hitting them. This altered his line of flight. He was almost at the end of the field by this time and here a large cottonwood barred his path. Banking steeply again, he missed the tree but the effort cost the machine the few feet of height it had so slowly gained. A wing tip struck the ground with a scrunch and the craft was wrecked in a smother of flying splinters and a cloud of Golden dust.

Hoy came out of the wreckage unscathed, and after shipping the remains of the Curtiss to Vancouver by freight, he returned to the coast by train, taking along the small pouch of mail which he had hoped to fly back. What became of the Curtiss is not known. It is regrettable that the remains of that mountain-flying Jenny were not looked after, and suitably embalmed as a reminder of a great event.

The C.A.F. Trans-Canada Flight of 1920

With the conquest of the Rockies an accomplished fact, the one great obstacle to the coast-to-coast flight was overcome. It is true that the great stretch of forest, lakes, and rocks that was northwestern Ontario remained a formidable barrier to land-based planes: but there was an obvious answer to that—the seaplane or the flying-boat.

The desire to make a trans-Canada flight was not motivated merely by an urge to conquer

space. Time, too, was a factor. To establish a mail service across the continent by air, if it were practicable, would be of the greatest value to the Canadian public. Perhaps even a passenger service would be feasible.

The fact is that trans-Canada air mail was flown between Halifax and Vancouver as early as 1920. Before I ventured to make a statement that most philatelists would flatly contradict, I considered it advisable to learn from experts just what does constitute air mail.

To be designated as official, flown covers must show suitable postage stamps thereon, duly cancelled by an authorized post office, showing date and place of origin. The addition of a *cachet* printed upon the envelope is desirable, although not compulsory. Furthermore, if indisputable proof can be shown that an envelope has been carried on an initial flight, such an envelope has definite claim as a first flight cover, even though it did not go through a post office for the usual official cancellation.

The story which follows is almost unknown.

The Canadian Air Board, which sponsored the flight, laid its plans carefully; and two preliminary flights were made over sections of the projected route.

On the first flight, two HS2L flying-boats left Halifax on July 17, G-CYAE piloted by Colonel R. Leckie, and G-CYAF in charge of Captain H. A. Wilson. Both reached their destination at Roberval, Que., at 6:50 P.M. on the 18th, in 9 hours and 2 minutes. The second flight from Halifax was by a single HS2L again piloted by Captain Wilson, who took off August 26 and reached Ottawa Air Station at 7:15 P.M. on August 28, in 14 hours and 7 minutes. This was the first occasion on which an airplane had been flown from the Atlantic seaboard to the nation's capital.

The final organization of the trans-Canada flight then got under way. Several machines were used, those for the western hops being sent on ahead by train. All pilots and ground crews were from the Canadian Air Force. The Air Force machines at that time had registration lettering similar to that of civilian aircraft, and to suggest a commercial rather than a military project the pilots wore civilian clothes.

The original plan was to use a Fairey aircraft to make a non-stop hop between Halifax, N.S., which had been specially fitted out in England and Winnipeg, Man. The Fairey was a seaplane by private enterprise and shipped to Newfoundland to be used on a non-stop trans-Atlantic flight, but this project was cancelled after the success of Alcock and Brown. In the light of what happened on the trans-Canada flight, it is just as well the Fairey did not head out over the ocean for ill luck dogged all its flights.

It had first to be flown from Montreal to Halifax. The pilot was Major Basil D. Hobbs, D.S.O., D.F.C., and his passenger was Lieutenant Colonel Robert Leckie, D.S.O., D.S.C., D.F.C., director of flying operations of the Air Board. After considerable delay, the two men set off from Montreal on October 4 for Halifax. West of Fredericton, N.B., they experienced engine trouble. At the same time they ran into heavy weather and were forced to alight on the St. John River, which was then in flood. However, they made repairs, doing so while the plane was moored under a bridge, and on October 5 they got away from Fredericton and reached Halifax, N.S., at 4:40 P.M.

Preparations were then made to fly to Winnipeg non-stop, but the seaplane could not be induced to take off with its heavy load. The weight was reduced and plans were changed to make Ottawa in a single flight.

On October 7, 1920, the ambitious trans-Canada flight finally began. The airmen took with them a number of letters from Mayor J. S. Parker of Halifax addressed to federal, provincial, and civic authorities *en route*. They were soon dropping the first mail at Saint John, N.B., as they flew over. Fifteen minutes later, as they proceeded northwest up the St. John River, trouble developed in earnest. A cowling covering the engine broke away, carrying with it a pump and oil pipes, and smothering the airmen in gas and oil. A forced landing was made on the river at once, and the heavily laden machine proved too much for the floats as it settled. In the resultant crack-up, the unfortunate Fairey became a total loss, though the two airmen were quickly rescued by boat.

Another machine was at once requisitioned from Halifax, and a service aircraft was despatched to the scene, piloted by the Commanding Officer of the air station, Major A. B. Shearer. The craft was G-CYAG, one of the old reliable single-engine HS2L flying-boats, well known in eastern Canada in forest fire patrols.

Hobbs and Leckie took off in the plane late in the afternoon, and arrived at Rivière du Loup at 10:10 P.M., alighting on the St. Lawrence River in darkness and a driving storm.

A large F3 twin-engined flying-boat had been dispatched from Montreal to Rivière du Loup, and on the 8th Captain H. A. Wilson turned it over to Hobbs and Leckie while he flew the HS2L back to Halifax. An additional crew member joined Hobbs at this point—an air engineer, Heath—and they set off for Ottawa at 6:15 A.M. After bucking 40-mile headwinds for most of the journey, they reached the capital at noon.

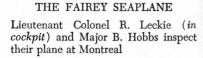

THE FAIREY SEAPLANE
Lieutenant Colonel R. Leckie (*in cockpit*) and Major B. Hobbs inspect their plane at Montreal

HS2L FLYING-BOAT

Here they were greeted by Colonel Redpath; and by Secretary J. A. Wilson and other officials of the Air Board.

The airmen now had 820 air miles behind them. The next morning they took on a fourth crew member, Captain G. O. Johnson, who was to be their navigator over the long and hitherto unflown route to Winnipeg.

Their course lay northwest up the long reaches of the Ottawa River to the mouth of the Mattawa, where they swung west to the city of North Bay on Lake Nipissing. The next objective was Sault Ste Marie, 250 miles west. They headed out over Lake Nipissing where it empties into the French River, then followed that waterway, with its many falls and rapids, until they reached its delta on Georgian Bay. From here onward their route kept closely along the northern shore of Lake Huron, bringing them to St. Joseph's Channel, thence to Hay Lake, and so at last to the "Soo" itself. They had left Ottawa at 8:30 A.M. and they arrived at Sault Ste Marie at 5:00 P.M.

After refuelling, the airmen taxied out to midstream on the St. Mary's River and took off with the intention of flying directly across Lake Superior during the night. But a heavy fog came up almost at once, so they alighted downstream, anchored their craft close inshore, and spent the night in it.

At sunrise on Sunday, October 10, the flying-boat was on its way to Kenora, Ont., which

was reached after an uneventful flight across Lake Superior. After refuelling, the flyers left Kenora late in the afternoon for Winnipeg, Man. Taking the Winnipeg River as their guide, they reached Lake Winnipeg, then turned due south to the mouth of the Red River. By this time darkness had settled and the river mists were beginning to rise, so a landing was made at Selkirk. Disaster was narrowly averted when they just missed a large dredge anchored in midstream.

Leckie and Hobbs then proceeded to Winnipeg by electric railway, taking along their precious mail bag so that it could be carried westward early the next morning by Captain Home-Hay. Next day Leckie and Hobbs returned to Selkirk and flew their big ship to Winnipeg, landing on the Red River at St. Vital. Later they went on by train to Vancouver to await the outcome of the flight.

Three different pilots were to make the western half of the flight, each in a wheel-equipped De Havilland 9, powered with a 400 h.p. Liberty motor. A spare machine was also on hand at Winnipeg in charge of Captain Pitt.

On October 9, the DH machine, G-CYAJ, assigned to the Moose Jaw–Calgary hop, left the St. Charles airdrome at Winnipeg, piloted by Captain C. W. Cudemore, with Sergeant Young as passenger. They reached the Saskatchewan city without incident.

At 4:52 A.M. on October 11, Captain Home-

TWIN-ENGINED F3 FLYING-
BOAT
On the Red River at Selkirk, Manitoba

Hay left Winnipeg for Moose Jaw, with the official passenger, Lieutenant Colonel Arthur Tylee, Commanding Officer of the C.A.F., on board. This departure was one incident of the trans-Canada flight at which I was present myself. I was employed by Canadian Aircraft at the time, and was one of the shivering spectators who watched that pre-dawn take-off. We were greatly impressed by the determination of the flyers to keep going. Nothing but the most threatening weather was allowed to delay them.

Things went well until the airmen were over Regina at 8:30 A.M., when engine trouble developed and they were forced to land near the Aerial Service Company's field just beyond the city. As the trouble could not be fixed at once, and the spare machine had been cracked up at Winnipeg, they had to send a rush telephone message to Cudemore at Moose Jaw to come to their aid. When Cudemore and Young reached Regina, Tylee hopped into the seat vacated by the Sergeant and was soon winging westward again, while Home-Hay and Young

remained to await repairs on the unlucky G-CYAN. Owing to this delay, Cudemore did not make the scheduled stop at Moose Jaw, thereby disappointing a large crowd of townsfolk.

On arrival at Calgary, a landing was made on the prairie near Bowness, where Captain G. A. Thompson was waiting with the De Havilland G-CYBF, which had been shipped from eastern Canada. By now the weather over the mountains was becoming alarming. Reports of storms and snow from all sections of the route prevented any flight attempt on the 12th. On the 13th, conditions were more promising and Thompson and Tylee got away at 11:55 A.M. to begin their flight to Vancouver. Following the beautiful Bow River Valley, they passed over Banff and on through the wide Kicking Horse Pass. They were sighted at Field at 1:30 P.M., at Golden 30 minutes later, and an hour and 25 minutes after bucking a hurricane wind they were passing over Revelstoke in fairly clear weather, having successfully negotiated

TYLEE AND HIS DH

Lieutenant Colonel A. Tylee stands beside the machine in which he flew from Calgary to Victoria

183

THE SELKIRKS   CAPTAIN G. A. THOMPSON

Photographed from the air by Lieutenant Colonel Tylee on October 13, 1920

the first aerial crossing of the Selkirks. Shortly after, dense clouds and heavy snowstorms closed down on them, and the flyers were forced to return to Revelstoke, where they made a good landing at Sam Crowe's ranch, three miles south of the city.

They were weather-bound all the next day. Dense storm clouds completely surrounded the valley in which Revelstoke lies, the ceiling reaching far down the sides of the great mountains. The delay was made pleasant by Mayor Walter Bews, who fêted the flyers at his home.

On the 15th a break in the overcast appeared to the west, high over Eagle Pass. Quickly the airmen bade their friends good-bye, and before noon they climbed out of the valley and vanished over the heights, just as the cloud ceiling settled down once again. They were spotted high over Sicamous at 12:12 P.M., and at last they were able to make a landing at Merritt at 1:45 P.M., after a violently rough and dangerous passage.

From then until the morning of the 17th, the weather would have brought profanity to the lips of a saint, much more two ambitious and impatient airmen. An attempt to carry on was made on the 16th but they were forced to return to Merritt an hour and a quarter after leaving, as dense clouds barred their path over the Coquihalla Pass.

At 7:50 A.M., on Sunday the 17th, although conditions were still adverse, they managed to get through the Coquihalla during a rift in the clouds. After a rough journey they sighted the Fraser River. Then followed a hectic flight, often only a few feet above the swift, swirling surface of the river, twisting and turning through the canyons with every bend; above the plane a cloud ceiling concealed the high tops of the steep cliffs.

When at last the storm was left behind the town of Agassiz appeared to starboard, and the flyers landed in a suitable field where additional fuel was soon forthcoming. With only a short delay they were off on the final hop. After speeding westward over the broad farm lands of the Fraser Valley, Thompson set the De Havilland down in a driving rain at 11:10 A.M. on the water-soaked centre field of the Brighouse Park race track near Richmond, on the outskirts of Vancouver. The great undertaking was ended.

The city of Vancouver honoured the airmen with a luncheon, but since little publicity had been given to the flight it aroused only mild interest among the public.

The total air distance flown from Halifax to Vancouver was calculated by the Air Board as 3,265 air miles; it was covered in a flying time of almost 45 hours, over a period of ten days. Considering the lateness of the year, especially for crossing the Rockies, and the fact that it was a pioneer effort, this was pretty good going.

Just to put a finishing touch to the flight, Thompson and Tylee flew from Vancouver across the Gulf of Georgia to Vancouver Island, landing at Uplands, near Victoria. Their stay was of short duration. The airmen and all the De Havillands later were returned to Camp Borden by train. The big F3 remained in Winnipeg where it was later used for aerial photography and forest patrol work.

A final word regarding the air mail carried on this notable trip may be of interest. As the planes proceeded westward, the small packet was increased by letters addressed to Mayor R. H. Gale from the mayors of Winnipeg, Regina, Moose Jaw, Calgary, and Revelstoke. A few were also sent to individual citizens, and in due course all the letters were received.

I became more and more curious to know what had become of these valuable envelopes and decided to try to find out if any had survived. At first the search seemed hopeless, but gradually I located each addressee (or the next of kin) and all in turn were questioned personally. Several had forgotten they had ever received such letters, and certainly could not remember where they had gone. Eventually I found one envelope, flown from Halifax to Vancouver, which bore all the pilots' signatures. The trail at long last led me to the Vancouver City Hall, and there Comptroller Frank Jones became interested. A thorough search of the filing vaults was made, and —wonder of wonders—the entire set of 6 letters received by Mayor Gale from eastern points was found intact. All the envelopes carry the signatures of the pilots concerned. These envelopes surely have a unique value for philatelists, but as they are not listed in any catalogues their monetary worth is not yet determined. The letters and envelopes were placed in the care of Major J. S. Matthews, city archivist of Vancouver.

As flying in Canada became more and more practical, experimental and record-breaking flights fell out of fashion, and the later twenties saw few purely amateur attempts. Two such flights deserve mention.

CALDWELL'S DISTANCE RECORD

Pilot C. S. Caldwell, whom we will meet again on a seal hunt, was the only Canadian to enter the so-called "On-to-Dayton Race," a publicity feature connected with the air meet held at Dayton, Ohio, in 1924. Caldwell flew his own plane, a Thomas Morse single-seater, fitted with an 80 h.p. Le Rhone rotary engine, and bearing Canadian registration G-CAEH.

In actual fact, the event was not a race at all, the performance of the various contestants being judged by points earned according to a set of rules which placed Caldwell's machine at a great disadvantage. Contestants were credited with one point for every ten miles over 500, with an additional ten points for every passenger carried. A point was deducted for every mile of cruising speed under 150 miles per hour, 100 points being the maximum allotted to each plane to start with. Five points were also deducted for each cubic inch of engine displacement (assumed to be equal to 10 horsepower) in excess of 10 h.p., the maximum number of points being 200. The rules governing the "race" were planned for aircraft of a more modern type than Caldwell's. As his was a single-seater with a low compression engine of no great horsepower, he did not stand a chance of winning. He flew 900 miles, one of the longest trips of any competitor —from Three Rivers, Que., to Dayton, Ohio, with stops at Ottawa, Camp Borden, and Detroit—but his total points did not gain him any prize.

A SEAPLANE FLIGHT ACROSS CANADA

In 1926, a wealthy American private pilot, J. Dalzell McKee, undertook an extensive flight over Canadian territory. He purchased a Douglas seaplane of the type that had earned a high reputation on the United States Army Air Service flight round the world in 1924, and enlisted the aid of a Canadian airman as pilot— Squadron Leader Earl Godfrey, who was attached to Headquarters Staff of the R.C.A.F. at Ottawa.

Starting from Montreal, the route was planned *via* northern lakes and rivers, following the St. Lawrence River to Ottawa, thence to Lake Nipissing, Sudbury, Longlac, Orient Bay, Sioux Lookout, Minaki, and Lac du Bonnet. From this point the course lay by Grand Rapids, and west along the North Saskatchewan River to Prince Albert, then to Wabamun Lake near Edmonton. The airman planned to fly over the Yellowhead Pass to the Fraser River, reaching the Pacific at the Jericho Beach Air Station, Vancouver. It was an ambitious endeavour and a private venture, with no prize or competition involved.

THE DOUGLAS SEAPLANE

Flown by Dalzell McKee and Squadron Leader Earl Godfrey over northern Saskatchewan

Courtesy Department of National Defence

Bad weather with vicious headwinds bucked the flyers on the first four days of their flight. Conditions improved during the latter part of their trip, but smoke-palls from many forest fires hung over great areas of British Columbia, making it very difficult for them to keep on their designated course.

The flight took 9 days, but the approximately 3,000 air miles were made in the flying time of 35 hours and 8 minutes. Not only were McKee and Godfrey the first airmen to cover so much of a trans-Canada air route, but they were also the first to make a direct non-stop flight across the Rockies between Edmonton and Vancouver.

Throughout the flight, the airmen were given the facilities available at the stations and bases of the Ontario Provincial Air Service and the Royal Canadian Air Force. In appreciation of this assistance, Mr. McKee established an endowment, with a trophy, to be awarded to the person giving the most meritorious service each year in the advancement of aviation in Canada.

It seems incredible now that it was to be another twenty-three years before the first non-stop trans-Canada flight was made. With Hoy's conquest of the Rockies in 1919, an actual coast-to-coast flight in 1920, and all the shorter flights over large sections of the nation by subsequent flyers, it is difficult to believe that the first non-stop flight between Halifax and Vancouver did not occur until January 1949.

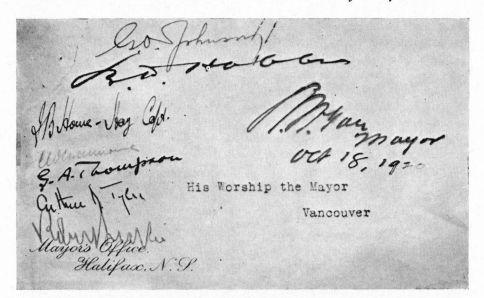

In the possession of the author is this original envelope which contained a letter from the Mayor of Halifax to the Mayor of Vancouver. It was flown over the entire distance of the trans-Canada flight made by personnel of the Canadian Air Force in 1920. The signatures of all the pilots who flew it, together with that of Mayor Gale of Vancouver, make the envelope a rare souvenir of that epic undertaking. One other similar envelope is known to exist and it is now in the care of the Vancouver Civic Archives.

22. Alaska, and round the world

NEW YORK TO NOME AND BACK, 1920

TODAY a chain of airfields extends along the Northwest Staging Route, which traverses the wilds of northern British Columbia, the Yukon, and Alaska. These airfields were built during the later years of World War II as part of a scheme to defend Alaska, to assist in the driving of the Japanese from the Aleutians, and to open up a practical, direct air route for lend-lease aid to Russia. Reams of paper have been used in writing about it since it went into operation— and it cost well over $58,000,000 to establish— but few people know that it really had its beginning back in 1920 when the Canadian and American governments decided to investigate the feasibility of establishing an air route into the far north.

The First Alaska Air Expedition planned a flight across the eastern states and the wide Canadian prairies, over the bristling peaks of the Rockies to the Pacific Coast, northward up the rugged coast and over the massive mountains of the Yukon and Alaska, to Nome—and a return journey over the same dangerous terrain.

Pilots and passengers who fly over the route today in a matter of hours owe much to those airmen who toiled many weary days on the ground as well as in the air to bring to a successful conclusion the first flights to the Yukon and Alaska and return.

The first knowledge the inhabitants of the Yukon had that such a project was being considered was when Captain H. T. Douglas of the U.S. Army Air Service* and Captain H. A. Le-Royer of the Canadian Air Force arrived at White Horse on June 18, 1920, to make arrangements for the preparation of a suitable landing-place for the machines. The two officers then left by steamer for Dawson and Alaskan points, where they made similar arrangements.

The aerial expedition consisted of eight men of the U.S. Army Air Service* with four De Havilland 4B's; each machine was powered with a twelve-cylinder 400 h.p. Liberty engine, carried 12 gallons of oil and fuel tanks with a 120-gallon capacity. The aircraft were numbered 1 to 4, and bore the U.S. Army Air Service insignia, and a wolf's head on their fuselages.

Captain St. Clair Streett, leader of the expedition, was pilot of No. 1 and with him was Sergeant Edmund Henriques. No. 2 carried First Lieutenant Clifford C. Nutt and Second Lieutenant Erik H. Nelson, the latter being engineering officer. (He later held the same position in connection with the United States round-the-world flight of 1924.) The crew of No. 3 were Second Lieutenant C. H. Crumrine and Sergeant James Long, while in No. 4 were Second Lieutenant Ross C. Kirkpatrick and Master Electrician Joseph E. English.

The flight began at Mitchell Field, New York, on July 15, and the four machines left within a

---

*United States military aviation originally was established in the form of Aero Squadrons attached to the U.S. Signal Corps, an arrangement which continued until June 4, 1920. From June 4, 1920 until July 2, 1926 the squadrons operated under the name of the U.S. Army Air Service, when their name was again changed to the U.S. Army Air Corps, continuing under this until June 20, 1941. From the latter date until September 18, 1947 they became the U.S. Army Air Force, and since September 1947 to the date of this publication they have flown under the name of the U.S. Air Force.

few minutes of each other. From then until their return, months later, the airmen waged a continuous battle with bad weather and primitive landing facilities.

They were barely on their way before they encountered heavy weather, and Streett was forced down for a while with engine trouble at Scranton, Pa. The three other machines reached Erie, Pa., in 5½ hours, on schedule. Many additional stops were forced upon them during their journeyings, but regulation stops were listed as follows: Erie, Pa.; Grand Rapids, Mich.; Winona, Minn.; Fargo and Portal, N.D.; Saskatoon, Sask.; Edmonton and Jasper, Alta.; Prince George and Hazelton, B.C.; Wrangell, Alaska; White Horse and Dawson, Y.T.; Fairbanks, Ruby, and finally Nome, all in Alaska. The return trip was planned to follow the same route.

After a number of delays, caused chiefly by storms, the four machines reached their port of entry at Portal, N.D., and on July 25, 1920, they crossed the border to Canada in open formation, reaching the McClelland airdrome on the outskirts of Saskatoon after a four and one-half hours' flight. During the airmen's short stay at the King George Hotel a congratulatory telegram was received by Captain Streett from the Canadian Air Board at Ottawa, extending the compliments and good wishes of the Canadian government.

On the morning of the 27th the expedition was on the wing to Edmonton, and after crossing Lake Manito and following the valley of the Battle River it was met several miles out of Edmonton by an aerial escort of two Curtiss JN4's, flown by Captain "Wop" May and Lieutenant George Gorman. With them as passengers were Acting Mayor James East and Alderman Charles Hepburn of Edmonton.

The United States airmen did not realize that they were supposed to follow their flying escorts, and when they were near Edmonton they went down to a landing in a likely field northeast of the city. May and Gorman followed them down and explained that crowds were expecting them at another field. After a good laugh the airmen

FIRST ALASKA AIR EXPEDITION
Left to right: Streett, Nutt, Nelson, Crumrine, Kirkpatrick, Henriques, Long, English

THE EXPEDITION AT EDMONTON
Work is in progress to fit the four De Havilland 4B's for the next stage of their flight

all took off, and landed a few minutes later at the field where the city fathers and a throng of citizens awaited their coming.

After a few days in Edmonton, spent in putting the De Havillands in good condition, the planes left for Jasper on July 31, but headwinds and heavy rain forced them to return, the first time they had had to back-track since leaving New York.

Conditions were good on August 1. After flying over the Pembina River country, where they could see that vast stretches of the forest were on fire, the airmen turned northward to the Athabaska. Going low over the small town of Pocahontas, they continued to Jasper Lake and Bride's Lake, to the convergence of the Athabaska and the Snaring rivers where a "landing-field" awaited their coming. There they were greeted by Colonel Maynard Rogers, superintendent of Jasper National Park.

The four craft left for Prince George on the 2nd and three of them reached their destination without incident, but Crumrine's machine blew a tire upon landing and nicked a blade of the propeller. The landing-area was much too small. Sergeant Long, his passenger, climbed out of the rear seat and slid along the fuselage towards the tail in an endeavour to slow the craft down as it taxied but the DH went up on its nose in the bush at the end of the clearing. Long went catapulting into the dense foliage, which broke his fall, so that he, at least, landed without injury.

Captain Streett had only just got away from the Jasper field when a burst oil pipe caused a minor fire and he returned without delay. The trouble was quickly remedied and he was on his way again by 1:00 P.M. following a different course from that of the others—by way of the Miette Valley over Lake Lucerne to the Yellowhead Pass. He then followed the Fraser Valley to Urling, thence to Prince George, where he arrived at dusk. The flyers already there had flares going to help guide him in, but a rainstorm was in full blast as he settled down to a landing. He made it, but the lower left wing was badly smashed.

The airmen had orders to remain together in

event of damage to any of the aircraft, so considerable time elapsed before repair work, in which, incidentally, a local cabinet-maker gave assistance, was completed on the No. 1 machine.

In the meantime, Captain Streett and Lieutenant Nutt had journeyed by train to Hazelton to size up the possible landing-grounds there. The outlook was melancholy beyond belief: such areas just did not exist. The only spot remotely suitable was on the farm of a Mr. Bierns, and at the time it was a thriving field of oats! But the owner rose to the occasion and had a wide swathe cut through his standing crop, then harrowed and rolled the stubble to a semblance of a runway. Thus did British Columbia's first northern "air field" come into being.

On August 13, everything was in readiness at Prince George for the airmen to complete the first east-to-west crossing of British Columbia by air. The four DH's took off just after 9:00 A.M., following the course of the Nechako River, later keeping the railroad in view until they reached Hazelton and the "Bierns Airport" at 12:15 P.M. Farm help and spectators alike had been pressed into service to assist in putting the runway into good condition. (One Indian who evidently didn't relish the job paused in his efforts to remark to Captain Streett, "You heap smart man, but heap damn fool.")

They left at 1:30 P.M. for Wrangell, equipped with maps which were completely lacking in accurate information for flying purposes. They were compelled to fly at an altitude of over 10,000 feet in order to avoid unmarked mountains which pierced the skies in that area. Dense clouds obscured their view of the ground for over two hours, before they at last recognized the Nass River. A short time later, Stewart Arm and Behm Canal were spotted, then Wrangell Island, and finally, Wrangell itself. They landed on Sergieff Island, where residents had kept a smoke-smudge going to guide them in. Kirkpatrick was first to set his machine down, 2 hours and 27 minutes out of Hazelton.

The landing area turned out to be a stretch of salt marsh grass, which in some places was a foot deep in water. At that, the airmen were fortunate, because no one had thought to tell them that the flats were innundated at high tide to a depth of 19 feet!

On the 16th, three of the aircraft left on the lap to White Horse. A nicked propeller had forced delay of Captain Streett's machine until the next day, when he too flew on to White Horse. The route took them from Wrangell, over the Stikine River Post and the famous Taku Glacier, on over Juneau and Skagway, across the mountains of the Coast Range, *via* White Pass, over Carcross, and so to White Horse. An historic flight indeed, this first one into the Yukon Territory, only 22 years after the sweating line of humanity had slowly struggled over the "Trail of '98," in the gold rush.

On the route traversed by the airmen between Hazelton and Juneau lies the area of the North American continent where the great stands of Sitka spruce grow; it was of interest to realize that their De Havilland aircraft were constructed of wood that once grew down below.

Dawson was putting on a big celebration on August 17 to mark the anniversary of the Klondike gold strike, and the American airmen were anxious to reach that point during the festivities. Machines Nos. 2 and 4, in charge of Nutt and Kirkpatrick, got away from White Horse and landed at Faulkner's Field west of Dawson on the evening of the 17th, where they were extended an official welcome by Commissioner George MacKenzie. Crumrine's machine blew a tire attempting a take-off at White Horse, so Streett remained behind with him until repairs were made. Early on the 18th, with the tire patched, the aircraft prepared to leave, and Streett was first away. As he circled above, watching his friend below, he saw Crumrine's machine swerve as it picked up speed on the ground, and he knew a tire had gone again. Rather than risk another landing, now that he was in the air, Streett flew on to Dawson expecting that the other machine would be along in a day or two. Imagine his surprise when Crumrine arrived not long after he himself had landed. It turned out that Crumrine and Long, with the aid of a local White Horse resident, Bert Peterson, had wound a hemp rope tightly around the rim of the wheel, then after replacing the outer casing of the tire, they had strapped it firmly in place with a lengthy fabric-web rein, taken from the harness of Peterson's horses. With this contraption the airmen made a successful take-off. They left the hospitality of Dawson on the 19th, reaching Fairbanks in Alaskan territory late in the afternoon. They were glad to be there, for repair parts which they badly needed were awaiting their arrival.

The expedition was away again, heading for Ruby on August 20, staying there over the week-end and leaving on the 23rd for the final lap which carried them to their goal at Nome. At 5:30 P.M. they brought their four machines to a stop on the old parade grounds at Fort Davis on the very threshold of the Bering Sea. Four thousand five hundred miles of flying lay behind them, accomplished in 55 hours of flying time, over a period of 40 days.

Even so, only half their job had been completed. Anxious to be on the move again, they did not loiter long by that northern shore. Three days later, at 3:30 P.M., they said good-bye to Nome on the start of their return trip, and for the first time set a course in a southerly direction. They were back at Ruby by dusk, and had reached Fairbanks by the 31st. There the four aircraft were given a thorough check in readiness for the great wilderness of forest and mountains which lay between them and Edmonton. They left Fairbanks for Dawson on September 3, arriving at their destination within a few minutes of each other, all decked out with souvenirs of their Alaska visit, including two husky puppies.

Early fall storms were beginning to brew, but they reached White Horse on the 4th. Then bad weather settled in with a vengeance. They left the Yukon town on the 5th, heading for Telegraph Creek, B.C., but violent winds and snowstorms forced them to return. Streett telegraphed from White Horse to Telegraph Creek, asking to be notified when the weather cleared. On the 8th, a good report came through and the four planes hopped off at once. Three were able to penetrate as far as Nahlin, B.C., about half way; then menacing weather conditions again beat them back to White Horse.

Kirkpatrick, becoming separated from the others, wound up four hours later at Wrangell. This was not due to miscalculation on his part, however, but to very fine navigation under the worst conditions. On the journey north he had met a young lady at Wrangell and admiration had been mutual, so instead of turning back to White Horse with the others, Kirkpatrick turned the adverse flying conditions into a happy break which enabled him to reach Wrangell and renew acquaintance. Whether a romance blos-

somed is not known. If it did, tragedy followed not long after, because in 1926 the young pilot lost his life in the United States during a forced landing in a storm.

On the 9th, the weather cleared sufficiently for the three DH's at White Horse to get away at last and fly to Telegraph Creek, where they came down to landings on a hay field of the Diamond C Ranch, twelve miles downstream on the Stikine River from the village itself. The undercarriage of Crumrine's machine suffered severe damage in the process, and the next day Captain Streett and Lieutenant Nutt left for Wrangell aboard the launch *Hazel H*, which happened to be upriver at the time. They returned with spare parts on the 15th, and made the necessary repairs on the disabled No. 3. Attempts were made to fly to Hazelton on the 17th and the 19th but dense cloud and snow forced them to return almost immediately.

Kirkpatrick had left Wrangell on the 16th, and as the weather was clear, he and English were requested to fly over the valley of the Unuk River, to keep a look-out for a party of men working on the Alaska–British Columbia boundary survey, who had not been heard from for some time. No sign of them was observed, but as the aircraft was leaving Wrangell the flyers met the Canadian Pacific steamer *Princess Alice*, just arriving from the south, so, circling close, they waved to the excited passengers lining her rails before setting their long course over the mountains for Hazelton.

The three De Havillands at Telegraph Creek

CRUMRINE'S MAKESHIFT TIRE REPAIR

were not so lucky. Weather continued impossible for flying until September 29; then they did at last get away and managed to reach Hazelton by 5:30 in the afternoon.

During the time they were earth-bound at Telegraph Creek, a modern version of the old-time beacon signals was used to relay weather reports from the telegraph office to the ranch. A man was stationed at a specified time each day on a high hill near the office, and another placed on a hilltop near the Diamond C Ranch. If favourable weather reports came in, the first man was to set off a charge of dynamite. This would be seen or heard by the fellow miles away, who in turn was to touch off a similar explosion which would tell Streett and his companions that it was safe to go.

Being so far in front of the others, Kirkpatrick felt he should not miss good flying weather while waiting for them to catch up to him, so he carried on, first to Prince George, then to Jasper and Edmonton, reaching Saskatoon on the same date the three other airmen flew into Hazelton.

Machines Nos. 1, 2, and 3 eventually reached Edmonton together on October 8, after beating their way through snow and storm most of the distance from Hazelton. On the 10th they left Edmonton, reaching Fargo, N.D., on the 11th. Kirkpatrick had crossed at the same point on October 1, his machine being the first to arrive back in the United States. He awaited the others along this stretch of the route. The flight officially ended on October 20 at Mitchell Field, Long Island, N.Y., at 1:37 P.M., when the four machines came in practically together.

The aircraft were much the worse for wear after their magnificent journey, but the men were in good condition and highly elated with the success of the project. They had flown 9,000 air miles, taking 112 hours of flying time, in the course of 97 adventurous days.

A week later the airmen were officially congratulated at Bolling Field, D.C., when General Pershing represented the United States government and extended his personal greetings as well.

In his report of the flight Captain Streett had written: "The air route to Alaska is not feasible until air fields are made along that route." It required many years for his words to become an actuality, but they did finally, when United Air Transport, Ltd., began charter flying to the

Yukon in 1934. They operated to White Horse from Edmonton, *via* air fields established at Fort St. John, Nelson Lake, and Watson Lake. A full-scale service went into effect. On July 7, 1937, the first scheduled air mail was flown from Edmonton to White Horse in a company machine, piloted by the firm's president, Grant W. G. McConachie. On August 20 the air-mail run was extended to Dawson.

In 1939 the Canadian government set aside an appropriation for a survey of the route already in use, with the object of establishing larger landing-fields. Not until the Canada–United States Permanent Joint Board of Defence was appointed in 1940 did actual construction work begin. Transport facilities for heavy equipment to the interior were non-existent because of the rough nature of the country, and it was February 1941 before the first caterpillar tractor train left Dawson Creek for Fort Nelson. The essential development was well under way when the United States entered the war against Japan. When the Alaska Highway was completed in November 1942, the different air fields became accessible to ground transport and the chain of huge landing-fields of the Northwest Staging Route came into everyday use.

First Round the World

All over the world from 1919 onward, space was being annihilated. Governments began to see the strategic value of air control; business glimpsed the value of air communication and swift passenger service; and the public generally experienced a thrill of expectancy as distances shrank and a world more fantastic than any Jules Verne had imagined seemed not so inconceivable. By 1924 the British had established flight routes to India, Australia, and South Africa: while the Dutch and French were busy with experimental flights to their respective colonies.

With trans-Atlantic routes already explored, the only formidable gap in a round-the-world route lay over Siberia to the north, or across the Pacific to the south. The simplest and safest way of closing this gap, since there were plenty of American naval craft available to act as surface bases, was to station ships along a route that followed the Aleutian chain from Alaska to Japan, and thence to the coast of China. The

THE *NEW ORLEANS*

east-west Atlantic crossing, which had not yet been made, could similarly be closed by stationing supply vessels at Iceland and Greenland points.

In 1924, Canada was twice visited by the aircraft of the United States Army Air Service in their famous flight round the world. It began at Seattle, Wash., on April 6 and ended at that city on September 28.

The expedition consisted of four two-seater Douglas Air Cruisers, each equipped with a 450 h.p. Liberty engine, 465 gallons of fuel, 30 gallons of oil, and 5 gallons of water reserve for the cooling system. The planes were fitted with pontoons. There were two men to each plane, Major F. L. Martin and Sergeant A. L. Harvey in the *Seattle*, Lieutenant Leigh Wade and Sergeant H. H. Ogden in the *Boston*, Lieutenants Erik H. Nelson and John Harding in the *New Orleans*, and Lieutenants Lowell H. Smith and L. P. Arnold in the *Chicago*.

The five-month route took the flyers up the west coast of North America to Alaska, across the Aleutian Islands to Yokohama, then to Shanghai, Karachi, Baghdad, Vienna, Paris, London, the Orkney Islands, Iceland, Greenland, Labrador, Nova Scotia, Maine, and thence across the United States to Seattle.

They suffered numerous accidents and delays. The first occurred on the flight up the British Columbia coast. Landing in a snowstorm at Prince Rupert, the *Seattle* smashed her pontoon struts. They were replaced on the spot by new ones, made of spruce wood procured from a saw-mill at Prince George, B.C. Bad weather delayed the take-off for some days, but the party was consoled with a civic banquet.

The next mishap occurred in Alaska. The *Seattle* was again the victim, fracturing the crankcase of her engine. A new engine was installed at Chignik, and Martin and Harvey set forth to rejoin the others, who had gone on to Dutch Harbor in the Aleutians. Again a snowstorm enveloped them, they struck a mountaintop and the machine was wrecked. The two men endured incredible hardships, but in sixteen days managed to get down the mountain and out to Port Moller. Three planes now carried on, the engines being replaced at certain stops. No serious trouble occurred until they were crossing the North Atlantic to Iceland. The *Boston*'s engine failed and she came down into high seas. The two men were rescued by a destroyer but the airplane sank.

The two machines then surviving were the first to conquer the Iceland–Greenland–Canada route. The next flight, from Greenland to Indian Harbour, Labrador, on August 31, was across 572 miles of ocean, and took just under seven hours. On September 3, the airmen reached Pictou, N.S., without further trouble. An R.C.A.F. plane flew out to escort them in, and on shore they found the men of the *Boston* waiting for them with a new plane, *Boston II*. The six Americans were greeted with enthusiasm by the Canadians.

At New York the three planes changed their pontoons for wheels, and proceeded to Seattle by easy stages arriving September 28.

Even with every assistance—replacement of engines *en route* and United States destroyers standing by—the risks and difficulties of the flight had been tremendous. It was calculated that the two planes that made the full trip had flown 26,345 miles, in 363 flying hours, over 175 days. America's highest aeronautical honour—the Collier Trophy—was awarded to the Army Air Service in recognition of this first aerial navigation of the world.

But here we leave the breath-taking span of oceans and continents to return to the interior of Canada, where bush flying had already been born, and where flights were to be made as epic in character as any the world had seen.

PART 6

NORTHWARD INTO THE BUSH AND SNOW, 1919-1929

23. Bush and arctic flights, 1919-1929

THE word "bush", like most popular terms, is used loosely by Canadian flyers and public alike. Fundamentally it refers to the areas of Canada covered with softwood forests, particularly the untamed areas of the Canadian Shield extending in a horseshoe around Hudson Bay, bounded on the west by the prairies and the Rocky Mountains, on the south by the Great Lakes–St. Lawrence system, on the east by the Atlantic, and on the north by the Barren Lands. But the word is also used to include all the more rugged features of all the Canadian provinces from Newfoundland to British Columbia, and even the barrens of the Northwest Territories.

A second meaning is attached to "bush flying": it is essentially of a practical nature. Pilots who flew into the bush had a job to do for which they were paid. By and large they had little concern with breaking records, making names for themselves, or even advancing the cause of aviation. They were flyers, pure and simple— although taken separately these adjectives don't quite apply. It is therefore impossible to point to a specific flight as the "first" bush flight, and it would be contrary to the spirit of bush flying to do so. All one can do is to pick out some stories that will tell what the earliest bush flights were like.

FROM HALIFAX TO LAC À LA TORTUE, 1919

When the war ended, a number of the HS2L machines used on the Atlantic patrol in 1918 had been dismantled and stored at Dartmouth, N.S. Two of these machines were purchased by the St. Maurice Fire Protection Association for forest patrol in the St. Maurice Valley in Quebec. The planes had to be taken a distance of 645 miles over Canadian and American territory to their new base at Lac à la Tortue. The St. Maurice Company engaged Captain Stuart Graham as pilot and William Kahre as engineer. An unofficial third crew member was Captain Graham's wife.

The party left Dartmouth, N.S., in the first plane, *La Vigilance,* at 2:25 P.M. on June 5, 1919, and made Saint John, N.B., on schedule, despite a fog. The mayor of Saint John gave them the keys of the city, and the owner of the local opera house invited them to a special performance of *Il Trovatore* These and other festivities delayed them until the 7th, when they departed for Eagle Lake in Maine. The next day they had a rough flight over mountainous and heavily forested country, and landed on Lake Témiscouata, Que., near the logging camp of the Fraser companies at Cabano.

The next stop, made the same day, was Rivière du Loup, on the St. Lawrence River. The take-off from this point was against a strong tide and heavy seas, and the flying-boat could not rise smartly enough to prevent Mrs. Graham, in the bow cockpit, from being drenched. Neither plane nor crew suffered any harm, but it was judged wiser to return and stay at the town for the night.

The next day they left Rivière du Loup in the afternoon, and with a tailwind to help them along they reached Three Rivers at 3:30 P.M. There they were greeted by Hon. J. A. Tession, to whom a letter was delivered from the Lieutenant-Governor of Nova Scotia for the Premier

TWO HS2L'S AT THEIR BASE AT GRAND'MÈRE

of Quebec. This was the first air mail flown between Nova Scotia and Quebec, and the envelope bore the following words, "Per Aerial Post," and was addressed to "Hon. Sir Lomer Gouin, K.C., C.M.G., Quebec."

Continuing from Three Rivers the same day, they carried on to their destination on Lac à la Tortue, landing near Grand'Mère at 8:15 P.M. Four days had been required to make the 645 air miles from Halifax, an actual flying time of 9 hours and 45 minutes, at the low flying speed of 66 m.p.h.

The three voyagers returned by ferry and train to Halifax, where the second flying-boat was waiting to be flown to Grand'Mère. At 2:00 P.M. on June 21, 1919, the three of them set off once again.

No matter how hard they tried, or what good intentions they had, early starts just didn't seem to be possible. Something would invariably occur to delay them, and rarely were they able to get on the wing before noon. They were 2 hours and 45 minutes reaching Saint John, bucking a strong head wind all the way. After landing there and refuelling, they were off again to make Woodstock, settling on the St. John River late in the day.

On the 22nd, their plans for an early start again went awry, this time because the 80 gallons of gasoline which were put into the tanks had to be poured in from four-gallon cans, a slow process. When they did take off from the river, they had a very close call. The sluggish flying-boat required a long run on the water before lifting free, and they came very close to a railway bridge before the craft was airborne. Graham kept the machine close to the water until the last moment, then pulled her sharp up, clearing the bridge and telegraph wires by a

very narrow margin. Once in the air, they carried on over the rugged country of Maine until a drop in the oil pressure made it imperative to land for a check-up. They came to rest on a large body of water which they found to be Long Lake. The oil trouble was quickly remedied and as they were close to a settler's home in the deep forest, they moored to the shore and made their way to the cabin.

The young woman who met them at the door was almost overcome by their extraordinary arrival, and their leather flying garb no doubt looked pretty strange, but after her first fright she asked them in to her two-room shack. Her large family stood around in wide-eyed silent interest, while the housewife served her guests cocoa made without either sugar or milk. Apparently this was regular fare, if not better than usual, for those hardy and probably desperately poor backwoods folk. Meanwhile the cabin was filling with people who said not a word, but just stood and stared at the flying trio as though they were visitors from another world—which they virtually were.

Leaving Long Lake, they reached Cabano on Lake Témiscouata, and refuelled before darkness set in; thus, for the first time, they were able to make an early start the following day. On the 23rd they were off at 7:00 A.M., arriving at Rivière du Loup to refuel, and landing that afternoon at Grand'Mère.

The numerous flights made by these two flying-boats during the following season constituted the first Canadian flights in connection with fire patrol. Their value in spotting fires and transporting fire rangers was soon recognized. One of the machines caused great excitement among the Indian population on the reserve at Senneterre, who immediately named it "Kitchi Chghee" (big duck).

Towards the end of the season, one of the machines became very difficult to handle in the air. Inspection revealed that the two sponsons (the separate buoyancy extensions outside the hull) had filled with water, and over 1,000 pounds was drained from them. How the plane managed to fly with a full load and the additional weight of the half-ton of water is a small miracle.

THE FIRST COMMERCIAL FLIGHT INTO THE NORTHLAND, 1920

The year following the Graham–Kahre flight from Lac à la Tortue, one of the first strictly commercial bush flights was made from Winnipeg. When Frank J. Stanley, a fur buyer and promoter from The Pas, Man., walked into Canadian Aircraft's down-town office and asked to be flown home, the staff were a bit staggered. The Pas lies on the edge of the Manitoba wilderness and the last two hundred miles or so of the route consist of bush, lakes, and muskeg—not the best landing-ground for a plane with wheels. That the plane could even be refuelled when necessary was by no means assured. However, the risks were accepted, and at 11:00 A.M. on October 15 Hector Dougall took his place at the controls, Stanley settled himself in his appointed place, and I stood at the propeller, ready to give it a flip.

Then the familiar litany of the day rang back and forth from mechanic to pilot.

"Switch off," says Ellis.

"Switch off," Dougall replies, to signify he has heard.

"Throttle open, air closed—Suck in!" I holler.

"Throttle open, air closed—Suck in," echoes Dougall, and I give the prop a swing to send a good rich mixture into the engine.

"Throttle closed—Contact!" yells mechanic Ellis.

"Throttle closed—Contact," Dougall yells back —a good stiff yank on the prop from me, and the engine breaks into an eager roar. I climb into my seat, we all buckle our belts, a ground helper pulls away the chocks from in front of the wheels, waves his hand to the pilot, dodges away, and we're off!

Our machine was one of the Avros bought from the British government after the close of World War I. It was a sturdy, reliable craft, converted from a two- to a three-seater and powered by a 110 h.p. Le Rhone rotary engine. Our aircraft, registration G-CABV, had served us faithfully on many exhibition flights throughout the prairies during the preceding months.

Ideal weather prevailed, and fifteen minutes out of Winnipeg we spied the smoke-haze rising up from the chimneys of Portage la Prairie. Another quarter-hour and we were past, with Gladstone our next objective.

Away to the north the vast stretch of Lake Manitoba met the rim of the horizon—an inspiring sight.

Over Gladstone a faulty spark plug caused engine trouble, so down we went in a long, lazy glide, to a landing in a ploughed field close to the town. Quickly fixing the trouble, we once more rose into the wind and turned toward the town of Dauphin, some 90 air miles away.

We climbed steadily to gain altitude to swing round the eastern fringe of Riding Mountain, then, an hour and twenty minutes out of Gladstone, we set our wheels down on another deeply ploughed field. From experience we preferred a rough landing near an available gasoline supply to a smooth landing a mile or two away from it. Autos packed with people arrived as though by magic, and the three of us were quickly whisked to the nearest garage to make arrangements for fuel.

Our next destination was the town of Swan River. We rounded Duck Mountain well to the east, and swung into a westerly course as the sun sank below the horizon.

Darkness was already settling over the northern farmlands as we made a rough landing in a stubble field at 6:30 P.M. about a mile east of Swan River. Securely staking the machine, we set out on foot to the hotel. To us that final mile,

WINNIPEG TO THE PAS
Frank Ellis (*left*) and Hector Dougall just after their 500-mile flight

FIRST AERIAL PHOTOGRAPH OF NORTHERN CANADA
Taken by the author on October 17, 1920, from 3,000 feet, looking westward

hoofed over a deeply rutted country-road in the darkness, was much more arduous than the three hundred miles we had just completed in the air.

Next morning the weather had switched completely. Dirty clouds were scudding from the north, the ceiling was low, visibility poor, and the temperature had dropped alarmingly. Rain squalls followed one after the other throughout the morning, and we despaired of getting away that day. By 2:30 P.M., however, things cleared a bit, so we decided to have a go at a take-off.

Our intention had been to fly directly from Swan River to The Pas, but the bad weather made us decide to go westward *via* the small village of Hudson Bay Junction, Sask., there to land or carry on as conditions would determine.

Our route of a hundred and nine miles carried us over the eastern end of the Porcupine Mountains. The air was desperately rough and visibility bad. Before we had been up fifteen minutes we plunged into a driving snow squall, and were flying blind. Fortunately we managed to keep our course, but several times we were obliged to alter our compass direction to avoid dense snow flurries, which forced us to fly close to the treetops to keep in visual contact with the ground.

During this hop we observed several large flights of Canada geese winging sturdily south-ward in grand formations before the wind. Usually geese fly high during migratory journeys. Those we encountered that day were keeping to a low altitude, just below the cloud layers, probably awaiting the time they would outfly the storm and rise to greater heights. It was a wonderful sight but not very encouraging to us, for their passing meant that the grip of winter was already binding the lakes and streams in the high north.

After an hour and a half of rough flying, we were relieved to spot the Junction. But our joy was short-lived. In ever-widening circles we swung around the settlement in search of a likely place to land, only to realize that a good spot did not exist! There were no areas under cultivation, only virgin bush and muskeg in every direction. But our fuel supply was diminishing and we would have a heavy headwind to buck if we attempted to try for The Pas, so we knew we had to sit down somewhere. We picked a muskeg near the village and prepared for the worst. We had swung into the wind which was so strong that it cut our landing-speed down to about 30 miles an hour; our wheels sank axle deep into coarse swamp-grass and mud and the machine stopped with a sickening jerk. The hefty ash undercarriage skid was all that lay between us and disaster. It took the full strain as the tail of the machine went up and the nose down, but it stood the gaff, and we settled back on an even keel.

The entire population of the village, about fifty all told, was quickly on the spot—a plane had never before been seen in that part of the world.

With about two hours of daylight left, we set about extricating our craft. The entire man-power of the village was quickly offered. Axes appeared like magic, and the men began clearing a path through the bush to near-by rising ground where a suitable stretch might be cleared for a take-off. The machine had been hauled to the partly prepared runway by the time darkness fell. During the evening we were guests of the village, and bed came late that night.

Our helpers rose early and worked all Sunday morning. By afternoon a fairly smooth although very short stretch was ready. Moreover, the weather was fine again.

If you have ever had the experience of running out of gas on a highway, walking back several miles to the nearest garage, and arriving only to find it closed for the day, you will understand our feelings when our request for fuel brought the reply, "Gas? Why, there isn't any here!" The village had no road connections with anywhere, and therefore boasted no automobiles and no gasoline to supply them. At this juncture, the local Chinese laundry and café owner heard of our plight, and with true northern hospitality offered us the entire supply of high-grade gaso-line which he had on hand for illuminating purposes. We almost hugged him when he pro-duced two four-gallon cans, full to the brim.

At 3:30 P.M. we climbed aboard. A number of the huskiest men held the machine back until our engine was going full out. At the "let go" signal, the craft surged forward and quickly picked up speed. The runway was desperately short. As we took to the air, our wheels slashed through the underbush beyond the cleared area. A steep climbing turn carried us away from a large stand of trees with only inches to spare.

As we gained altitude, we circled the village to wave a last farewell to our friends, whose up-turned faces and waving arms conveyed their congratulations on our successful escape.

At 3,000 feet we levelled off, heading north-east. As far as the eye could see in all directions lay a jumbled mass of land and water, the browns, greens, and blues forming the gigantic pieces of a vast and intricate jigsaw puzzle. Low in the sky, the waning sun cast its glittering reflection on a myriad of lakes to westward. Once, far below, we spotted the faint "V" ripples cut by the bow of a tiny canoe, as its solitary occupant paddled across the surface of an un-named lake.

A strong tailwind helped to push us along at a fast clip, and the 87 miles between the Junc-tion and The Pas slipped under our wings in 42 minutes. The townsfolk of The Pas had re-ceived word of our coming by telegraph. All Saturday and Sunday they had waited im-patiently for us to appear.

THE FIRST "AREOPLANE"

Left: The three-seater Avro, the *Thunderbird*. *Right*: Opening sentences of *The Pas Weekly Herald*'s account of the exploit

After circling the town, Dougall chose a landing-site behind The Pas Lumber Company's yard, and at 4:30 P.M. we settled to a safe landing, although the cattle and stumps that dotted the area gave us a few anxious moments. Mayor Stitt and his councillors were quickly on hand to extend an official welcome, and to congratulate Mr. Stanley on his initiative in becoming the first passenger to be flown into the northland. A banquet in honour of the event was held the following day.

Our machine stirred up great interest, since very few of the townspeople had ever seen an airplane before. Indians came from miles around to view the machine, but few of them would come up close to it. One elderly Cree, however, examined the machine very minutely. When I asked him what puzzled him, he asked, "How Thunder Bird stay up in sky?" His choice of a name for the machine impressed us at once and we had her christened with some ceremony. (The Thunder Bird of Indian lore is a mighty bird of immense wingspread. Flying at tremendous heights, it is rarely seen by human eyes, but the flashing of its eyes causes the lightning, and the echoes from the sound of its great pinions make the thunder.)

During the week that we spent at the outpost, we carried many passengers aloft on short sightseeing hops. The first to go up was Ruth Taylor; the second, Mr. McKay, manager of the Hudson's Bay post at Grand Rapids.

As winter then set in, the machine was dismantled and shipped back to Winnipeg by ground transportation, a rather ignominious experience for a staunch aircraft to undergo.

Although the *Thunder Bird* has long since passed into that mythical graveyard which swallows up aged elephants—and aircraft—the thunder of her engine over the silent northern lakes was to have its echoes multiplied by the thousands in the years to come.

24. Rene and Vic: and the Viking

OCCASIONALLY, even today, when chatting with a young airman, I detect a note of pride, even affection, in his voice when he refers to his machine. And yet, compared with the days before the thirties there is something lacking that we once knew: a kind of personal affinity—you might even call it love—between an airman and the craft to which he entrusted his life. For all I know it might have worked both ways: certainly there were some machines in the early days that seemed to have a personality and even a soul of their own.

I was not personally acquainted with the illustrious twins *Rene* and *Vic*, or with the Viking G-CAEB: but their stories have come to me and none have a better right to a place in the history of Canadian aviation.

THE *Rene* AND THE *Vic*

In 1921 it would have taken a bold imagination to predict the great pipeline that would one day convey crude oil from Norman Wells, in the Northwest Territories, across the mountains to White Horse, in the Yukon. This gigantic project was actually undertaken and completed by the United States and Canadian governments during World War II. When Imperial Oil Limited pioneered flying in the Northwest Territories in 1921, Fort Norman had only the one small Discovery Well, situated on Bear Island in the Mackenzie River, fifty miles downstream from Fort Norman.

Charlie Taylor, Imperial Oil manager at Edmonton, was responsible for influencing head office to purchase two large monoplanes in order to send a company man to Fort Norman long before the spring break-up would allow travel in the ordinary way by water. The planes were German Junkers for which a New York firm had obtained the agency. As they were of all-metal construction and were adaptable to wheels, skis, or pontoons, they seemed well fitted for the work ahead. Each craft was equipped with a 175 h.p. engine, and could seat five people, or carry an equivalent weight.

Captain W. R. ("Wop") May and Lieutenant George W. Gorman of Edmonton were engaged as pilots, and S. ("Pete") Derbyshire as engineer. Towards the end of November 1920, the three men went by train to New York to fly the machines back. Winter had set in when the long cold flight was made back to Edmonton. May's plane arrived on January 5 in a temperature of 50° below zero; Gorman's was delayed several weeks by damage suffered when icy conditions forced him down at Brandon, Man. The flight of the two machines across two-thirds of the North American continent in mid-winter drew little publicity or tribute, but it was a great achievement.

"Wop" May severed his connection with the venture on his return to Edmonton. Lieutenant Elmer G. Fullerton and a mechanic, William Hill, were engaged.

A christening ceremony followed; Mrs. McQueen, wife of the vice-president of Imperial Oil Limited, bestowed the name of *Rene* on machine G-CADQ, and that of *Vic* on G-CADP. The machines were then flown north to Peace River Crossing, where a landing-site, hangar,

CHRISTENING THE TWO
IMPERIAL OIL JUNKERS
Left to right, standing: Charles Taylor,
Mrs. McQueen, T. Draper, A. M.
McQueen, Dick Myhers; *kneeling*: G.
Gorman, "Wop" May

and living-quarters had been established. The engines were given a complete check and the wheel undergear was replaced with skis. As fuel could not be obtained between Peace River and Great Slave Lake, a cache was established between those points. On March 22 the two Junkers left their home base, each carrying 25 four-gallon cans of gasoline and additional oil. With little incident they reached Upper Hay River post, 190 miles to the north, landing on the frozen river. They stored the fuel with the Hudson's Bay Company, and flew back to Peace River the same day.

The company man whom they were to fly to Fort Norman was W. Waddell, a federal land surveyor. At almost the last minute an additional passenger was included. Sergeant Hubert ("Nitchie") Thorne, R.C.M.P., had just completed an eight-week trek from Fort Providence to Edmonton, escorting an Indian murderer to jail. The officer's request to be flown back to his station at Fort Simpson was granted by Imperial Oil authorities.

With Lieutenant Gorman in command, the expedition left Peace River Crossing on its attempt to reach Fort Norman at 9:00 A.M., March 24, 1921, carrying equipment and emergency rations for 10 days. After contending with headwinds and stormy weather, both machines made good landings near the Hudson's Bay post at Fort Vermilion, where a royal welcome was extended to the crews during a storm-bound stop-

over of three days. They took off again at 2:00 P.M. on the 27th, and made the 200 miles to Hay River post in just under 3 hours. As they had been able to replenish their fuel supply at Vermilion they bypassed Upper Hay River post where their cache of gasoline was stored.

Landings were tricky, in deep, powdered snow. During this hop the first aerial photographs ever made in the Northwest Territories were taken, when the magnificent Louise and Alexandra Falls on the Hay River were snapped, white and immobile in their ice masses. The next day the flyers left Hay River post to battle their way through a heavy snowstorm over Wrigley Harbour before reaching Fort Providence, where they made hazardous landings in 3 feet of dry snow. After refuelling, three separate attempts were required before the airmen could lift their machines clear from the engulfing snow to push on to Fort Simpson, 165 air miles away. They were over the Fort buildings in an hour and forty minutes, but were dismayed to find that their intended landing-place, the surface of the Mackenzie River, was a mass of jagged ice blocks. Sergeant Thorne, in Gorman's machine, suggested a landing-spot near the Hudson's Bay post, so down went Gorman with the *Rene*. She had barely rested her weight on her skis when they plunged into a frozen drift. The landing-gear was wrenched away, and a ski and the propeller were shattered.

Fullerton in the *Vic* was more fortunate—a

204

mere matter of the undercarriage being buried in the snow. Later, Fullerton flew the *Vic* out light to a near-by "snye"—a narrow channel off the main river—where a good landing was made.

With the *Rene* out of commission it was decided that Fullerton, taking Waddell and Hill, should carry on to Fort Norman in the *Vic*, but engine trouble developed due to excessive carbon deposits formed by the low-grade fuel they had obtained *en route*. So plans were changed. The *Vic*'s propeller and a ski were removed and fitted to the *Rene*. Gorman then prepared to vanquish the final hop. First he flew the *Rene* out light to the snye, landing alongside the *Vic*. Then, with Waddell, Hill, and equipment aboard, he attempted to take off. But before the craft was fully airborne, deep drifts again intervened. Another ski was broken, and worst of all, the propeller too.

But the resourcefulness of Canadian airmen is justly renowned. The pilots conferred with Mr. Godsell of the Hudson's Bay post and concluded that propellers might be made. Good oak sleigh-boards were available, and the Roman Catholic mission had a fine workshop which was willingly offered by Father Decoux. Walter Johnson, one of the employees at the post, was an expert cabinet-maker and possessed a complete set of wood-working tools. Mechanic Bill Hill thoroughly understood the making of propellers, so the two men rolled up their sleeves and went to work.

The one unbroken propeller blade was used as a model. Numerous templates of tin were cut to shape for use in forming the new propellers, the planks were laid out, and with babiche glue, made at the post from the hide and hoofs of a moose, the laminations were glued together and clamped tightly in place. Everyone at the post assisted to the best of his ability, but the work, which occupied two weary weeks, was done mainly by Hill and Johnson. In the meantime Fullerton, Derbyshire, and Waddell had been overhauling the *Vic*'s motor, working out in the open in sub-zero temperatures. By the time the propellers had been fabricated, the trio had repaired the damaged skis and the slightly twisted wing of the *Rene*.

The first propeller turned out by Hill and Johnson was made completely of oak, and was given a thin coat of red paint. It was fitted to the *Vic*, Fullerton's machine, and stood up well in a severe ground test at full engine speed.

Hopes and enthusiasm were realized to the full when, on April 15, Fullerton took the Junkers up for a thorough work-out in the air. The prop functioned perfectly without any sign of vibration, and a critical inspection after the flight revealed no flaws whatever.

As there were not enough oak sleigh-boards to form the blocks for two propellers, the No. 2 prop had alternate planks of oak and birch. When finished, it was actually a closer reproduction of the original prop than was the No. 1. It was fitted to the *Rene*, Gorman's craft, and found to be satisfactory in a test flight on April 20.

By this time—late April—Fort Norman could not be reached safely in ski-equipped planes, so a return to Peace River was planned at once. The spring break-up was imminent, and it was imperative to get away before melting ice and snow might leave the planes stranded again.

In the pre-dawn hours of the 24th, an excited Indian rushed to the Mission to announce that the ice was going out on the Mackenzie. The two machines were parked on a channel of the river, and this ice, too, might begin to move out without a moment's notice. The airmen pulled on their clothes and raced down to their machines. Bad luck still dogged them. As Gorman attempted to lift the *Rene* clear, the tail broke through the snow crust and suffered serious damage. A team of oxen was pressed into ser-

LOUISE FALLS IN WINTER
Taken on March 27, 1921

WILLIAM HILL AT WORK
At the Mission Workshop at Fort Simpson

the rest of the party should return to Peace River in the *Vic.*

Equipment was accordingly back-packed to the *Vic* during the next day and at 8:00 A.M. on the 26th Fullerton took off with Waddell beside him as navigator, and Gorman and Hill in the cabin as passengers. The new propeller cut the northern air with a smooth, steady rhythm and took them in a straight, 500-mile, non-stop flight to Peace River in 6 hours.

When the airmen flew over their home base they were dismayed to see bare ground showing everywhere through melting snow, so a note was dropped to the caretaker stating that they would land 15 miles to the north on Little Bear Lake, which was still ice-bound. They asked that wheels and fuel be dispatched as soon as possible. They were surprised to see another Junkers aircraft, identical with their own but wheel-equipped, sitting right on their own field.

Fifteen minutes later, when Fullerton set the machine down to a safe landing on the soft ice of Little Bear Lake, a fuel check showed that they had only two quarts left!

Soon the drone of an airplane engine was heard, and down came the monoplane they had just seen. They learned that it belonged to a New York agent for Junkers, who had flown to Edmonton in the hope of finding buyers.

And thus, with wheels fitted, tanks refilled, and skis stored aboard, the *Vic* returned to her home base.

Upon arrival at Peace River, the faithful "home-made" propeller was removed and a new factory-made one fitted. For some time the No. 1 prop lay in the hangar, then it finally vanished, and where it went no one seemed to know.

Although the *Vic* had returned without accomplishing her mission, Imperial Oil authorities were still anxious to send representatives to Fort Norman as quickly as possible. New propellers were at once ordered for both machines, together with parts for the *Rene* and two sets of laminated wooden pontoons to convert the two Junkers into waterborne craft. On May 27, the *Vic* took off on a second attempt to reach the oil well at Fort Norman. Fullerton was pilot, and with him were Hill, Waddell, and an additional passenger, Theo Link, the company's geologist. Fully loaded, the plane weighed 4,300 pounds.

By now spring was in the air. Smoke from distant forest fires was observed during the two

vice and the machine was hauled to high ground, there to await repairs.

By this time the ice at the downstream end of the snye was heaving and breaking away in great chunks. It was hurriedly decided that Fullerton should fly the *Vic* out light to a frozen lake 5 miles away. At the last moment Jack Cameron, a trapper, was asked to go along as guide. So vital was every second of time that Jack piled aboard the machine with his snowshoes still on. When the machine reached the end of the unbroken surface of the snye ice, Fullerton lifted it clear. The heels of the skis trailed the heaving, grinding ice blocks for over a hundred feet! Certain death lay in that jumbled maelstrom below, but the home-made prop bit cleanly into the air and the *Vic* and her venturesome crew climbed steadily.

A safe landing was made on the snow-covered lake a few minutes later, but it took the men three and a half hours of weary trudging through bush and snowdrifts to get back to the post. There it was decided that Derbyshire should remain at Fort Simpson until parts could be shipped north for the *Rene's* repair, and that

hours they were on the wing to Fort Vermilion, where a landing was made on the Peace River near the Lamson-Hubbard store.

The next morning, when flying directly over the treacherous rapids known as the Vermilion chutes, trouble developed. The single exhaust-stack, projecting upwards directly in front of the pilot's cockpit, burst wide open. Back to Vermilion they went, and Waddell, Link, and most of the supplies were put ashore. Fullerton and Hill then returned to Peace River Crossing, where a welding job was done. The next day they flew back to Vermilion to take aboard the waiting trio and baggage, and by 5:50 P.M. of the same day the voyagers again had reached the chutes.

From Point Providence on the Peace River the airmen flew across a wilderness of stunted trees, muskeg, and lakes to Slave River. Swinging north, they made good time, helped by a tailwind. Smith Rapids slipped by, 2,000 feet below. To river travellers, these rapids are a terrible menace and many lives have been lost in them; the waters of the Slave River swirl and boil past a maze of rocks and islands for some 18 miles, during which they make a drop of 109 feet. During 1920, Theo Link and a party had taken 8 days travelling by water to negotiate that 18-mile stretch, and a loaded barge had been lost during the process.

At 8:03 P.M. the airmen landed near the Hudson's Bay post at Fort Smith, where an excited crowd came down to the shore to welcome them. The previous year it had taken Link and his companions exactly one month to make the trip between Vermilion and Fort Smith which the airmen had just completed in 2 hours and 40 minutes!

On the 31st, a strong north wind brought with it a downpour of rain, and it was not until 2:30 P.M. that the flyers left Fort Smith behind. Following the general course of the river to reach Great Slave Lake, they were able to cut across the many winding loops which make river travel so slow. For example, the Grand Detour is a loop where the distance between two bends is only two miles across, yet river traffic must travel sixteen miles to cover that distance. As they approached Great Slave Lake the flyers were greatly relieved to find that the ice had already gone. Later, they learned that this was the earliest date in history on which the lake had been open for navigation. The airmen alighted on a small channel at the mouth of the river at Hay River post, and the Hudson's Bay Company put them up for the night. Mrs. Vail of the Anglican Mission set out such sumptuous meals that the airmen admitted they were loath to continue the journey. However, they tore themselves away shortly after 2:00 P.M. the following day, June 1. They passed over Wrigley Harbour during a snowstorm, and at 3:25 P.M. Fullerton circled low over Fort Providence to give the lonely people there a friendly wave.

Massive piles of ice blocks, 50 feet high, girded the banks of the Mackenzie, but the river was free. The break-up of the ice held no terrors for the flyers now, since they were equipped with pontoons, and by 4:00 P.M. they

THE CRIPPLED *VIC*
Kept from sinking by a small scow

CLOSE SHAVE
Left: Vic's shattered pontoon. *Right*: The bent and battered right wing after its removal from the fuselage

were being warmly welcomed by their old friends at Fort Simpson.

June 2 dawned a perfect day. With happy expectations the airmen set off at 1:51 P.M. on the final lap of their journey to Fort Norman. A bad radiator leak developed *en route*, necessitating a landing on the Mackenzie, and two hours went by with their craft moored to a large ice-block ashore, while Bill Hill did a repair job.

Wrigley was passed at 3:52 P.M. and Fullerton went down to less than a hundred feet, circling the place at about 100 m.p.h. Excited whites and Indians rushed from the few scattered dwellings to gape up at their first airplane. From this point, the pilot began to gain altitude, going up to 4,000 feet so that all aboard could gain a good view of the surrounding country with its maze of hills, lakes, and watercourses. The air was very clear, and the huge stretch of Great Bear Lake, still solidly ice-bound, could be seen to the east. The gorge through the Franklin Mountains was also distinct, although both of these immensities were a great many miles away.

At 5:20 P.M., the flyers were over Fort Norman, and Fullerton put the nose of the machine down. The air was still at the time, and the surface of the river was glassy, a condition which makes judgment difficult, and landing often dangerous. The pilot set the craft down expertly enough, and except for the usual impact, everything seemed fine. The travellers were astonished to find, when the Junkers came to a stop, that the right wing was resting on the surface of the water and the left wing was high in the air, while the right pontoon floated behind, badly shattered. The crew quickly scrambled out on the left wing as the right began to sink. Fortunately, the machine did not

turn turtle; the river bottom was only 10 feet below, and there the tip of the right wing stuck.

Help was soon forthcoming and eventually the *Vic* was pulled towards the bank, but her right wing was badly battered on the rocky bottom. After great effort, a small scow was pushed under her, to which she was firmly lashed. In that manner the lame duck was later steered downstream some fifty miles to the site of the Discovery Well on Bear Island, 100 miles south of the Arctic Circle, to make a decidedly unimpressive arrival at her final destination.

Gorman eventually came along from Edmonton by water transportation, bringing the various parts required to put both machines into flying trim.

At last, early in August, the *Vic* was repaired by Hill and Derbyshire. Fullerton successfully tested her, taking up A. W. Harris, superintendent of all Imperial Oil Arctic operations. Then the geologist, Mr. Link, and the surveyor, Mr. Waddell, were taken on an extensive flight. In one hour's time they were flown over territory which had taken Link's party the entire 1920 season to cover by foot and canoe. The geologist was strongly impressed with the possibilities of aerial geographic and photographic survey. Chiefly through his recommendation, Imperial Oil Limited contracted with Western Canada Airways Ltd. to map 1,000 miles of Alberta foothill country in 1929.

Bill Hill had already gone south by steamer to Fort Simpson to complete the work on the *Rene*, and the personnel on the *Vic* when she left Bear Island for Fort Simpson were Fullerton, Gorman, Derbyshire, Mr. McKinnon of Imperial Oil, and Chester A. Bloom, reporter for the Calgary *Herald*.

The Arctic sun was setting in a low arc towards the northwest when the party took off on August 6. The rays cast pink lights and purple shadows on the distant peaks of the Rocky Mountains, far up the winding valley of the Carcajou River.

The *Vic*'s party remained at Simpson until the *Rene* was finally put into flying shape. Then, on August 21, the machines took off within a few minutes of each other to start their long flight back to Peace River Crossing.

The return trip was disastrous for the *Rene*. When alighting on the choppy waters of the Peace River a pontoon struck a submerged log and the craft capsized. The occupants were rescued by a police launch, but the *Rene* was done for. Although she was salvaged, she never flew again.

In 1922 the *Vic* became the property of the Railway Employees' Investment and Industrial Association of Hazelton, B.C. Major G. A. Thompson was engaged to pilot the craft—the same Thompson who made the hazardous flight across the Rockies described in chapter 21. In May 1922, Tommy flew the *Vic* from Edmonton to Henry House, B.C., and after a short stay, carried on to Hazelton. L. S. Bell, president of the association, was his passenger. It was the first occasion on which a Canadian registered aircraft had flown across central British Columbia. At Hazelton the aircraft was equipped with floats, which enabled it to make many flights into the wilderness, carrying hunting and prospecting parties to uncharted waters of the vast hinterland of British Columbia.

Flying into such wild territory in that period was an extremely risky business. The entire country is a mass of mountains and if a forced landing or a crash had occurred in that area in 1922, the prospect of passengers and pilot getting out alive would have been slim indeed. On one

occasion in more recent years an aircraft lost in that country lay unseen for five years, despite an extensive air search. The wreckage was finally discovered accidentally by a hunter and an Indian guide on a high and remote mountain top. The eleven occupants of the plane had perished.

As the 1922 season wore on, the operational costs of the *Vic* became excessive, and the next year she was hauled up on the banks of the Skeena River. Here she lay for six long years, a forlorn hulk. Then she was bought by R. F. Corless and shipped to Prince George, B.C. The engine was given an expert overhauling, but air authorities refused to renew the licence on account of age and deterioration. The son of her new owner was so enraged with one government official for punching holes in her thinning shell with a screwdriver that he dealt the inspector a blow that landed him in the icy waters of Taber Lake. A few days later, no machine was to be found. She had flown off, unlicensed and unseen. During the summer she was heard flying to and from a mysterious, unregistered gold mine operated by her owners, becoming more or less a "ghost ship" to local inhabitants. This furtive existence went on until September 20, 1929, when, in landing on Stuart Lake, she received serious damage to her weathered pontoons and wooden engine-bearers from the heavy waves. She was pulled ashore. Before her owners returned the following spring, vandals had smashed the wings and shattered all the instruments.

At this stage a brief moment of recognition came to the old plane. The Junkers firm got wind of her plight and asked that she be shipped back to their factory in Germany, offering to rebuild her completely, install a new engine and propeller free of cost, and then ship her back to New York. But her owners considered her

FIVE PIONEERS
Left to right: Elmer Fullerton and George Gorman, pilots; William Hill and Pete Derbyshire, engineers; Walter Johnson, carpenter

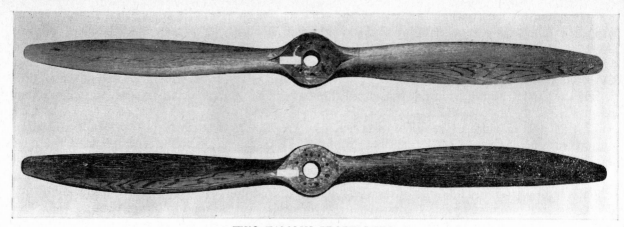

TWO FAMOUS PROPELLERS
The hand-made "sleigh-board" propellers

too far gone; they reluctantly declined the offer, and the *Vic* never flew again.

The engine was sold to an up-coast mining venture where it was rebuilt as an air compressor—a humble ending for a creation that had lorded it over the northern wilds and waters on so many adventurous missions.

In the years which followed, the story of the home-made propellers that were made at Fort Simpson for the *Rene* and the *Vic* gradually came into the limelight through numerous articles, many of them inaccurate. Efforts were made to locate the No. 1 propeller, but apparently it had vanished. The historical value of both props increased, and they appeared to have become the most famous propellers in the history of Canadian flying.

Concerted efforts to find them were finally made by the National Research Council at Ottawa and it was learned that the No. 2 prop, the one made of birch and oak which had been fitted to the *Rene*, had remained at Fort Simpson in the possession of the Roman Catholic Mission. Negotiations culminated in 1938, when Father Cathy, O.M.I., donated the prop to the Aeronautical Museum of the National Research Council at Ottawa.

J. H. Parkin, showed both faith and foresight when he had the glass display case made for the museum: he ordered it built to hold two propellers. In a letter to me in 1941, he asked me to let him know if any news ever filtered through to me which might lead to the discovery of prop No. 1. But waiting for news to drift in is a slow way to get results, so I instituted a search.

The first investigation was too easy to be true. I learned from a newspaper that the original home-made prop taken from the *Vic* was on view in an Edmonton garage. It turned out to be a factory-made propeller, which had seen service on a DH Moth. Months slipped by, while flyers of the two planes were traced and written to. A letter from Bill Hill in April 1942 roused my hopes when he said he thought that the prop had gone to the Standard Oil Company at New York, to be placed in the company museum. But Standard Oil replied that they had

MRS. MARION GORMAN

neither the propeller nor any museum. Exhaustive enquiries at the Imperial Oil Limited offices at Edmonton and Toronto also brought negative results.

In August 1943, an interview with R. F. Corless, last owner of the *Vic*, brought a real thrill. The propeller was resting on the rafters of a barn near Fort St. James, and could be sent by motor freight to Vancouver. When, after weeks of suspense, it arrived, it proved to be another shop-made job, although it had seen service on the good old *Vic*. (It was later presented to the Hudson's Bay Company by Mr. Corless, and is now on display in their museum at Winnipeg.)

A letter from Johnson (co-worker in making the propeller), who was finally located at Waterways working for the Hudson's Bay Company, suggested that the famous prop went to England, to be exhibited at the British Empire Exposition at Wembley in 1924. So off went letters to London, war or no war. After a long interval, an authoritative letter came back—they had no knowledge of any propeller from Canada.

Then in 1945 came a break. During correspondence I learned that the librarian of the Edmonton *Journal* was a cousin of Mrs. Marion Gorman, widow of the pilot. (Her husband had died in January 1942.) After that, the five-year search quickly moved towards a happy conclusion. Mrs. Gorman and her family had the propeller, which they cherished for sentimental reasons. But they generously agreed to donate it to the Aeronautical Museum at Ottawa. There the two propellers are now on view, as sound as when they were first put together with moose glue. Side by side, they are a permanent reminder of the resourcefulness of Canadian airmen.

VIKING OF THE SKIES

This is the story of a staunch aircraft, a Vickers Viking Mark IV amphibious flying-boat, which, like the Viking ships of old, set out with a valiant crew on missions which carried it into unexplored regions where white men were unknown.

Built in England, and well powered with a 450 h.p. Napier Lion engine, the Viking had been sold to the Laurentide Air Services in 1922. Her registration letters, G-CAEB, were seen by thousands of people throughout Canada. She served for ten years rich in accomplishment, before she crashed to a fiery death in Pacific waters.

In August 1924, EB first earned a name for herself by a 900-mile flight over a period of 12 days. The expedition was made for the purpose of paying treaty money to the Indians on the numerous reserves in Northern Ontario. With Roy S. Grandy at the controls and B. McClatchey as engineer, the journey began at Remi Lake, Ont. The airmen followed the course of

THE VICKERS VIKING MARK IV
Near Telegraph Creek, British Columbia, 1925

the Albany River to Fort Albany on James Bay, flew up the coast to Attawapiskat, and turned south again to Moose Factory. The return to railhead was by way of the Moose, Mattagami, and Groundhog rivers. Mr. Aurie, the official of the Department of Indian Affairs who went on the flight, expressed the greatest admiration for the crew, the machine, and the expeditious manner in which the route was covered. To have travelled the usual way—by canoe—would have taken many weeks.

In 1925, an American mining syndicate hired the machine from the Laurentide Air Services for a trip into the wilds of northern British Columbia. The syndicate signed on J. Scott Williams as pilot and Jack Caldwell as co-pilot and engineer. The amphibian was shipped from Montreal to Prince Rupert, B.C., from whence the airmen set off for Wrangell, Alaska, on June 1. At Wrangell, they took on the members of the prospecting party and flew them, in two relays, up the Stikine River to Telegraph Creek, B.C. From that inland point, men and supplies were then flown into the Dease Lake area, where a base was established.

During the summer, the Viking flew to some of the most remote parts of northern British Columbia, penetrating areas which had never before been seen by white men, and probably by few Indians. Parties of men were set down and picked up again after they had had an opportunity to prospect in the vicinity.

The entire area is wild in the extreme, with great peaks 10,000 feet high thrusting massive barriers into the sky; yet the flying was carried out without a single mishap or forced landing, which speaks highly for the skill of the crew and the reliability of their Viking ship. The longest journey undertaken from the main base carried the airplane beyond the northern border of British Columbia, 150 miles into the Yukon Territory, following the Dease and Francis rivers to the Hudson's Bay post on Francis Lake. The latitude of that outpost is between 61° and 62°N., and a glance at the map will prove that these airmen well earned a reputation as flying explorers of the north.

Trips were also made to the Hudson's Bay Company post on the upper Liard River. Many of the spots at which other landings were made were nothing but blanks on any available maps —and still are, as a matter of fact.

Flying and prospecting were carried on for six weeks and a total of 95 hours' flying was logged for the entire venture, which was the first mineral exploration mission by air in northern and western Canada.

EB was dismantled at Vancouver and shipped back to Sault Ste Marie, where she was overhauled in the shops of the Ontario Provincial Air Service.

In 1926, she passed into the ownership of a group of investors at Calgary, Alta., known as the Northern Syndicate Limited, whose object was the exploration for minerals in a blank area on the map of the Northwest Territories east of Great Slave Lake.

The story behind the formation of the Northern Syndicate Limited reads more like fiction than fact. One day a grizzled prospector arrived in Calgary and went calling upon a number of the city's outstanding business men. He showed them rich samples of gold which he claimed he had found in the Northwest Territories. He would not say where, but promised to reveal the place to them for a price. It would not be too hard to locate from the air, as he had marked it with a rough cross cut in the bush by the side of a large lake. He added that he had left his Indian wife there to keep an eye on his find.

COMPLETE OVERHAUL
Vachon rigged this derrick to move the engine

THE VIKING ON CALDWELL
LAKE

However, before plans had been completed to obtain the Vickers Viking, the prospector became drunk one night in a Calgary dive, was struck on the head with a beer bottle during a fight, and suffered a fractured skull. He made a partial recovery, but his memory was affected. He could no longer give any coherent description of where his gold find was located; and he could not be taken north as a guide.

Plans went forward just the same. The syndicate engaged Jack Caldwell as pilot, and Irenée Vachon as engineer—a mechanic who had much service with aircraft engines in the employ of the Napier Company in England.

The dismantled machine was sent by flat-car to Edmonton, where the Napier engine was given a complete overhaul by Vachon and Caldwell. It is of interest to note that, although there were no air-mail or air-freight facilities across the Atlantic or even across Canada in those days, when an order for parts for the engine was sent by wire to the Napier works in England, the required items were in the engineer's hands at Edmonton just 16 days from the dispatch of the telegram.

Repairing the motor took a month. Then the entire machine was sent by railway to Lac la Biche, 127 miles north of Edmonton, arriving June 16, where, out in the open beside the lake, assembly of the machine began. Facilities for handling the Vickers and for lifting the engine were lacking, so Vachon rigged up a derrick, using part of a near-by tree. In 6 days the job was done.

The first test flight was made on June 22, and the next day Caldwell and Vachon set off for Fort Fitzgerald, passing along the route that the Imperial Oil planes, the *Rene* and the *Vic*, had followed five years before. At Fitzgerald, the airmen joined the prospecting party, which

had come up on the Hudson's Bay Company's river steamer. On June 30, the establishment of caches began. The main camp site was on the sandy beach of a large lake (which they named Caldwell Lake), at 61° 30'N., 107° 30'W. It was 1,300 feet above sea level, and actually was on the height of land in that vicinity where on one side of the watershed the run-off flows southeast to Hudson Bay, and on the other side north to the Arctic sea.

Through most of July, the search for the prospector's cross was carried on. Flights were made into the Barren Lands and surrounding country. On several occasions the flyers penetrated over 200 miles into the rocky wastes north of the lake, an area where landmarks of a defined character do not exist. They had to take every precaution with regard to directions and flying time, to avoid getting lost.

Search as they would, no trace was found of the cross, nor was there any sign of any human habitation. If the prospector told the truth, his Indian wife must have perished, for her camp was never discovered.

By the beginning of August, the nights were turning cold and thin ice was forming on the lakes, so the search was halted and the entire party flew back to Edmonton. Here the prospecting party said good-bye to the airmen, who had received orders to fly on to the R.C.A.F. station at High River, south of Calgary. They landed there on September 4, 1926—the first time, incidentally, an amphibian aircraft had accomplished a "ground-landing" at any point in Canada.

Until the fall of 1928, EB lay dismantled in a hangar at High River, and as disuse means deterioration, it appeared that her days were numbered. Then along came B. Lundy of Cal-

THE METAL PROPELLER AND NAPIER ENGINE
Being tested at Sudbury in 1924

gary and Vancouver, who purchased her from the syndicate, and had her hauled by truck and trailer to the foothill city, later shipping her to Vancouver. Much work had to be done on her after two idle years, but when, in the spring of 1929, she was test flown at the R.C.A.F. air station at Jericho Beach, her performance was considered excellent.

Then while her new pilot, Mr. Van der Byl, was being given a refresher course, misfortune arrived in large doses. The doses in this case were of castor oil, which was the specified lubricant for the massive Napier engine. The supply which Mr. Lundy had acquired must have been in storage for years in some corner of a warehouse. While in flight several miles from the air station, over Howe Sound near Bowen Island, the motor began to act up and soon became almost red-hot. A landing was quickly made on English Bay. When the engine was stripped in the hangar later on, the cause of the trouble was easy to see. The castor oil had formed into a thick, gooey mess, and every bearing had melted from frictional heat. The fine engine seemed totally beyond repair.

Until February 1932, the Viking hid herself away in storage in Vancouver; then an enterprising Vancouverite, Captain Fred Clarke, purchased her from Mr. Lundy. Boeings Limited, of Vancouver, were given the job of repairing

the hull and wings, the latter being completely stripped and re-covered.

The motor was dismantled down to the last nut and bolt, and Captain Clarke's brother, H. E. Clarke, an engineer and a master craftsman at his trade, went to work on it. He fashioned new bearings throughout and made a number of new aluminum cylinder heads and divers other items. When at last the motor was tested in the Vickers, its twelve cylinders turned in better revs and gave more power than they had done before. Several flights were made during August 1932. On one trip, with a full passenger list, the airplane climbed easily and smoothly to 10,000 feet.

Captain Clarke intended to send EB on prospecting trips into the Yukon in the summer of 1933, but fate decreed differently. On September 16, 1932, when 2,000 feet above the Strait of Georgia, with Gilbert Jenkins at the controls, Bill Bolton as engineer, and everything going along serenely—crack! A gasoline feed pipe to one of the carburetors snapped, fuel sprayed over the red-hot exhaust pipes, and in an instant the motor was a mass of flames. Three passengers aboard at the time—Mrs. Reece, Captain Landheim, and Floyd Kurtz—received an unexpected thrill as the pilot put the flaming amphibian into an almost straight dive; no one breathed freely again until the plane alighted safely on the water. Small boats from shore soon rescued all aboard.

Although the engine was later salvaged, the Vickers became a total loss. She had gone to her Valhalla in a blaze of glory, like a true Viking ship.

One or two notes conclude EB's history. The propeller which had arrived with the engine from England was three-bladed and made entirely of metal; it had been donated as an experiment by the Leitner-Watt Co. It was soon discarded in favour of a more suitable, four-bladed, wooden propeller, and the original metal prop is now in the Aeronautical Museum of the National Research Council.

In the Museum, too, is the Viking's engine. A brief, matter-of-fact legend, "Napier Lion aircraft engine, No. 24634, originally fitted to Vickers 'Viking' aircraft, Mark IV, G-CAEB," describes one of Canada's most famous airplane motors, but it does not convey to visitors the fact that they are looking at one of the sturdiest engines ever to cleave Canadian skies.

25. Over the icefields

SINCE the middle of the seventeenth century the sealing industry has been one of the primary assets of Newfoundland, but for close to two hundred years the methods used in the search and capture of seals underwent only minor changes.

Sealing is a hardy business. Men from all parts of the island gather at St. John's in the late spring, trekking in from outlying points by any means of transport available to them. They are all well aware of the dangers ahead.

As a matter of fact, the actual capture and killing of a hooded seal is not a risky job, but the journey to and from the sealing grounds is exceedingly hazardous. Ships often have had to fight for days, butting through massive icefields, in terrific blizzards and icy seas. Many a fine ship and her crew have gone to the bottom of the Atlantic as a consequence.

Spotting the seal herds on the ice-floes was for many years mainly a matter of luck. Throughout the daylight hours, crew members took turns aloft in the crow's-nests, scanning the surrounding ice with powerful glasses, but it was as easy to overlook a herd as it was to spot it, and yearly catches varied considerably.

Once located, the seals could be caught and killed only on the icefield. Every available man had to go overside, each armed with a long iron-hooked pole. The method of taking the defenceless creatures is cruel, for they are simply beaten to death by the thousands. Then the sculps, as the bodies are called, are dragged by the hooked killing-pole to the ship's side. The skins are stripped from the carcasses, and the liver and blubber are removed to be taken aboard ship.

THE SEAL PATROL

Towards the end of 1920, Major Sidney Cotton and Captain Bennett of the Aerial Survey Company of St. John's persuaded the owners of several sealing ships to let them show what they could do in spotting seal from the air. But the severe winter conditions were too great a handicap. Radiators and engines froze, carburetors would not function properly, ski undergear had to be designed and tested. As a consequence, the 1921 season was almost over before the airmen had an opportunity to test their plan. Several flights were made, one circling out over the Atlantic for 350 miles, but as it was almost all open water by this time, no seal herds were spotted.

In March 1922, the two machines owned by the company flew out over the icefields. Major Cotton spotted a great herd, but over a week slipped by before negotiations could be completed with the shipowners. By that time winds and tides had completely changed the picture and the opportunity was lost.

All the skippers were strongly against the suggestion that aircraft should assist them. Having been at the work from childhood, sailing to the icefields year after year, it was natural for them to feel that the sealing industry should be conducted only in the way they knew.

In 1923, a completely different idea was evolved by a Newfoundland pilot, Captain Roy S. Grandy, a former member of the R.A.F., who had been a sealing skipper himself. Job Brothers and Company, Ltd., in conjunction with Bowring Brothers, both shipowners of St. John's, were prevailed upon to buy a suitable airplane for the use of the fleet. Grandy purchased from the A. V. Roe Co. Ltd. in England an 80 h.p. two-seater Baby Avro, which originally had been supplied to Sir Ernest Shackleton.

The tiny craft was shipped to Newfoundland, accompanied by an English pilot. Then the Avro set out for the sealing grounds on board one of the company ships, the S.S. *Neptune*. But the antagonism of skippers and crews of the sealing fleet had not diminished. The Avro was never unleashed from her perch aboard the *Neptune*, and the English pilot earned his pay the way the others did—as a sealer.

In the winter of 1923, Laurentide Air Services, Ltd. commissioned Grandy, who had been flying for them in Quebec during that year, to visit St. John's with a view to purchasing one or more of the Aerial Survey Services aircraft, as that company had by then abandoned operations. While there, Grandy prevailed upon the fleet-owners to let him try to find seals from the air during the coming season. So when the fleet sailed in March 1924, the Baby Avro was again in attendance, this time on board the S.S. *Eagle*, whose skipper was more co-operative. Fully rigged, the aircraft was lashed to a platform aft of the ship, ready to be lowered overside. Below decks were Grandy and one of Cotton's ablest mechanics, Engineer H. E. Wallis.

When the fleet reached the sealing grounds, the huge main herd could not be spotted from the ships. When Grandy urged that the airplane be tried out, the unenthusiastic crew carefully lowered it on to a suitable icefield.

A runway of about 150 feet by 30 feet was smoothed off, and Grandy took off, carrying with him as observer Jabez Winsor, whose title aboard ship was Master Watch. The flight lasted only 35 minutes, but it was long enough to break tradition. The airmen spotted a herd of some 125,000 seals, and a fine catch resulted.

One other flight was attempted. The Avro had been designed for use with wheels, skis, or floats, and the extra equipment was aboard ship. When the *Eagle* was in open water, the floats were fitted to the plane and an effort was made to get her off the water, but she vibrated so badly that the flight was postponed. That night the ice closed in, and there was no further opportunity to fly.

During the 1925 season Grandy was not available as a pilot, and the Baby acquired a new boss in C. S. (Jack) Caldwell, who was both engineer and pilot of his craft. He had been previously employed by Laurentide Air Services.

Again the tiny Avro went out aboard the *Eagle*, but despite the success of the previous year the masters were very reluctant to make use of her. The sailors believed that their own knowledge, gained over years of bitter experience, was a match for any airplane.

Luckily for Caldwell, observations from the crow's-nests again brought negative results, and he at last prevailed upon the skipper to put him and his machine to work. So the *Eagle* was tied alongside a large floe and the ski-equipped plane

THE S.S. *EAGLE*

Left: Among the icefields in the seal-hunting area   *Right*: The Baby Avro on its platform aft

ROY GRANDY (*right*) AND H. E. WALLIS
Beside the Avro on board the S.S. *Eagle*

JACK CALDWELL

was swung out on to the rough surface. With little ado, Caldwell took off. He saw immediately that the fleet was proceeding in the wrong direction. A new course was set, and a few days later, on a second flight, Caldwell spotted the edge of the main herd. Continuing on a triangular course of about 70 miles, he was able to give the fleet information which enabled every ship to obtain a full catch.

Flying of this nature was hazardous in the extreme; fog could spring up in a few minutes to envelop the entire fleet. If that had happened, the airman's chances of returning alive would have been just about nil. The skipper of the *Eagle* kept a good smoke going from the stack of the ship as a guide for the flyer, and it was of great assistance.

During 1926 and 1927, Caldwell again contracted to fly the Baby, and all the members of the fleet welcomed his efforts. Numerous observation flights were made and results were good. It was noticed that Bowring Brothers' ships, which made full use of Caldwell's reports, were the ones which returned with the greatest catches.

During the summer of 1927 Caldwell received a new machine, a two-seater Avro Avian, powered with a Cirrus air-cooled engine, capable of a flying range of 500 miles. Its British registry was G-EAQP.

Arrangements for the 1928 season were quite elaborate. During the summer, land bases were selected along the coasts of Newfoundland and Labrador, the Gulf of St. Lawrence, and Anticosti Island. Caches of 600 gallons of fuel were established at each of these points. The purpose of the observation was not only to assure a good

catch, but also to learn more about the migratory habits of the seals. Very little was known of their movements, although they had been hunted for hundreds of years.

Before a single vessel left port that season, Caldwell made an extensive reconnaissance flight, and, following his advice, the captains of the ships were able to proceed directly to the most profitable areas. Not only were capacity catches made by all the ships, but several were able to return to the sealing grounds for a second catch; this had never happened before.

During the 1929 season Caldwell was not available as a pilot, so the aerial spotting was taken over by Alex Harvey, formerly of the Ontario Provincial Air Service. Plans for the 1929 seal spotting were made, but during one of his early missions Harvey was trapped aloft in a dense fog. In making an emergency landing near St. Anthony, Nfld., the Avian was wrecked, although the airman was uninjured.

By now the shipowners had gained much useful knowledge of the habits of seals, and as the depression of the thirties came on, they decided to dispense with the airplane service. In less than a decade a unique phase in the history of flying in Newfoundland had begun and ended.

Airports Remote, 1922

Hundreds of miles beyond the Arctic Circle, along a flat stretch of shore on the barren southern coast of Ellesmere Island, stakes were driven in 1922 to mark the site of what was then the world's most northerly air strip, 830 miles from the North Pole.

This was part of the work of the Canadian

Government Arctic Expedition of 1922, in connection with the establishment of Royal Canadian Mounted Police posts on the islands northeast of the mainland. Information had seeped through to Ottawa that foreign trading companies were operating illegally in that area, and the protection of the law for traders, natives, and game had to be assured. The murder of a white man by natives, about a year before, was also proof that the police were needed in those high latitudes.

The officer in charge of the 1922 expedition was J. D. Craig, and, thanks largely to the efforts and foresight of J. A. Wilson of the Air Board and O. S. Finnie, Director of the Northwest Territories, the personnel included Major R. A. Logan of the Canadian Air Force. He was a qualified surveyor, and his job was to investigate flying conditions and stake out possible sites, particularly on the northern tip of Baffin Land, and on Devon and Ellesmere islands. Captain J. E. Bernier was in charge of the government ship, *Arctic*, and the party included 9 Mounted Police under Inspector Wilcox, a ship's crew of 5 officers and 20 men, L. O. Brown of the Geodetic Survey (who also acted as official meteorologist), Dr. L. D. Livingstone (of Aklavik agricultural fame) as medical officer, W. H. Grant of the C.N.R. staff as secretary, and G. H. Valiquette, official photographer—a total of 42 men.

The cargo included hundreds of tons of coal and lumber and 75 tons of miscellaneous supplies suitable for a stay of possibly two years. The party left Quebec harbour on July 18 and two weeks later Captain Bernier identified the Greenland coast still some 45 miles away, sighting the far-away heights of 4,710-foot Mount Umelik in the early afternoon. Beyond could

THE AVRO AVIAN

vaguely be seen the vast, overwhelming ice-cap which spreads over almost all of the great island. The going was slow, because of the ice about them, and on August 8 they were obliged to stop, being completely hemmed in.

By August 15, however, the ship was forging steadily across Baffin Bay, with the intention of reaching Pond Inlet, but that area was still solidly ice-bound, and they were forced to heave to several miles from their destination. Here they were visited by Mr. Caron of the Arctic Gold Exploration Company and Inspector Joy of the R.C.M.P., who were brought to the ship in an Eskimo whale-boat. It was decided that the *Arctic* should go on to Ellesmere Island and call at Pond Inlet on the return.

Ellesmere had been unpopulated for years, and since experience had proved that white men's lives in the north often depended on the presence and help of natives, a group of Eskimos with their dogs and belongings was taken on board. They were hired to stay on Ellesmere for a period of one year. While aboard ship, they lived in their skin *tupiks* (tents) on the foredeck.

After battling up and down the coast of Ellesmere Island through heavy floe-ice, the expedition selected a small harbour on Smith Island—they named it Craig Harbour—at which the unloading of supplies could proceed, and a party went ashore to select a site for the buildings. On that lonely, barren spot, the police post began to take shape, and by August 28 the building was fast nearing completion and all the supplies were ashore.

Meanwhile, Logan had surveyed the shore at the head of the harbour and had staked out a landing-strip. Its position is 76° 10′N. latitude, and 81° 20′W. longitude, or, roughly, 2,200 miles straight north of Toronto, Ont. An official ceremony marked the opening of this, Canada's most northerly air field to date, and a Canadian Air Force flag was raised on a pole in the presence of a "crowd" of three Eskimos and the photographer!

Time was precious, for thick ice was beginning to form on the ocean surface at night. As Captain Bernier realized that to delay was to court disaster, departure was speeded up. With little ado, Inspector Wilcox and his band of six men said good-bye to the ship's company, and climbing into their small boat, they rowed slowly away towards shore, as a heavy snow

THE *ARCTIC*

Left: At anchor in Pond Inlet, September 6, 1922.   *Right*: **The northern part of the route taken to Ellesmere Island**

squall swept out of the north and hid the ship from view.

Heading south at reduced speed through thick icefields, Captain Bernier passed to the east coast of Devon Island, rounded Cape Warrender, and anchored in Dundas Harbour, where the ship lay to for 24 hours. The call was too short to enable Major Logan to make the scheduled survey at that point.

Attempts were then made to reach Pond Inlet again, first by a route to the west of Bylot Island, into Navy Board Inlet, and thence by Elipse Sound. This proved to be impossible, as solid ice stretched right across between the shore of Bylot and Baffin Island, so the ship was forced to retrace her course, going back around Bylot and following down the eastern coastline.

When reached, Pond Inlet was found to be still bound solid with ice. As the weather was bad, Captain Bernier sought shelter in Albert Harbour on Baffin Island, some eleven miles from the Hudson's Bay post at Pond Inlet. While at anchor there, the Hudson's Bay supply ship *Bayeskimo* arrived, and on September 6, both vessels moved out and were able to force their way to Pond Inlet. The *Arctic* spent three days there, landing the supplies and material for building the police post. Major Logan scouted around and discovered ample space for a large air field, on a site southwest of the Salmon River, a few miles west of the post. Several short but serviceable runways were also selected just south of the post itself. All sites were duly surveyed, staked out, and written down in reports.

It is regrettable that funds did not permit the Air Board to send a small aircraft with this expedition. However, the reports and recommendations of Major Logan were of great importance to the Air Board when an aerial expedition was finally sent to Hudson Strait in 1927.

In that high latitude, winter sets in when only the first crisp touch of fall is in the air along Canada's southern border. Knowing this all too well, Captain Bernier had no desire to be caught napping, and be forced to winter in the Arctic. The work of unloading was speeded up, every member of the ship's personnel helping with a will. The skipper himself handled the winch levers, to relieve the winchman for other duties.

By 4:00 P.M. on September 7 the work was completed and the anchor was weighed immediately. The Mounties who were to be left at Pond Inlet went ashore, and with a salute of three long blasts of her whistle, the *Arctic* moved out of the inlet for the return journey south. An hour later, the inlet filled solid with ice.

With the exception of a brief stop at Godhavn, Greenland, the *Arctic* pushed straight on, buffeting cold winds and heavy seas all the way, until she docked at Quebec harbour on October 2.

An aftermath of the flag-raising ceremony at Ellesmere belongs to this story. Upon Major

Logan's return to Ottawa, the source from which the flag had been procured asked for its return, or payment in full by the airman. As he had left the flag waving proudly in the wind over the Arctic wastes, the Major was a little nonplussed. The official record closes here, but it became well known that Logan told the flag-seekers of a much warmer and distant place to go than the Arctic, although he knew full well they would not find the Air Force bunting flying down there.

Eleven years afterwards, during the summer of 1933, Major Logan returned to the north in charge of supplies on the Pan American Airways vessel, the *Jelling*, which was the floating base used by Colonel and Mrs. Lindbergh on their flights over Labrador, Baffin Island, Greenland, and other areas, during their aerial circumnavigation of the North Atlantic. On one lengthy flight, Major Logan accompanied the American airman along the west coast of Greenland, where they spotted a number of likely sites for possible future commercial bases. This knowledge was utilized by both the Canadian and United States governments when it became necessary during World War II to establish bases between Canada and the British Isles. Serving then with the R.C.A.F., the Major spent three months during the early war years helping to re-locate the places he and Colonel Lindbergh had spotted from the air in 1933.

UNITED STATES MACMILLAN EXPEDITION

Who were the first flyers in the Canadian Arctic? This question has given rise to many incorrect answers, some of which have been accepted as true. Careful research reveals the following facts.

In the summer of 1925, the United States MacMillan Expedition, sponsored by the National Geographic Society, went north in two ships, the S.S. *Peary* and the S.S. *Bowdoin*, under Lieutenant Commander D. B. MacMillan and Lieutenant Commander E. F. McDonald, Jr., respectively. Aboard one of the ships were three (dismantled) single-engined Loening amphibian aircraft, NA-1, NA-2, and NA-3. The Naval Air Unit which went along to operate the aircraft consisted of eight men under Lieutenant Commander Richard Byrd.

From a strip of beach at Etah, on the west coast of Greenland, exploratory flying was conducted over portions of Greenland and of Ellesmere Island, which is, of course, Canadian territory. A number of trips were made across Smith Sound to Ellesmere, the first on August 8, 1925, when the NA-2 and the NA-3, in charge of pilots Schur and Reber, completed a circular flight from Etah. Four other members went along, Lieutenant Commander Byrd, mechanic Rocheville, and photographers Gayer and Williams. These six men were the first to look down from the air on Canada's High Arctic regions. The first landing by aircraft in any Arctic area of Canada occurred on August 12, when the NA-1, flown by Lieutenant Commander Byrd and Floyd Bennett, and the NA-3, with Reber at the controls and Lieutenant Commander McDonald aboard, made a short stop-

ELLESMERE ISLAND
Major Logan displaying the Canadian Air Force flag at Craig Harbour, 1922

THE LOENING AMPHIBIAN NA-2
On the beach at Etah, northern Greenland, 1925

MOUNTAINS ON ELLESMERE ISLAND
An historic photograph, taken from the NA-3, 5,000 feet above Smith Sound

over on the sea in Hayes Fiord on the east coast of Ellesmere.

An interesting feature of this expedition was the use made of short-wave radio communication. Lieutenant Commander McDonald had prevailed upon naval authorities to install a short-wave radio set aboard the *Peary*. During the expedition's stay in Arctic seas, it was able to keep in almost daily contact with United States headquarters, and on numerous occasions the short-wave messages were received by U.S.S. *Seattle*, then cruising near Tasmania.

An example of such broadcasts over the short-wave hook-up was made from the *Peary* while anchored at Etah. Eskimo singing was sent over the air lanes, to be picked up by Lieutenant Schnell's short-wave set aboard *Seattle*, which was then lying at anchor off the coast of Tasmania, almost directly opposite Etah, on the other side of the world! This was the real commencement of successful short-wave radio communication over vast distances.

THE HUDSON STRAIT AIR EXPEDITION AND SURVEY, 1927–1928

The aerial survey of Hudson Strait in northeastern Canada, in 1927–1928, may be regarded as the culmination of some three hundred years of search and discovery in that formidable area. During a year of voluntary exile, airmen under the Canadian government made observations and collected a body of authentic data which made the passage of those waters much less hazardous and arduous. The knowledge gained of weather and ice conditions made possible

safer sea and air transport in that area in World War II.

The immediate purpose of the aerial survey, however, was to assist in finding a northern water route for the export of Canadian grain to Europe—a route which would be many miles shorter than that *via* the Great Lakes and the St. Lawrence River. Hudson Strait, between the northeast part of Hudson Bay and the Atlantic, was the most difficult part of the passage. In 1927, therefore, when the Hudson Bay Railway was being extended to the Manitoba shores of the Bay, the Hudson Strait Expedition was also at work.

Three men were appointed as an Advisory Board: the late Major N. B. McLean of the Department of Marine, E. B. Jost of the Department of Railways and Canals, and Group Captain J. Stanley Scott, M.C., A.F.C. of the R.C.A.F. who was later replaced by the late Wing Commander J. Lindsay Gordon, D.F.C.

Equipment and stores for three sub-Arctic centres were shipped to Halifax; buildings, all prefabricated, comprised six hangars, six dwellings, and three each of store, radio, and blubber houses, the last being for food storage for the native Eskimos. Machines included six new Fokker Universal aircraft, a two-seater DH Moth biplane, three Fordson tractors, three 30-foot motor launches, and three complete radio stations with motor-driven generators and 150-foot, steel masts. Supplies for the three bases consisted of 326 drums of gasoline, 45 gallons to the drum; 38 oil drums of similar capacity; 1,000 gallons of Ethyl airplane spirits; 160 tons

PILOTS OF THE 1927–1928 HUDSON STRAIT SURVEY EXPEDITION
Left to right: Leitch, Ashton, Coghill, Lawrence, Carr-Harris, Lewis

of hard coal packed in sacks; and sacks of sand and cement for the foundations of buildings. In addition to all this were the useful beds and bedding, stoves, and dishes; two dories, a skiff, guns, ammunition, clothing, medical supplies, and meteorological instruments; and sufficient food for 16 months. So wisely did the Board, in co-operation with J. R. O'Malley, secretary of the expedition, anticipate every contingency, that when the bases were established in the north it was found no essential item had been overlooked.

Two ships were commissioned to carry men and material north, and on July 17, 1927, they sailed out of Halifax together. The 44 members who comprised the personnel of the expedition were aboard the Canadian government ice-breaker, C.G.S. *Stanley*, commanded by Captain John Hearn. Aboard the second ship, a freighter, the S.S. *Larch*, under Captain W. J. Balcom, were 57 men of the construction crew together with 2,585 tons of general cargo and 2,700 tons

of coal. After bucking violent weather, and battling heavy ice-floes for ten days, both craft reached Port Burwell at the northeast point of Ungava Bay.

As base "A" was to be established in that locality, four members of the expedition were left ashore at the Hudson Bay post there, while C.G.S. *Stanley* and the *Larch* continued westward to find sites for the other two bases. On August 1 they anchored at Eric Cove near Cape Digges, were welcomed by another Company manager, and directed by him to a spot on the southwest shore of Nottingham Island. An aerial inspection of the island in the Moth seemed to confirm this choice, and unloading of equipment began. It proved to be a tough job, but all hands went at it with a will. By the time everything had been placed ashore on the 17th, derricks had been rigged, buildings were in process of completion, the radio masts were already in position, and before the *Larch* left Nottingham Island to establish the third and final base, radio contact was established with Ottawa. Hon. Charles Dunning, then Minister of Railways and Canals, sent back a radio message in which he announced that Fort Churchill on Hudson Bay had finally been selected as the northern terminal for the Hudson Bay Railway.

It had been hoped to set up the third base along the northern shore of the strait, but aerial observations showed it to be a low, rock-bound, barren area, quite unsuitable for year-round operations. On the 24th, both ships anchored at Wakeham Bay while Major McLean conferred with the Hudson's Bay Company manager and some Roman Catholic missionaries who lived there. They learned that a good site was available at the east end of the bay, and as a short flight in the Moth showed nothing better in the vicinity, the job of unloading supplies and material began at once. Every item of equipment had to be trans-shipped by launch and scow, yet by September 8 everything was ashore and buildings were nearing completion.

Shortly before that date the dainty but sturdy little Moth came to a sad ending. She had been moored well off shore; when the wind suddenly freshened to gale proportions, heavy seas made it impossible to reach her. She was left to battle it out alone through the dark hours of the night. Tough as she was, the huge waves beat her

under, and the next morning she was a total loss.

The flights of the Moth seaplane were the first to be seen in Canada's eastern mainland Arctic area, and the Eskimos were greatly interested in the machine. They seemed to have none of the fear and awe which many Indians showed when they viewed an airplane for the first time.

C.G.S. *Stanley* and the *Larch* now sailed back to Port Burwell. The party left there six weeks before had selected a site near by, Fox Harbour, in the northwest arm of Burwell Harbour. Time was pressing, so instead of erecting living quarters, Major McLean's party rented an old Moravian Mission building close by. Base "A" was ready by October 28.

When all construction was completed, the two ships set off on the return voyage south, taking with them the workmen and, regrettably, Major McLean, who had been taken ill at Wakeham Bay and could not carry on.

The three bases, now under the command of Squadron Leader Lawrence, knuckled down to the work ahead. In spite of bad weather, the airmen were ready to begin actual survey flights by November 7, by which time installation had been completed at all three radio stations for sending out meteorological information to each other, to planes in flight, and to Ottawa.

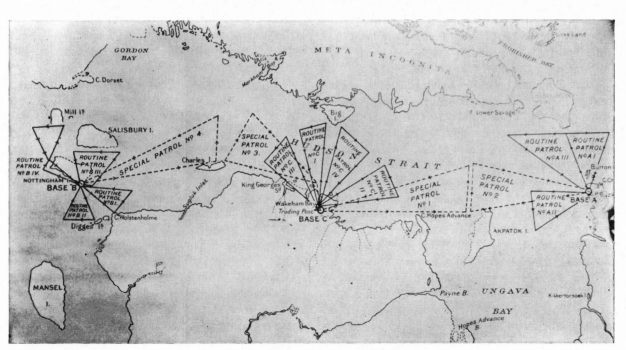

CHART OF ROUTINE AND SPECIAL PATROLS
Hudson Strait Expedition 1927–1928

THE *SPIRIT OF THE VALLEY OF THE MOON*

The name given to the De Havilland Moth used by the Hudson Strait Expedition
Left: Squadron Leader T. A. Lawrence
Right: Flight Lieutenant A. A. Leitch

The first signs of ice during the winter of 1927–1928 were observed from a base "B" plane flying south from Nottingham Island on November 16, when large masses were seen afloat miles to the south in Hudson Bay. Fog, strange Arctic vapours, wind, and snow prevented flying operations on a great many days, but at every opportunity, often under hazardous conditions, air patrols and reconnaissances were made.

An adventure which befell the airmen from base "A" on one routine patrol is an example of the dangers of the work.

Just before noon on February 17, Flying Officer "Jaggs" Lewis, Flight Sergeant Terry, and an Eskimo named Bobby Anakatok took off from base "A" on a northerly course across the entrance of Hudson Strait, with the coast of Baffin Land as their objective. Their turning-point was a promontory on Resolution Island, but when they reached it, visibility became almost zero, and they had to set a compass course for Burwell. As if that were not trouble enough, the engine began to misbehave, losing revs and vibrating badly. After flying this way for some time, Lewis thought he glimpsed the Button Islands on the port side. He carried on for about fifteen minutes; then the topography of the land led him to believe he was far off his course, and near Akpatok Island in the centre of Ungava Bay. Being comparatively sure of his position—he had steered a somewhat exaggerated course after leaving Resolution so as to be

sure of flying into Ungava Bay instead of over the Atlantic—the pilot set a course which he thought would carry the plane direct to Burwell.

The airmen flew on westward until 3:00 P.M. while the engine became rougher and rougher. At last, with fuel almost exhausted and no land yet in sight, it became necessary to put the machine down on the ice. This Lewis managed to do without a crack-up, although not without some damage to the propeller and the skis. The surface was a mass of huge hummocks, with drifts over twenty feet high, and cracks and crevasses everywhere.

The men stayed with their machine all night. In the morning a heavy snow was falling, and they were unable to check their position. Concluding that they were down on Ungava Bay, they struck out on foot in an easterly direction, lugging along their life-raft, with their emergency kit and compass stowed aboard. They travelled all that day, until the condition of the ice made further progress impossible; then spent the frigid night in a snow hut, expertly fashioned by Bobby.

The 19th broke clear, and from a high hummock they glimpsed clouds and what seemed to be open water to the east, while to the west, in the direction from which they had journeyed, were high banked clouds and what appeared to be a high contoured skyline. They decided to back-track, and on the afternoon of the 20th the irregular, cloud-packed horizon to the west was definitely seen to be land. For the next four days the weather was thick, and they were forced to rely entirely upon compass readings.

The coastline was seen again on the 24th, and rather than lose it the trio travelled all that day and the following night, finally reaching land on the afternoon of the 25th. Later they learned that it was the Labrador coast, in the vicinity of Kamaktorvik. They had been nine days afoot on the Atlantic ice! The temperature was then showing 25° below zero.

During the early part of their trek, they crossed open-water leads on their rubber life-raft, but as their strength began to give out they abandoned the raft, together with items of clothing and even some of their precious rations.

On one stretch they were without food for 24 hours, until Bobby Anakatok managed to bag a walrus, parts of which they ate raw. After they had left their raft behind, they used ice-pans to ferry themselves across the numerous leads of open water.

Once on land, they set off in a northerly direction. They sighted no living person until the afternoon of the 28th, when they met an Eskimo hunter and his wife, with a komatik (sleigh) and dog team. With Bobby acting as interpreter, arrangements were made with the hunter to lead them to Port Burwell.

Meanwhile, the men at base "A" had called on the machines at the other bases for aid. The flights which were taken in immediate response were in themselves a great achievement. From Nottingham Island, Flight Lieutenant Leitch flew 590 miles, and from Wakeham Bay, Squadron Leader Lawrence flew 340 miles, both over unknown terrain, with visibility almost zero, and with unreliable compass readings.

At base "A" the weather now became too stormy to permit flying for about a week. When the planes did succeed in getting aloft, the

INSPECTION BEFORE TAKE-OFF

The vital emergency kit laid out before a search flight for the missing Fokker. *Facing camera, left to right:* Flying Officer Carr-Harris, Flight Lieutenant Leitch, Squadron Leader Lawrence, G. Valiquette, the photographer

225

MANHANDLING OVER ROUGH ICE
Three komatiks protect the skis and tail skid from damage

searchers centred on Ungava Bay, misled by Pilot Lewis' radio message just before his forced landing. Over 1,500 miles were covered by the planes, but since they were searching in the wrong area, they had no chance of locating the lost men.

For twelve days every effort was made, and willing assistance was given by Royal Canadian Mounted Police, by Mr. Ford of the Hudson's Bay Company post at Port Burwell, and by Captain Bennett and members of the C.G.M.M. steamer *Canadian Raider*, which was wintering at Port Burwell. Eskimos had been dispatched to points on Ungava Bay and gas caches were being established when the missing party finally trudged in at midnight on March 1, frostbitten, hungry, and desperately weary, but otherwise in surprisingly good condition.

All aircraft when on a lengthy patrol carried one of the local Eskimos. It was this precautionary measure which saved the lives of Lewis and Terry, for without the aid of Bobby Anakatok the two R.C.A.F. men would in all probability have perished. The abandoned aircraft was never seen again.

Considerably less serious, but trying enough, was the experience of Lieutenant Leitch, Corporal F. J. Ewart, and Constable McInnis, R.C.M.P., when returning from a patrol flight to base "B." A dense black haze, such as only the north can "cook up," gradually obscured land, sea, and sky. Flying past their destination, the

airmen were forced to alight on the open water, where they rode out the night among heaving and groaning ice-floes. Although threatened with destruction at every moment, the airplane managed to survive the long ordeal.

On another occasion, Flying Officer Carr-Harris was flying out from base "C" when his engine went dead over rocky terrain. He was

34.2° BELOW
Registered on the thermometer on January 21, 1928

BASE OF THE HUDSON STRAIT
EXPEDITION AT WAKEHAM BAY

The buildings at the upper left are the
Hudson's Bay Company and Revillion
Frères posts

able to set the float-equipped craft down on a small lake, and as it was summertime, he experienced no hardship; but several long and lonely days elapsed before he and his machine were located and rescued.

The Fokker aircraft used were all of the cabin type, but the pilots' cockpits were open. This made flying rugged for the airmen during the winter months. Hands and faces were frequently frostbitten, and the doctors stationed at each base were kept quite busy, apart from the meteorological duties which they also performed.

During the expedition's year of duty, over 200 routine patrols were flown successfully, together with a considerable number of special patrols, and non-patrol flights. Throughout the winter, daily measurements were taken at each base of the depth of sea-ice. The greatest depth was registered at Nottingham Island, where a maximum of 54 inches was noted on May 10, and again on June 4, 1928.

The lowest temperature recorded was also at base "B," when the mercury twice dropped to 34.2° below zero, on January 21 and February 24. This temperature, while not extremely low, was made cruelly severe by the bitter winds

that howl across the barren land much of the time, averaging 25 miles an hour. The most violent wind recorded was at base "C," when gusts during a sou'wester reached 70 miles an hour on August 23, 1928.

One major outcome of the expedition's work was the establishment of three Direction Finding Stations along the shores of Hudson Strait, which enabled ships to navigate with more security than previously. So many inaccuracies were found in the maps and charts used by the expedition that R. G. Madill, assistant magnetician of the Dominion Observatory, and his assistant, Charles Rump, were, at the request of their department, sent north on C.G.S. *Montcalm* when she set out in the fall of 1928 to bring back the personnel of the expedition. Before the ship returned south, the two men had performed work of the greatest value in correcting some of these inaccuracies.

Some of the personnel travelled south aboard C.G.S. *Voyageur*, and others on the Hudson's Bay Company's S.S. *Nascopie*, but the majority went on C.G.S. *Montcalm*. When *Montcalm* reached Quebec on November 14, she was welcomed tumultuously with blasts from the whistles of all the shipping in the harbour.

26. The bush pilot–and his engineer

It would be only too easy to grow sentimental in describing the exploits of the Canadian bush pilots of the late twenties, for in what other country has a like story been written in the sky and on the ground? Elsewhere individuals have made more historic single flights, and certainly no Lindberghs nor Mollisons nor Earharts, no Costes nor de Pinedos, have emerged from the Canadian bush. Yet, wherever snow falls or water freezes, flyers the world over are indebted to our airmen; and their total contribution, in terms of aggregate hours of flying and accumulated variety of experience, has no parallel in the aviation history of any other country.

From the days of the *voyageurs* and the *coureurs de bois*, the Canadian northland has lured adventurers with its rich rewards of fur or minerals. The revolution in northern transport created by the airplane staggers the imagination. A point in the hinterland that could be reached only by weeks of paddling, tracking, and portaging, or on foot behind a dog sled in winter, could, in the late twenties, be reached in hours. An inaccessible point such as Trout Lake Post in the heart of Ontario's Patricia District was now as close by air to Winnipeg as Ottawa was to Montreal by train. At first only an occasional wealthy man or a mining syndicate took advantage of this fact. But as the possibilities of air transport began to sink in, the prices dropped, and as the reliability of the bush pilot became known, more and more people availed themselves of plane transport, till finally the prospector, the trapper, and even the Indians accepted it in their stride.

Much has been made of the bush pilot in romantic fiction, and a popular picture has emerged of an irresponsible daredevil who would cheerfully fly through the gates of hell and back—even if the gates were shut. Nothing could be farther from the truth. Like pilots the world over these men did the job that was set before them, but the nature of the job was unique. Every dollar earned by commercial companies was won by good organization and by good flying, which assuredly did not include any recklessness on the part of their pilots. It should be remembered, too, that the Customs and other taxes on aircraft imported into Canada increased the cost of their machines nearly 50 per cent. As the majority of aircraft came from outside sources, the total cost of a plane was in itself a big reason for entrusting it only to pilots of good judgment.

An aviator's first concern in flying over new territory is, where can I land? The answer to this in the Canadian bush is simple enough— anywhere. But having answered this, we must hasten to add that anywhere is nowhere. Equipped with pontoons in summer, or skis in winter, the experienced pilot can find a smooth spot of water, ice, or snow within gliding distance of almost any place in the Canadian hinterland where he may find himself in trouble. But once down, if engine trouble, or fuel shortage, or disability, or any other of a dozen possibilities makes a take-off impossible, the pilot and his mechanic find themselves nowhere, with only emergency rations, stout hearts, and resourceful minds to get them out.

But this is only a minor aspect of the bush pilot's problem. A major one is finding his way. In the twenties and early thirties most of Canada's rugged hinterland was unmapped, except in the broadest sense. Worse still, such maps as existed were frequently inaccurate in such details as they did show, so that they could cause more trouble than if they had been left blank. More typically, in the early days the only map a pilot had to go by had been sketched on on the back of an envelope by someone who had made the trip by dog team or canoe. As he neared the magnetic pole—and "near" might be 1,000 miles—and sometimes in the presence of large bodies of ore, his compass became a mere ornament in the cockpit. There was the sun to reckon by in the long flying hours of the summer, but in winter good visual flying time was desperately short. Caught in the air under conditions of poor visibility, a good pilot flew by the seat of his pants: that is, by physical reaction to the stability of his machine, the angle of pressure of the seat against his behind imparting the exact flight position of the aircraft so that instinctively he kept the airplane on an even keel.

If landmarks were lost, and an airman was obliged to go down to a landing to wait out the weather, continued low ceiling or zero visibility could hold him up for days. Then too, in the fall and spring of the year, when changing temperatures made the north country unfit for landings or take-offs either with floats or skis, to be caught away from base was bad medicine, to be avoided at all costs.

Yet all of these hazards were merely routine obstacles or nuisances. A bush flight was never made without a reason, though the "reasons" were not always too rational.

"They want fresh eggs in at Cache Lake. Can you fly in a couple of crates?"

"Why not?"

"We want to get our outfit in to Cedar River by the 8th, and there's still no sign of a break-up."

"Sure thing—take you in tomorrow."

"What about our canoe?"

"Don't give it a thought: we'll fasten it under a wing or between the pontoons."

"Doc Smith says that the baby in at Sturgeon needs cow's milk."

"What makes you think we're running a dairy?"

"Well, we're getting a cow in on Tuesday night's train."

"What do you want us to do—fly the cow in?"

"We'll pay you for it—whatever it's worth."

"A cow? A live cow? Sure we can take her, if Doc will give her a hypo."

So went the story. No matter what was requested, if the plane could hold it, in it went and off the pilot flew, to make delivery. As larger and more powerful aircraft came into the picture, so the volume and size of freight increased.

"We need a portable saw-mill up at Woman Lake."

"Fine, when do we start? We fly anything."

A missionary bishop wants to arrange for a round trip to northern missions; a Hudson's Bay post needs urgent supplies; a trapper wants to be flown into his territory in the fall—dogs, equipment and all—and what's more, he wants to make sure he'll be picked up again in the spring.

Finally, as air freighting throughout the country took on a new aspect, practically anything needed in the north and in outlying districts could be whipped in by air, and it could never get there any other way. Today, teams of horses and oxen, giant tractors, and similar heavy equipment all go in by air, and if the machinery is too big or too heavy for one trip, or cannot be got through even the wide door of the machine, there's nothing to it—just cut 'em up with a welding torch, pack 'em, and weld 'em together again at their destination as good as new!

This is, perhaps, as good a place as any to speak for the almost forgotten man—the engineer, or mechanic, as he used to be called. Nearly always a bush flight was made by a team: pilot and mechanic. Usually the mechanic could fly the plane at a pinch; often he was a pilot temporarily without a machine, and sometime a qualified pilot who preferred the responsibility of keeping the engine going to that of piloting the machine. A good mechanic was often the difference between death and survival; a poor mechanic could ruin a routine flight.

FREDERICK J. STEVENSON

Pilot and mechanic: one was no good without the other.

As the twenties moved into the thirties, flying in the high north came into its own, and winter trips became routine. The successful maintenance of aero engines under sub-zero conditions is one of the epic accomplishments of flying anywhere in the world. The engineers who serviced the engines under such temperatures, completely exposed to the weather, were as doughty a bunch of pioneers as ever earned a place in history.

Few laymen know that when those early airplanes in the north came to a safe winter's landing at some lonely outpost, miles from communication with the outside world, it was the mechanic's chore to see that the engine did not freeze up, and that it would be in readiness to break into song again when required, without too much waste of time. At the end of any flights in near zero weather it was imperative for the engineer—and pilot too—to get out of the machine as quickly as possible after landing, and, just as soon as the engine was stopped, to open the tap which drained the oil from the motor in order to allow the hot oil to run out before the

whole engine cooled off and froze into a solid mass. Many a time, if there happened to be a thick layer of snow covering the ground, the oil was drained out directly on to the surface of the snow. The cold quickly congealed the oil into a solid mass which the engineer would then gather up like a small log, and lug off to a nearby cabin—if there was one. In the cabin he would melt the oil into a can; when it was needed again, it was heated almost to boiling point before being put back in the engine. Woe betide the engineer who returned the hot oil to a frozen crankcase, only to find something else needed doing before the engine could start. There would be nothing for it but to drain out the oil again quickly, or serious trouble was in store. When engines did solidify through excessively low temperatures, the flyers erected a rough tent around the nose of the machine to enclose the engine, then with gasoline blowtorches or small stoves heated the motor back into mobility. This heartbreaking task was carried on for many years by engineers who kept aircraft in the air in the north during the winter. It was not until 1938 that the invention of the oil dilution technique solved the difficulty.

Fuel was another problem in the early days. Gasoline had not reached the stage of perfection it has today. Early pilots and mechanics, when purchasing fuel for the motors, simply bought "gasoline," and high octane as used in modern aircraft engines was a fuel beyond even wishful thinking. Right up through the years to well into the thirties, no good mechanic would think of pouring fuel into an airplane's tank without first straining it through a suitable chamois.

FRED STEVENSON AND THE HARMON TROPHY

All this having been said about bush flying in general, the interesting question comes to mind: who in particular was the outstanding airman of the lot? Here it is not so much a question of who was first, as of who stayed with it the longest and did the most. Many bush pilots would dismiss the whole question with an indifferent shrug of the shoulders. Others would toss off dozens of names in quick succession. As an historian I would be completely on the spot if the selection had not already been made by an authority any airman would be willing to accept: La Ligue Internationale des Aviateurs. This organization was formed in 1926 in Paris, France, by an American pioneer balloonist and airman, Clifford B. Harmon. Shortly afterwards the league established the Harmon Trophy as a memorial to the members of the Lafayette Escadrille, the famous air squadron of American volunteers which fought under the French flag during World War I. The award is international, symbolized by a four-foot bronze trophy and individual medals, and is conferred annually on the pilot of any country whose flying is judged to have been the most outstanding in the advancement of aeronautics in the year. The only Canadian who has so far won this honour is Frederick J. Stevenson of Winnipeg, who was Canada's outstanding pilot in 1927.

"Steve" was born of Irish stock at Parry Sound, Ont., on December 2, 1896, but grew up chiefly in Winnipeg. He was not quite nineteen when he left Wesley College to enlist in the 196th University Battalion at Winnipeg. In 1917, he transferred to the Royal Flying Corps, trained in England as a fighter pilot, and returned to France. By the end of the war he was flying one of Britain's fastest, single-seater fighters, the tricky, two-gun *Dolphin*, a machine of which little was ever heard, as the war ended shortly after it had been placed in service.

Like all fighter pilots, Steve brushed past death many times, and at the close of the war he was fast reaching a high score in enemy aircraft destroyed—18 airplanes and three observation balloons. He came out of the war unscathed, with the rank of Captain, wearing on his tunic the ribbons of the Distinguished Flying Cross and the Croix de Guerre.

After the war Stevenson served first as one of the R.A.F. pilots entrusted with the job of ferrying between London and Paris the flock of diplomats and other officials connected with the Peace Conference. Then a call went out for volunteers from the R.A.F. to serve as flying instructors in Russia, and Steve immediately signed up.

The Russian training candidates proved to be fine specimens of manhood, but not very adept at flying, and few were allowed to attempt solo flights until the week before their R.A.F. instructors were recalled to England. For his service to the Russian government, Stevenson received the Order of St. Stanislas.

Back in Canada in 1920, Steve was taken on as pilot by the Canadian Aircraft Company, then beginning their second year of air operations at Winnipeg; and it was there that I met him. Like everyone else, I found Steve likeable and unassuming, and a grand companion. We flew together many hundreds of miles over the wide prairies of Manitoba and Saskatchewan, filling contracts with towns and villages for exhibitions of aerobatic flying, or taking up passengers at $10 a flip.

We carried on in a staunch, three-seater Avro, bearing Canadian registration G-CAAR, until August 25, at Fort Frances, Ont., when we met our match in one of many unsuitable landing places. We had taken up a number of passengers, and then went aloft to do our usual programme stunts. After going through all the aerial gyrations which an expert pilot can conjure up, Steve stuck the nose of the machine down in a tight spiral, from some 2,000 feet, preparatory to making a landing.

The engine was of the rotary type, a Le Rhone, and one of its failings, unless one babied it along, was to foul up with oil if the pilot allowed it to rotate at low speed for any length of time without giving the throttle a full blast to clear the cylinders. Steve "blasted" the throttle several times as we came down. We were approaching our landing over a maze of telegraph wires on the main-line railway and as he straightened out, Steve gave the engine the gun to speed us safely over, but the motor failed to take it! With a dud engine, little flying speed, and the wires coming fast, Steve banked away from them, still expecting the engine to give song at any moment. It didn't; the nose of the Avro suddenly snapped down, and instantly we

THE CRASH AT FORT FRANCES
In which the author and Stevenson were involved

WESTERN CANADA AIRWAYS LIMITED

The men responsible for the flight to Port Churchill, 1927. *Left to right*: "Rod" Ross, Bernt Balchen, Al Cheesman, Fred **Stevenson**

were in a spin. We smacked the ground with a hefty impact from 200 feet. Seconds later I was picking myself up from the wreckage of the right wing, bruised, but miraculously whole.

Steve was less fortunate; he was tightly jammed in the débris, and out cold. Examination shortly after revealed a dislocated hip, broken jaw, and fractured ankle, but after a lengthy spell in hospital he made a full recovery. These injuries, the first he had ever suffered in his flying career, put a damper on his piloting for a considerable time.

In 1924, Steve joined the newly organized Ontario Provincial Air Service, to fly on forest patrol from bases established at Sault Ste Marie, Sioux Lookout, Sudbury, and elsewhere. After two years with that service he handed in his resignation and left to join "Doc" Oaks and the famous Western Canada Airways Ltd.

In March 1927, at the request of the federal Department of Railways and Canals, Western Canada Airways Ltd. contracted for the transportation of eight tons of material and equip-

ment and a crew of fourteen men from Cache Lake at Stop 423 on the Hudson Bay Railway to Fort Churchill on Hudson Bay, before the spring break-up. Two Fokker Universal aircraft were put into service for the work; they were cabin machines, but with open cockpits. J. R. ("Rod") Ross was in charge of field operations, Steve and Bernt Balchen were pilots, Al Cheesman was the mechanic. (Balchen later became famous for his Arctic, Antarctic, and Atlantic flights, and Cheesman rose to fame as one of Sir Hubert Wilkins' pilots during the search for the Russian airmen lost in the Arctic in 1937.) After the machines had been flown from New York to Hudson, Ont., they took off on March 23, and the following day reached their base of operations at Stop 423, on the Hudson Bay Railway.

Trouble was experienced with the ski-equipped undergear when the metal tubes between the axles and the skis buckled under the rough usage they received, but the difficulty was overcome by using chunks of wood cut from bridge timbers 12 by 12 inches. Then, as the engines had been specially prepared for winter flying conditions, warmer temperatures in April caused overheating and much consequent strain on the pilots, as well as on the motors.

On one trip, when Steve was returning alone from Churchill, the main oil pipe broke and he was forced down 75 miles short of Stop 423. Landing safely on a small frozen lake, he spent the night in the machine with the wolves prowling and howling around outside. Next morning he spotted a recent trail in the snow, and followed it to an Indian trapper's cabin, where he was able to make arrangements to be taken to his base three days distant by dog sled. But the Indian's team was in poor condition and it soon became apparent that Steve would have to "hoof it," in spite of a limp left as a souvenir from the Fort Frances crash.

Balchen and Cheesman meanwhile had flown out to search for Steve when he did not show up from Churchill on schedule. They found his machine abandoned, without oil, and filled her up, but without discovering the broken pipe. Although Cheesman had had only a few hours as a qualified pilot, he took the plane off the ice and set a course for camp. On the way, the engine caught fire through the continued oil leakage. Cheesman made a quick emergency landing, put out the flames, and after repairing the break, flew back to base without further trouble. So it was that when Steve came in, after hoofing it through the bush for three days, weary, hungry, and slightly frostbitten, the first thing that met his eyes was the machine he had left behind!

By the middle of April, workmen and material

THE HARMON INTERNATIONAL TROPHY

had all been safely flown in to Churchill. On the 20th the Fokkers and their four crewmen headed south from a lake where the ice was already under water from spring thaws. Caught by a storm on their way to Norway House, they spent three very uncomfortable days, sitting in the machines most of the time, grounded on a small, frost-bound lake, until the weather brightened and they were off. When they arrived at Hudson they landed safely on the surface of a frozen lake, which by the very next day was too soft for use.

A realization of the value of the work these airmen accomplished can better be conveyed by a quotation from the report on the undertaking issued by the government in the outline of civil air operations for 1927: "Notwithstanding the distance from any prepared base and the severe winter conditions, the operation was an

STEVENSON'S FATAL CRASH
The remains of the Fokker at The Pas, January 5, 1928

THE HARMON INTERNATIONAL TROPHY
MEDALLION

unqualified success; the men and equipment were landed at Fort Churchill. The decision during 1927 as to the selection of Fort Churchill as the ocean terminus of the Hudson Bay Railway was made possible by these flights. There has been no more brilliant operation in the history of commercial aviation."

In August 1927, R. J. Jowsey, who was planning to open up the Sherritt-Gordon mine, inquired whether Western Canada Airways could contract to fly machinery and equipment—thirty tons altogether—from The Pas, Man., to the site of his property. Oaks accepted the challenge and Stevenson was the pilot who did the job.

That contract to fly heavy and bulky machinery into a remote area actually marked the beginning of freighting by air in Canada on a big scale. Today freight traffic by air has reached amazing proportions, and has opened up properties in the north, which, inaccessible by any other means, might have remained undeveloped forever.

Steve continued flying for Western Canada Airways during the remainder of 1927, but on January 5, 1928, destiny cut short a fine flying career during a simple test flight at The Pas. Through some undetermined cause, the Fokker aircraft went into a spin at a low height, crashing with terrific impact on one of the town's streets. Steve was killed instantly, and Canada lost one of her best airmen in the first fatal flying accident to occur in the northland.

Later in the same year, the city of Winnipeg and the municipality of St. James opened a large joint airport west of Winnipeg. At the opening ceremony, Steve's mother and father unveiled a marble plaque which bore the inscription: "This aerodrome is named Stevenson Aerodrome in dedication to the late Captain F. J. Stevenson of Winnipeg, Canada's Premier Commercial Pilot." Today it is well known as Stevenson Field.

Steve's exploits had already come to the notice of La Ligue Internationale des Aviateurs, and from nominations sent in by twenty-two member nations, Stevenson's name was selected as winner of the Harmon International Trophy for 1927.

Residents of Winnipeg or visitors with a few hours to spend in the Manitoba capital, who wish to offer silent homage to a great bush pilot, should make a brief pilgrimage to Brookside Cemetery where, in a quiet spot, an airman sleeps, untroubled by the whine of enemy bullets or the cough of a faulty motor.

On the plain cross which marks his resting place, this final message can be read, expressive in its poignant simplicity:

Captain F. J. Stevenson, D.F.C., Croix de Guerre. Royal Air Force. Faithful unto death.

27. Commercial flying hits its stride, 1924-1929

THE year 1924 might be designated as the date when commercial flying into the rugged interior began to look like a paying proposition to the more imaginative, yet hard-headed, type of business man. Bush pilots were logging long hours in the air, and newly formed air transport companies began to pile up impressive totals of passengers, parcels, and freight deliveries. As their reputation for reliability and resourcefulness grew, more and more was demanded of them, until by the end of the twenties there were pilots who were ready to fly anywhere, and customers sufficiently confident in their ability to ask them to do just that.

Here it is only fitting to pay tribute to the small group of pioneer Canadian investors who had the foresight and the insight to back the early air transport lines. At a time when Canadian finance generally leaned backwards in its caution to avoid speculative investment in new enterprises it took a good deal of courage and imagination for financiers—of whom a typical example was the late James A. Richardson of Winnipeg—to invest heavily in the new air transport companies.

The earliest Canadian flying companies whose aircraft penetrated the high northern latitudes of Canada were: Western Canada Airways, Canadian Airways, Yukon Airways and Exploration Company, Dominion Explorers, Northern Aerial Minerals Exploration, Prospectors Airways, Tredwell Yukon Company, and Consolidated Mining and Smelting Company of Canada. To the pilots and the ground personnel of these eight companies, the northland owes a great debt.

Except for the flying done by the Hudson Strait Expedition during 1927–1928, and by Stevenson and Balchen to Churchill on Hudson Bay in March 1927, little winter flying had yet been seen in the far north. The arduous conditions caused by extremely low temperatures, high winds, storms, and low visibility, had to be met and overcome.

One of the most unreasonable requests perhaps ever made came to the office of Great Western Airways of Calgary, whose manager, Fred McCall, accepted a contract to fly 300 quarts of nitro-glycerine from Shelby, Mont., to Calgary, Alta., where the explosive was urgently needed for use in "shooting" a well in the Turner Valley oil field. McCall flew to the Montana town on February 22, 1928, and took aboard Charles B. Stainaker of the Independent Eastern Torpedo Company, with 100 quarts of nitro and a dozen sticks of dynamite. After bucking severe headwinds, McCall landed at the air field used by the Calgary Aero Club. He was so short of fuel that his engine stopped after landing and he was unable to taxi to the hangar until more gas was brought. The dangerous cargo was then transferred to a truck which carried it off to the oil fields. Two more trips were made by McCall to fetch the rest of the nitro, 100 quarts at a time. On each occasion when he returned to the Calgary field, all personnel in the vicinity shook in their shoes. Everyone knew what he was carrying, and breathed with relief as the nitro disappeared down the highway in its delivery truck.

PRIVATE AND GOVERNMENT FLYING

A major event of 1924 was the organization of the Ontario Provincial Air Service. The first bases were established at Sudbury and Sioux Lookout, with a total of 13 HS2L flying-boats, 16 licensed pilots, and 19 licensed air engineers. A tremendous amount of flying was carried on right from the start. In 1924, 2,595 flying hours were logged, covering approximately 170,000 air miles. During the season a total of 597 forest fires was reported from the air.

Later in 1924 the main base was established at Sault Ste Marie, with numerous sub-bases throughout the northern area of the province, and the number of aircraft was increased to 19, with a corresponding increase in personnel.

In 1924 Chief Pilot Roy Maxwell flew Hon. James Lyons, Minister of Lands and Forests, on a lengthy trip, which carried them from Remi Lake down the Mattagami River to Moose Factory on James Bay. From that point they flew to Rupert House and Eastmain River, returning from those Hudson's Bay Company posts to Sault Ste Marie, with a few intermediate stops.

The work of the bush pilots of the Ontario Provincial Air Service was both strenuous and risky, as may be judged from their log books. For instance, the biggest day's flying by an individual pilot in 1924 was chalked up on July 5 by C. J. Clayton, who spent 10 hours and 40 minutes in the air on forest fire patrol. The highest total for any one week was established by G. A. Thompson with 38 hours and 15 minutes. In 1925, J. O. Leach flew 9 hours and 30 minutes on August 24, and T. B. Tully recorded the best weekly time with 49 hours and 5 minutes. In 1926, one pilot corralled both records, when C. A. ("Duke") Schiller flew 12 hours and 10 minutes on August 4, and was also credited with 52 hours and 15 minutes in the air during a seven-day period. This was a brilliant group of pilots.

Once the Ontario government had organized its own air service, the Laurentide Air Services greatly curtailed theirs. They retained only four aircraft, but with these they did excellent work in establishing a passenger and freight service to the Rouyn gold fields in Quebec, operating from bases at Larder Lake and Haileybury. The year 1924 was a bumper one for the company: their planes carried 1,004 passengers, 78,800 pounds of express and freight, and over 15,000 letters and telegrams.

To the Laurentide Air Services belongs the credit of establishing the first regular air-mail

ONTARIO PROVINCIAL AIR SERVICE BASE, SAULT STE MARIE
Nine HS2L flying-boats during the winter of 1925–1926

AIRMEN OF THE ONTARIO PROVINCIAL AIR SERVICE

Back row, left to right: J. O. Leach, H. C. Foley, T. B. Tully, W. Lyons, T. G. M. Stephens, H. A. Oaks, P. J. Moloney, W. H. Ptolemy. *Centre row*: J. C. Ruse, R. C. Guest, C. A. Schiller. *Front row*: H. J. B. Bain, F. J. Stevenson, J. R. Ross, W. R. Maxwell, C. J. Clayton, G. A. Thompson, R. Vachon

route within Canada. The first flight was on September 11, 1924, and special stickers were sold by the company for the mail flown.

This was the beginning of flying activities in the gold fields of Quebec and Ontario. Other flying organizations invaded the area in later years and the traffic grew to considerable proportions. Only one other firm however, began flying out of Haileybury and Rouyn during 1925. This was Northern Air Services, Limited, which was formed by its chief pilot, B. W. Broatch, who had been a pilot with Laurentide Air Services. By the end of the year his company had transported 503 passengers, 22,580 pounds of freight, and 1,030 pounds of mail. Regular air-mail service by this company began on June 27, 1925, between Rouyn and Haileybury.

The year 1925 was a singularly good one from many angles. Commercial aircraft in Canada logged 255,826 air miles, totalling over 4,091 flying hours, and in all that flying there was not one accident which meant death or injury to anyone aboard.

The R.C.A.F. did not suffer the same reverses in its operations as did private organizations in the years following World War I, and after the Force was fully established, increase in flying was steady, chiefly for departments of the government. Most of its patrol aircraft were equipped with radio, which greatly assisted their work in forestry, survey, photography, and other projects.

The construction in Canada of JN4 Curtiss training planes and a few machines of other types had risen to considerable proportions between 1915 and 1918, but this enterprise

A VICKERS VIKING OVER JAMES BAY, 1924

ended with the war. It was 1923 before the work was taken up again by a Canadian firm, this time Canadian Vickers, Limited, of Montreal. Their first contract, which came from the Canadian government, was for the construction of five single-seater Avro patrol aircraft for the High River Station in Alberta.

In 1924 a design department was added under W. T. Reid which set to work immediately on the new Vickers Vedette. The model was tested in the wind tunnel at the University of Toronto, the first complete model to be tested in this manner in Canada. The machine then went into production and the finished job very successfully passed its tests. The R.C.A.F. later made use of many of these serviceable, single-engined, three-seater flying-boats. For many years they operated efficiently from coastal waters, and from countless lakes and rivers across Canada.

Until 1928 Vickers were the only designers and constructors of aircraft in Canada.

By 1926, Laurentide Air Services had discontinued flying operations, but the J. V. Elliot Air Service of Toronto began to put aircraft to work in the Rouyn area. Elliot also began operations from another base at Rolling Portage, Ont., to serve the newly discovered Red Lake gold fields. During 1926, the first season of operation, 587 passengers, 2,000 pounds of freight, and 800 pounds of air mail were carried. Jack Elliot flew one of the machines in the Red Lake area, and was pilot of the first air-mail flight of his company on March 6, 1926. Covers flown on that flight vary in value according to the cancellations they bear, up to a maximum of $25 each.

In 1926, several other flying companies which were to influence the development of aviation in Canada had their beginning. Of these, Patricia Airways and Exploration, Limited is the most important, not so much for its accomplishments in its two years of existence, but because of the men responsible for its formation.

The Company was organized by H. A. ("Doc") Oaks and G. A. Thompson, pilots with Ontario Provincial Air Service. These two men sold some claims which they had staked in the Red Lake gold fields, the money going towards the purchase of a Curtiss Lark aircraft, fitted with a Wright Whirlwind engine. The machine could be fitted equally well with wheels, floats, or skis. Sammy Tomlinson was their engineer.

With the Lark and a leased HS2L flying-boat, Patricia Airways flew a total of 217 hours in 1926, carrying 259 passengers, 14,000 pounds of freight, and 3,000 pounds of mail. The inaugural air-mail flight took place on June 27, 1926, when "Doc" Oaks flew the Lark from Sioux Lookout to Red and Woman lakes.

On one occasion, a mining engineer en route from Winnipeg to New York, by Canadian National Railways was picked up at Minaki, Ont., and flown to a claim 120 miles north of the railroad. After examining the claim, he was flown to Sioux Lookout, where he caught the same train he had left four and one-half hours before.

During 1927, the second and last year that Patricia Airways operated, two additional pilots were taken on, J. R. ("Rod") Ross and W. N. Cummings.

In July and August alone of 1927, the Lark

"DOC" OAKS (right) WITH HIS ENGINEER
Just before the inaugural air-mail flight, June 27, 1926

was flown over 12,000 miles, and during the entire season, aircraft of the company flew 47,787 miles, carrying over 67,600 pounds of freight. The fares charged by the various companies flying in northern Ontario and Quebec during that period averaged around $1.50 per passenger mile. Freight and express rates depended on many conditions, but they ran from 9 to 25 cents per pound, depending chiefly on the distance flown.

Another important company started in 1926. The success with which "Doc" Oaks had operated Patricia Airways led him to believe that flying had a great future throughout the Canadian north, so he looked about for someone who would not only listen to his plans, but would back them up with cash. His search ended when he laid his suggestions before James A. Richardson, the Winnipeg financier. The outcome was the incorporation of Western Canada Airways, Ltd., in November 1926, with Oaks as manager.

The first aircraft purchased was a Fokker Universal, a five-place cabin-type, but with the pilot's cockpit not enclosed. On Christmas Day Oaks flew the plane in several hops from Teterboro, N.J., to Sioux Lookout, Ont. Before the year had ended, Oaks had flown the Fokker 31 hours on commercial work, and had carried 18 passengers and some 850 pounds of mail before New Year's Day of 1927.

This first aircraft, which bore registration G-CAFU, was the forerunner of a fleet of various types which Western Canada Airways owned in the four years of its varied operations.

In 1926, Dominion Aerial Exploration Company, formed by H. S. Quigley in the autumn

A SCHRECK FLYING-BOAT
Similar to the one in which Count de Lesseps was lost

of 1922, was taken over by the newly incorporated Canadian Airways Limited. From the main base at Three Rivers, Que., operations spread out till the name of the company became a household word in eastern Canada. At first Canadian Airways owned only four HS2L flying-boats. Their record for 1926, between May 1 and November, was comparable with that of other flying companies in Canada. They carried 537 passengers and 18,000 pounds of freight, but only 100 pounds of mail.

Originally, Canadian Airways Limited had been one of a group of eastern flying companies which became absorbed into the Aviation Corporation of Canada. On November 30, 1929, the latter organization joined forces with Western Canada Airways, and the union in its entirety adopted the name of Canadian Airways Limited, as being more representative of their country-wide activities. The well-known Canada goose emblem came to Canadian Airways from Western Canada Airways, which had originated the design for use on its aircraft.

Other flying firms in business during 1924–1926 included the Compagnie Aérienne Franco-Canadienne at Gaspé, Que., the Fairchild Air Transport Co. Ltd., and Fairchild Aerial Surveys Company (of Canada) Ltd. Brock and Weymouth of Canada, Ltd., a branch of an American company, was also at work in photography contracts in Quebec.

The Compagnie Aérienne Franco-Canadienne had the big job of mapping the entire Gaspé peninsula from the air, under contract with the Quebec Provincial government.

THE OLD AND THE NEW
A dog team and a Fokker Universal at the Western Canada Airways Office at Pine Ridge, Ontario, January 1927

THE *QUEEN OF THE YUKON*

Operations were conducted from a main base established at Gaspé Basin and a sub-base at Goose Lake, using five Schreck flying-boats, four of them powered with 180 h.p. Hispano-Suiza engines and one with a Napier Jupiter of 420 h.p. This was the company that persuaded the famous Count de Lesseps who had commanded a French flying squadron during World War I to serve both as an executive and as one of their pilots.

On October 18, 1927, during a routine photographic flight along the Gaspé coastline, de Lesseps encountered a storm of terrific suddenness and ferocity. As the hours slipped by and no word came of his safety, company officials were reluctantly forced to conclude that he was down at sea, a dread which became a reality when parts of his battered flying-boat were found cast ashore on the Gaspé coastline.

It will never be known why the Count and his engineer, Theodor Chickenko, were obliged to come down into the open waters, but this they did, somewhere in the vicinity of the fishing hamlet of Ste Félicité, a short distance from Mont Joli. Residents there reported seeing the machine, apparently in difficulties. Twice the pilot endeavoured to gain sufficient altitude to get over Notre Dame mountain, and reach the calmer waters of the lake near Val Brillant. Unable to make it, he was forced to turn back and his machine was quickly lost to sight.

On December 4, almost seven weeks later, the body of the pilot was found on the shore of Clambank Cove, Port au Port Peninsula, Nfld., but that of the engineer was never found.

Eastern Canada accounted for most but not

all of the commercial flying during this early period. At Regina, Roland Groome was still flying and managing Aerial Service Company, Ltd., and on the coast, Pacific Airways, Ltd., under the pilot-management of Don R. MacLaren, and Dominion Airways, Ltd., under E. C. W. Dobbin, were both going strong.

A number of individual pilots also owned and flew their own machines on commercial enterprises. For example, in 1926, Bert McConnell, operating in Ontario, flew a total of 75 hours on passenger and advertising missions, and R. G. Trenholme of Windsor, Ont., flew some 20 hours on passenger trips.

Before 1927 only one Canadian-owned airplane had invaded the Yukon Territory, and only three Canadian planes had penetrated into the Northwest Territories. The solitudes beyond the Arctic Circle had not yet been vanquished by Canadian flyers, but the end of this seclusion was in sight.

COMMERCIAL FLYING IN THE YUKON

In 1927, commercial flying came to the Yukon with the incorporation of Yukon Airways and Exploration Company, Limited, with headquarters at White Horse.

The firm was organized with a capital of $50,000, and a Ryan monoplane, fitted with a 200 h.p. Wright Whirlwind radial engine, was purchased. It had a payload of 1,200 pounds or seating for four people, besides the pilot. Charles Lindbergh had flown an exact counterpart of the machine on his famous flight across the North Atlantic.

David (Andy) Cruickshank was selected to fly the Ryan, and a splendid choice he proved to be. Andy was born in England of Scottish parents. During World War I, he served as a

THE CRUICKSHANKS
On their honeymoon voyage to Skagway

THE *QUEEN OF THE YUKON* AT REST

A tent-hanger protecting the engine. Mrs. Cruickshank (*left*) and her husband

pilot with the R.F.C. and the R.A.F., chiefly in the 84th Squadron. After the war he came to Canada and joined the R.C.M.P. He served five years with the force, much of the time in the Yukon. He resigned in 1924 to take up flying again, and he barnstormed in California, Utah, and Nevada for almost three years.

When Yukon Airways bought their plane, they made Cruickshank general manager of the company and pilot of the machine, which they named *Queen of the Yukon*.

The journey from California to Vancouver was done in a leisurely manner, stops being made at many towns, where paying passengers were taken up for short flights. Over 1,000 persons had been flown before Cruickshank landed the machine at Lansdowne Park race track, near Vancouver, on September 27, 1927. He was accompanied by Clyde G. Wann, the company president; W. A. Mundy, manufacturer's pilot, Mrs. Mundy and their 2½ year-old son; and J. E. Smith, engineer. The Ryan was dismantled at Vancouver and shipped to Skagway, Alaska, where it arrived on October 21.

The voyage to the north was also a honey-moon trip, as Esmée Bulkley of Vancouver had become the bride of the airman on October 17.

At Skagway the *Queen of the Yukon* was put ashore, and assembly began. Facilities for a take-off were extremely limited as there was no suitable large area. As a makeshift, a rough wooden skid road was selected for use as a runway, which meant that once the craft became airborne, a safe landing would hardly be possible. On October 25, 1927, with Mrs. Cruickshank, Clyde Wann, and J. E. Smith aboard, Cruickshank made a fine getaway from the improvised air strip, taking off directly over the sea. Right from the start, lowering clouds presented a serious menace to flying. Knowing this, the airman had endeavoured to persuade his wife to travel to White Horse by train, a suggestion she flatly refused, preferring to be with her husband and share the risks.

Once in the air it was impossible to gain immediately sufficient height to clear the many peaks which surround Skagway, so Cruickshank set off, steering a winding course through gullies and valleys, until an altitude of 12,000 feet had been reached. Twenty minutes were required to gain that level, and a number of times they had narrow escapes from destruction as jagged, snow-clad bulwarks loomed up in their path. At last, with height and safety assured, Andy set a course for White Horse, over White Pass, and reached his destination in an hour and ten minutes.

The day after, the flight continued to Dawson.

Beginning on November 11, 1927, an air-mail service was established over the 500-mile stretch between White Horse, Dawson, and Mayo. The entire distance was covered in just under 4½ hours; whereas by dog team, in good weather, it had required from 12 to 14 days.

Mrs. Cruickshank accompanied her husband on the majority of the hazardous flights he made throughout that winter. The aircraft did not have ski landing-gear, but wheels, which certainly tested Andy's prowess as a pilot and testified to his wife's faith in her husband's skill.

On the first air-mail trip, the journey was not broken at Dawson *en route* from White Horse to Mayo, and to Mrs. Cruickshank fell the task of heaving the mail-bag from a window of the cabin, as the plane flew low over the Dawson field.

All mail was flown under permission of the postal authorities, but they gave no financial aid and the venture was purely a commercial undertaking. Letters flown carried regulation postage, and in addition Yukon Airways and Exploration Company was permitted to make an extra charge of 25 cents per letter; issuing "stickers" of their own design which were affixed to each piece of mail. A single specimen envelope of the first air mail flown from White Horse on November 11, 1927, to any point, is valued in stamp catalogues at $35, and if the envelope bears Cruickshank's signature its value increases $5.

The *Queen of the Yukon* had a brilliant career until wrecked at Mayo during a landing on April 5, 1929, on what was to have been a pioneer flight from Dawson to Aklavik. The pilot was killed, and all that remained of the aircraft, G-CARR, was a pile of shattered parts.

Andy Cruickshank had left the Yukon company in 1928, and joined Western Canada Airways. One of his first assignments with them was the piloting of aircraft into the Red Lake gold fields area of Ontario.

On July 5, 1932, while on a flight in the service of Canadian Airways Limited, Cruickshank became enveloped in a dense fog between Cameron Bay and Fort Rae, in the Northwest Territories. That country is hilly and densely wooded, and the Fokker aircraft Andy was flying crashed into a forested hillside, killing him and his two engineers, Horace W. Torrie and Harry King.

Mrs. Cruickshank had not been the first woman to fly in the northland. The Italian airship *Norge*, having flown across the North Pole from Spitzbergen with Captain Nobile, Roald Amundsen, Lincoln Ellsworth, and thirteen crew members aboard, landed at Teller, Alaska, on May 14, 1926. There the *Norge* was dismantled and, later, shipped back to Italy. The three leaders, anxious to reach the United States as quickly as possible, arranged with a flying company at Fairbanks, Alaska, to fly them to White Horse, Y.T., so that they could take train to Skagway and then travel by boat to the States. A pilot named Bennett undertook to deliver a large batch of movie films, taken on the Arctic flight, to railhead at White Horse in a Waco two-seater biplane, in order to get them into the hands of film syndicates in the United States. He made the trip to White Horse safely, and is reported to have received $8,000 in payment. When returning to Fairbanks, he ran into foul weather on May 22 in the vicinity of Fort Selkirk, in the Yukon. Being a good airman, he put his small craft down in a field near by to wait for better weather. His arrival caused great excitement among whites and Indians

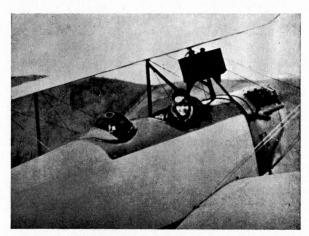

PAULINE WHITE AND BENNETT,
MAY 23, 1926

THE BARREN LANDS

This photograph shows clearly the absence of landmarks that makes flying in the north so dangerous if visibility is poor

alike, as his was the first aircraft ever seen at Fort Selkirk.

Teaching at the grade school in Selkirk was Pauline White, a girl born in Dawson of American parents. When she learned that Bennett intended leaving for Dawson the following morning, she asked him if he would take her along as a passenger, and he agreed. At 6:38 P.M. on the 23rd they set off from Selkirk, and after a brief enforced landing en route, caused by ignition trouble, reached Dawson. Residents of the place quickly gathered about the plane on its arrival, and were considerably astonished when out of the passenger seat popped Pauline, who was well known to most of them. She seems to have been the first woman to fly in that northern Canadian territory.

Over the Barren Lands

The rocky vastness known as the Canadian Barren Lands, which stretches over much of the eastern portion of the Northwest Territories, is almost devoid of habitation. No large trees exist and vegetation is sparse in the extreme, consisting chiefly of stunted willows, grasses, lichens, and moss. Thousands of square miles of this area were represented by blanks on maps of the north country when airplanes first flew over the Barrens.

In 1928, an exploratory flight was made, starting from Winnipeg, covering portions of the Barrens about which practically no knowledge existed at the time. The object of this flight was to obtain information about the conditions under which prospecting parties were working. These parties consisted of employees of Dominion Explorers, Limited, Northern Aerial Minerals Exploration, Cyril Knight Prospecting, Nipissing Mines, and other organizations, all of whom had reached the territory by ground transportation.

It was, of course, realized that before flying operations could really get under way in the north, supplies and fuel had to be shipped in. For this purpose, Dominion Explorers and Northern Aerial Minerals chartered schooners to carry cargoes into Hudson Bay. Once gasoline and oil were available at Churchill and Baker Lake, aircraft of the companies began to establish caches at many other outlying points. Thus the aerial invasion of the high north began.

Dominion Explorers sponsored the 1928 flight, using a Fokker seaplane, G-CASK, chartered from Western Canada Airways, and

PASSENGERS AND CREW
Left to right: Lieutenant Colonel C. D. H. MacAlpine, "Punch" Dickins, Richard Pearce, W. B. Nadin

piloted by C. H. ("Punch") Dickins, with William Nadin as engineer. The two passengers were Lieutenant Colonel C. D. H. MacAlpine, president of Dominion Explorers, and Richard Pearce, editor of the *Northern Miner*.

On August 28, 1928, Dickins left Winnipeg with his passengers. On the first day he reached the Hudson's Bay Company post at Norway House, at the northern end of Lake Winnipeg. The journey continued next day to Jackfish Isles, where the Royal Canadian Air Force already had a base on the Nelson River. A stop-over of two days was forced upon them by a dense blanket of smoke from forest fires, which spread over the surrounding country in a gloomy, acrid pall. When the airmen took off on the 31st, dense fog on the shore of Hudson Bay caused them to turn in at Churchill. On September 1 they arrived at Mistake Bay, and on the day after they reached Baker Lake. By this time they had left the tree-line behind and were well into the Barren Lands.

Only one airplane had been seen at Baker Lake before the arrival of SK, and, by coincidence, it had arrived just the day before. A pilot for Northern Aerial Minerals Exploration, "Duke" Schiller, had landed in a Fokker Super Universal seaplane, in which he was ferrying supplies to Baker Lake to establish a fuel cache.

On September 3, the G-CASK flew from Baker Lake across 840 miles of the Barrens—the first crossing by air of this desolate territory. Before the party landed at Stony Rapids on the Cat River, near the eastern end of Lake Athabaska, they had viewed from the air thousands of square miles which white men had never seen before. Roughly, the route followed was west from Baker Lake to Aberdeen Lake, then south along Dubawnt Lake and the Dubawnt River to Stony Rapids.

From Stony Rapids a number of side trips were made to unmapped lakes. There was a gasoline cache at Stony, but rather than deplete it altogether for a trip to Fort Smith, only sufficient fuel was put in the tanks to see the machine through. The route proved to be much longer than anticipated, because a heavy smoke-haze compelled Dickins to follow the shoreline of Lake Athabaska until the Slave River was reached.

With fuel running low, the flyers were obliged to make an emergency landing thirty miles upstream on the Slave River from Fort Fitzgerald. How soon help might come along, they could not know, so they first tried floating downstream in the machine, but it proved unmanageable, and that idea was abandoned. They then set about constructing a raft, and had it completed and just launched when round the bend came a welcome sight—the *Northland Echo*. Gasoline was procured from a supply she had aboard, and after only an hour's delay, the flying explorers were airborne. They reached Fitzgerald without further delay, and later went on to Fort Smith. From that Hudson's Bay post they flew back to Winnipeg, *via* Fort Chipewyan, Stony Rapids, Reindeer Lake, Cold Lake, and The Pas, making numerous stops and side excursions *en route*.

The flight ended officially 12 days after its start from Winnipeg, having accomplished a journey that by other modes of travel would have required eighteen months. The log of the flight showed a flying time of 37 hours and a total air distance of 3,965 miles.

THE R.C.A.F. FLIES THE MAIL

Although the Canadian Air Force had carried letters across Canada as far back as 1920 (chapter 21), it was not until a flight was made by Squadron Leader Earl Godfrey in

1928 that officially stamped air mail was carried on such a journey. What is more, no single aircraft, R.C.A.F. or commercial, had previously made the trip in as short a time as Godfrey.

Sponsored by the R.C.A.F., and using established bases across the country, the flight was scheduled to be made in six hops, only three of them for refuelling. The route was not a direct one. The machine was a Fairchild seaplane, and for safety it had to follow a northern course, by way of the many waterways and lakes of that region.

With Sergeant Major M. Graham as his companion and engineer, Squadron Leader Godfrey left the Rockcliffe Air Station at Ottawa on the morning of September 5, 1928. He flew to the Ontario Provincial Air Service base at Long Lake, non-stop, in 7 hours and 45 minutes. After refuelling, he set off again on a flight of 4 hours and 55 minutes which took him to Lac du Bonnet, north of Winnipeg. To fly from Ottawa to Lake Winnipeg the same day was an achievement.

On the 6th, Godfrey and Graham reached Larder Lake, Sask., with another 7 hours and 30 minutes of flying behind them; then they pushed on to Edmonton, landing on the Saskatchewan River 3 hours and 45 minutes later.

The vision of a quick trip to the Pacific was spoiled by stormy weather, and the flight could not continue until the 8th. On that date the airmen flew to Wabamun Lake, twenty miles out from Edmonton, where refuelling facilities awaited them, and, leaving at 10:30 A.M., they made a direct hop to Vancouver, arriving at the Jericho Beach Air Station at 6:00 P.M. During much of their journey over the mountains visibility was poor, owing to smoke. Once Godfrey had picked up the course of

THE R.C.A.F. FAIRCHILD
In which the first official air mail was carried

the Thompson River, he followed it, and later he followed the Fraser, keeping a hundred feet or less above the water to avoid the thickening smoke-pall.

Godfrey gave great credit to Sir Frederick Stupart, Director of the Government Meteorological Services, for the weather reports which the airmen found to be remarkably accurate.

The mail carried on this memorable flight had a special, two-line description printed on the envelopes, just beneath the regular 5-cent postal stamp. It read:— "Plane left Ottawa 7 a.m. Sept. 5—arrived Vancouver 6 p.m. Sept. 8. Insure your parcels at the post office." On each cover there is a special *cachet*: "Special— Trans-Canada Flight—Ottawa—to Vancouver— September, 1928." Approximately 550 letters were flown, and all were back-stamped with a circular cancellation, marked "Vancouver, Sept. 8, 6:30 p.m. 1928. B.C." Today a single specimen is valued at $20, and at $30 if signed by the pilot.

A whole book could be written containing nothing but the history of the flying done by and for the mining syndicates, and it would make fascinating reading. Some few of these experiences, typical of many, are presented in the next few pages.

Two flights made in August 1928 were of a pioneer kind. A. H. Farrington left Winnipeg on August 12 in a Western Canada Airways Fokker seaplane, on a flight which carried him by stages to Norway House, Jackfish Isles, Deer Lake, Churchill, and finally Eskimo Point, half-way up the west coast of Hudson Bay. He was to bring out mining men operating in that area. By the time Farrington had

WESTERN CANADA AIRWAYS FOKKER
Used by Dominion Explorers

reached Winnipeg on the return flight, his log book showed a mileage of well over 1,000 miles for the trip.

Later the same month, T. M. ("Pat") Reid, flying a Loening amphibian aircraft G-CATM, owned by Northern Aerial Minerals Exploration, made the first aerial coastal circumnavigation of Hudson Bay. The purpose of the flight was to transport supplies and obtain reports from prospecting crews at various points. On the first lap, Reid carried a party of men to Chesterfield Inlet on Hudson Bay. Leaving them at Baker Lake, he flew south to Moose Factory on James Bay, then north again up the east coast to Richmond Gulf, stopping at Rupert House, Fort George, Eastmain, and Great Whale River. He finally flew back to Winnipeg, arriving there in mid-September.

For a week in August 1929, Reid had occasion to search for the Loening machine he had used previously, when it disappeared in the Barrens on a flight between Baker Lake and Wager Inlet, with "Duke" Schiller, Tom Creighton, and Jack Humble aboard. Reid finally spotted the three men trudging across the Barrens, miles from their machine. They were tired and hungry, but none the worse for their experience, although they had been walking for fifteen days towards their Baker Lake base. They had come down when their machine had run out of fuel. It was later located, refuelled, and flown out without difficulty.

The first midwinter flights into Hudson Bay by a private company became history in January 1929. Two N.A.M.E. aircraft flown by "Pat" Reid and "Doc" Oaks, with engineers Murray and Mews, went in from their base at

DOWN IN THE BARREN LANDS
The Loening Amphibian, out of fuel in August 1929

Sioux Lookout, *via* Longlac, Remi Lake, Moose Factory, and up the east coast to Richmond Gulf. The trip was made to fetch out thirteen prospectors who had been caught in the freeze-up. A missionary and his bride were passengers from Cochrane to Rupert House. On the way north, the flyers were forced down by a blizzard, and the party was obliged to spend three cheerless days and nights in the cabins of the grounded aircraft, seven miles off shore on the sea-ice, until the storm blew itself out and they were able to proceed. They then carried on to Richmond Gulf, picked up their men, and returned to Sioux Lookout without further incident.

The Barrens was first invaded by air in winter when two ski-equipped Fairchilds owned by Dominion Explorers were flown in by Stan McMillan and Charles Sutton during March and April in 1929. The expedition was sponsored by the Lindsley prospecting organization, which subsidized the formation of Dominion Explorers. Sutton left Ottawa on March 26, 1929, flying CF-AAO, which, incidentally, was the first commercial aircraft in Canada to be fitted with a radio transmitting and receiving set. With him as passengers were Lieutenant Colonel MacAlpine, president of Dominion Explorers, Alex Milne, engineer, J. C. Rogers, geologist, Monty Priske, wireless operator, and Edward Harrison, Fox Films movie cameraman. From the capital, the flyers went to Missanabie the first day, and on the next flew the entire distance to Lac du Bonnet, Man., with one short stop at Orient Bay to refuel.

The second aircraft, G-CACZ, joined them at Lac du Bonnet, coming up from Winnipeg. McMillan had with him Major Robert Baker of Dominion Explorers, going in to Baker Lake, and Norman Pearce of the *Northern Miner*. On the flight to Norway House on the 29th, both planes were engulfed in a fierce blizzard. Sutton managed to reach Berens River post; the other machine was forced down on the ice of Lake Winnipeg, where its occupants spent a chilly night in the cabin.

The next morning, the flyers had great difficulty in getting their engines started, because of the low temperature, but both machines went on to Norway House, and then to Churchill. By this time, landing-gears were

WINTER ON HUDSON BAY

Left: The two aircraft at Richmond Gulf, January 1929. *Right*: "Pat" Reid, "Doc" Oaks, Kel Mews

showing the strain of contacts with broken and hummocky ice, and required repairs before the journey could continue to Tavani, 300 miles up the coast. On the trip to Tavani on April 2, a letter was dropped to the Mission at Eskimo Point as the planes flew over. The main base of Dominion Explorers had been established at Tavani, and a number of prospectors were wintering there; they were overjoyed to receive mail and news from outside.

The flyers reached Baker Lake post on April 4, after dropping a sack of mail at the Hudson's Bay post at Chesterfield Inlet as they went by.

Sutton's machine carried a sun compass as part of its navigational equipment. It was the only one in use in Canada. Only three such instruments existed at that date, the other two being in the possession of Admiral Byrd and Sir Hubert Wilkins.

The intention had been to fly across country to Lake Athabaska after company business had been completed at Baker Lake, but when endeavouring to establish a gas cache at distant Dubawnt Lake, both planes just avoided getting lost in the great white waste, where land and sky merge into one featureless void. The only thing they could do was to fly in an easterly direction until they hit the coast of Hudson Bay, from which they were able to find Baker Lake again without much difficulty. In this long flight they used up all the fuel intended for the cache, and so had nothing to show for an exhausting flight of some 900 miles.

The idea of crossing the Barrens was then abandoned, and both aircraft returned south over the route they had taken "going in." The two craft parted company at Thicket Portage, Man. Colonel MacAlpine and the others in Sutton's machine went on a lengthy flight to Fort Resolution, then finally returned to The Pas. McMillan managed to fly back to base at Winnipeg, although his aircraft had suffered structural damage.

Snow conditions on the ground present quite a hazard in the far north. The continual winds blow the snow into sharp-edged drifts, which freeze as hard as cement. Shovels are useless; only an axe will cut into them if skis become stuck. The strain on McMillan's aircraft from so many rough landings did considerable damage, cracking struts, breaking tail skids, and even causing structural trouble in the wings. At times it had been a toss-up whether he could make Winnipeg or not.

About 8,000 miles were flown by the two planes. It was a fine undertaking and fulfilled its purpose of gaining first-hand information about winter flying conditions in the north.

ONWARD TO ARCTIC SEAS

The first plane to fly across the eastern Canadian Arctic Circle belonged either to Northern Aerial Minerals Exploration or Dominion Explorers, since both these companies were operating out of Baker Lake during the summer of 1929. The following entries are from the diary of the manager of the Hudson's Bay Company post at Repulse Bay:

WINTER IN THE BARREN LANDS
Left: The two Fairchild aircraft at Baker Lake, Northwest Territories *Right*: J. C. Rogers, Lieutenant Colonel C. D. H. MacAlpine, Charles Sutton, Alex Milne, Edward Harrison

"1929, August, Sunday, 25th. Wind from N.W. Sunny. . . . About 5 p.m. an airplane belonging to Dominion Explorers was sighted and landed in a few minutes. Messrs. Rogers, Sutton and Wheelers [sic] were aboard."

"1929, August, Monday, 26th. Slight wind from north west veering at intervals. The D.E. party accompanied by Mr. McHardy of this post left by plane for Committee Bay. Returned in about three hours after nice trip. Committee Bay is packed with ice. The plane left shortly afterwards for Baker Lake."

The post at Repulse Bay is practically on the Arctic Circle, and Committee Bay is beyond it, some thirty or forty miles due north, over the strip of country which connects Melville Peninsula with the mainland.

The first pilots to reach Coppermine on the Arctic Coast in Coronation Gulf were N.A.M.E. men, "Pat" Reid and Jimmy Vance. They flew from Baker Lake to Bathurst Inlet, *via* Beechey Lake, on August 31, 1929, and the next day flew along the coastline to Coppermine.

In the western Arctic, the first pilot to reach the "Circle" was C. H. ("Punch") Dickins, flying a Western Canada Airways Fokker aircraft. In January 1929, the Company arranged a schedule of ten winter trips to be flown by Dickins between Waterways and Fort Simpson. As a considerable volume of mail was waiting to go north from Fort McMurray by dog team, the postal authorities took advantage of the opportunity to dispatch it north by air.

The first trip began on January 23. On March 6, Dickins extended his northern journey by going on to the Hudson's Bay Company post at Fort Good Hope with mail. Fort Good Hope, situated on the Mackenzie River, only a few miles below the Arctic Circle, was the farthest point a plane had reached in that area to that date. The following day the pilot set out for Edmonton, carrying a full load of baled furs, the first time a valuable cargo of this kind had been flown out of the north. The temperature at the time was 54° below zero. On July 1, 1929, Dickins landed a float-equipped plane on the Mackenzie River at Aklavik, thus becoming the first airman to reach any point by air along the western Canadian Arctic coastline.

Another notable flight was a journey made by W. Leigh Brintnell, who was manager of Western Canada Airways at the time. He set off from Winnipeg on August 5, 1929, on an inspection tour, flying a Fokker seaplane. Before he again set foot to the Manitoba capital he had flown some 9,000 miles. He flew first to Edmonton and then to Fort McMurray, where, upon instructions from James A. Richardson, president of W.C.A., he picked up two prospectors, G. A. LaBine and Charles St. Paul, who wished to go to Great Bear Lake. These men later staked the claims on Echo Bay which developed into the world-famous Eldorado Mine. The discovery of these deposits of radium ore put mining on the map in that area. The mine, on Great Bear Lake, became the world's largest known source of radium, and U-235-Uranium, the base mineral from which the first atom bombs were made.

When LaBine and his partner discovered radium, it was worth $70,000 per gramme, world control then being in the hands of a company in the Belgian Congo. With the Canadian discovery the price dropped to $25,000 per gramme, breaking the hold of the Belgian radium monopoly. Only 300 grammes of commercial radium existed in the world in 1930. A few years after the find in Canada, 800 grammes were available to aid in the fight against cancer.

During this work, Brintnell circumnavigated Great Bear Lake by air. He then flew on to Aklavik, taking as his passenger O. S. Finnie, Director of the Northwest Territories. From Aklavik Brintnell made the first crossing of the northern Rockies by air, during a non-stop flight of 780 miles to Dawson, Yukon, over some of the wildest and least inhabited country of the continent.

The long flight back to Winnipeg was made in easy stages by way of White Horse, Skagway, Prince Rupert, Prince George, Edmonton, and The Pas. In 94 hours of flying time, Brintnell had covered a distance equivalent to about one-third of the circumference of the earth.

A flight such as Brintnell's showed how, even in 1929, before jet planes and superspeed, the new medium of travel had altered the former conceptions of time and distance.

SOME FACTS AND FIGURES

The prosperity of the boom years of 1927 to 1929 was reflected in the great development in Canadian aviation. The amount of commercial flying increased, new uses for airplane services were developed, and new production companies were started.

The following table shows the progress made in commercial flying:

	1927	1928	1929
Reg'd Aircraft	67	293	445
Pilots	43	258	349
Engineers	74	131	212
Commercial Flights	16,748	75,285	144,143
Passengers Flown	18,932	74,669	124,751
Air Freight (pounds)	1,098,346	2,404,682	3,903,908

Air mail also became a more regular feature of commercial flights. In 1929 the total weight carried under government contract was 430,636 pounds, compared with 277,184 pounds in 1928, and approximately 15,000 pounds during 1927.

During this period, Royal Canadian Air Force operations and government civil operations also increased. In 1927, a government

aircraft began experiments in the control of forest insect pests. The first effort was made in June against the spruce budworm, which was destroying trees on Cape Breton Island. In July, aerial dusting for the control of wheat stem rust began in Manitoba at Graysville, Portage la Prairie, and Morden. The work was originally planned and directed by the Forest Insect Division of the Entomological Branch and the Dominion Rust Research Laboratory, in conjunction with Royal Canadian Air Force machines and personnel. It developed to great proportions and resulted in saving millions of trees and thousands of acres of crops.

By the end of the 1920's the aircraft manufacturing industry in Canada had become established on a firm footing. Canadian Vickers Limited had been first in the field, beginning in 1923. In 1927, the Ottawa Car Company arranged with the Consolidated Aircraft Company of Buffalo, N.Y., for the assembling and servicing of the latter's products in Canada, and two years later they also obtained Canadian manufacturing rights for the Avro Avian, under contract with the A. V. Roe Co. Ltd. of England. In 1927 a branch office and factory at Montreal were opened by the Wright Aeronautical Corporation of Patterson, N.Y., for assembling and servicing their engines in Canada.

De Havilland Aircraft of Canada, Limited, began operations in a small plant at Mount Dennis, near Toronto, in 1928, confining itself at first to assembling and servicing aircraft shipped from England. During that year, 62

FIRST CARGO FROM THE NORTHLAND
C. H. "Punch" Dickins (*wearing gloves*) and Mickey Sutherland his mechanic (*kneeling*) with fur bales from Fort Good Hope at Edmonton, March 1929

two-seater Gypsy Moth aircraft arrived and were rigged and tested at Toronto before being shipped throughout Canada for use chiefly in connection with the newly formed light airplane clubs sponsored by the federal government. By the end of the next year the company had enlarged its plant and produced 192 Moths of its own manufacture.

The first Moth assembled in 1928 was allotted Canadian registration G-CAJU, and was presented to the Toronto Flying Club by Sir Charles Wakefield. E. L. Capreol, De Havillands' first test pilot in Canada, put her through her original flying tests at Toronto in 1928, and hundreds of Toronto's civilian pilots received their initial flying training seated at the controls of faithful old JU. During her span of active flying, which included trips from coast to coast, that first Moth flew over 3,000 hours. The "old girl" gave good service up until 1940, when she was struck off strength on account of age. When last heard of she was being used by the club as a rigging trainer.

The Reid Aircraft Company was incorporated in 1928, with W. T. Reid as president, and established an "airpark" of 242 acres at Cartierville, on the outskirts of Montreal. The company designed and built the Reid Rambler, a light, two-seater, all-metal, fabric-covered aircraft, with folding wings and other novel features. The prototype was first flown on October 1, 1928. In January 1929, a merger was effected with the Curtiss Aeroplane and Motor Company of New York, and the company became the Curtiss-Reid Aircraft Company Limited.

In 1929, Fairchild Aircraft Limited built a large factory at Montreal, and construction of various types of their design began in 1930.

Aircraft construction also got under way at the Pacific Coast with the formation of the Boeing Aircraft of Canada, Limited, in conjunction with the Hoffar-Beeching Shipyards at Vancouver.

In 1928–1929, the number of aircraft engine companies in Canada increased to four. They engaged only in the distribution and servicing of engines manufactured in Britain and the United States. Canadian Wright Limited was reorganized as Aero Engines of Canada Limited, and became agents not only for Wright engines but also for the engines of the

Bristol Aeroplane Company, Limited, England. The other firms were Armstrong-Siddeley Motors, Limited, English engine manufacturers, who opened a branch at Ottawa; Pratt and Whitney Aircraft Company, Limited, a branch of the American company at Longueuil, Que.; and Bellanca Aircraft of Canada, Limited, at Montreal. Because of the rapid progress in flying, the facilities which the different companies were able to offer became available none too soon.

At the end of the 1920's bush flying had become an established feature of Canadian business life, especially in the fields of mining exploration and development. Governments, both federal and provincial, had discovered a useful and speedy method of conducting surveys, transporting mail, protecting forests and fighting fires, and of servicing remote communities.

Some of the handicaps that faced the first bush pilots were disappearing; others had only emerged. In the settled areas of the north, from which many bush outfits operated, private and public airports came into operation.

On the other hand, though much winter flying had been done in the southern bush, airmen were just beginning to feel their way into the high north, and innumerable problems had yet to be faced.

Except for the flying carried out by the Hudson Strait Expedition during 1927–1928, and by Stevenson and Balchen to Churchill on Hudson Bay in March 1927, little winter flying had yet been seen in the far north. However, operations were carried out by commercial companies in the winter of 1926–1927 from bases close to the railway, and were gradually extended into the far north in the following three years despite the arduous conditions caused by extremely low temperatures, high winds, storms, and low visibility. Once they had started, the indomitable bush pilots and engineers met and valiantly overcame the innumerable exigencies that cropped up. During the months of "open-water" flying in 1928, aircraft in the high north operated on floats. Later, when fuel caches were established at distant points, and winter flying was carried on in ski-equipped machines, the utter isolation of the northlanders became a thing of the past.

NOTHING in the history of Canadian aviation has justified the introduction of the airplane more than the possibilities it opened up for alleviation of human distress in isolated areas, some of which were almost inaccessible by other means of travel. Such "mercy flights" were far from uncommon, although many of them were never reported in the pages of the daily press: in one year alone Canadian Airways Limited flew a total of 120. East, west, north, and beyond to the Arctic rim, the bush pilot and his versatile mechanic flew on missions of mercy with a singleness of purpose and a disregard for their own safety and convenience that greatly increased the respect they had earned in routine commercial flights.

To tell every story of these flights would fill a dozen books; here we must be content with a few selected samples.

S.O.S. EAST: THE CREW OF THE *Bremen*

In 1928 the famous Junkers plane, the *Bremen*, making the first east-west crossing of the Atlantic, crash-landed on Greenly Island, in the Strait of Belle Isle off the northern tip of Newfoundland. Two Fairchild ski-equipped aircraft were dispatched by Canadian Transcontinental Airways Limited from their base at Lake Ste Agnes, Que., with orders to fly the 700 miles to Greenly Island to rescue the three men in the Junkers. The first machine, flown by "Duke" Schiller, with Dr. Louis Cuisinier aboard, left at 10:45 A.M. on April 14, 1928, and spent the night at Seven Islands. On the following day they continued to Greenly, landing on the sea-ice between the island and Blanc Sablon on the coast. The entire flight was made through blinding snow-storms. The second Fairchild, which also made a safe landing, was flown by Romeo Vachon; it had on board representatives of the Canadian and American press and a movie cameraman.

It was first decided to repair the *Bremen*, and Schiller flew out to Ste Agnes for repair parts brought by air from New York. When the Junkers had been repaired, the flyers tried to take off, but the space available for a runway was too short, and the machine was again damaged. The *Bremen* was left on Greenly, and the pilots flown to New York.

S.O.S. WEST: THE INJURED SUPERINTENDENT

At the other end of the continent a very different type of emergency arose. In September 1929 the Consolidated Mining and Smelting company requested Western Canada Airways to send a plane to Burns Lake in central British Columbia, to pick up the superintendent of their Emerald Mine, H. C. Hughes, who had been severely mauled by a grizzly. Infection had set in and his condition was so serious that the company wished him to be flown to Vancouver at once for treatment. A pilot, Walter Gilbert, hopped 260 miles from Stewart, B.C., to Burns Lake on the afternoon of September 10 in 2 hours and 10 minutes. The next day the injured superintendent and a nurse were taken aboard and Gilbert set off for Vancouver. With brief stops at Quesnel and Bridge River, he covered the 520 air miles in 6 hours and 40 minutes. A waiting ambulance rushed the stricken man to the Vancouver General Hospital. Eventually he made a full recovery, although his scalp had been torn completely off his head by the infuriated bear.

S.O.S. NORTH: DIPHTHERIA AT LITTLE RED RIVER

Trudging out of the tiny settlement of Little Red River, 600 miles north of Edmonton, the musher and his team of dogs soon became dots in a great white wilderness. They were making for the nearest telegraph station, to summon help to combat an outbreak of diphtheria. The only doctor was stationed at Fort Vermilion, fifty miles away, and his supply of antitoxin was inadequate and far from fresh. Unless a new supply could be procured, he knew an epidemic would result. A Hudson's Bay Company man was dead and other residents were already afflicted with the disease.

So the crack musher of Little Red River set out for the railhead at Peace River, some 300 miles away, with a message to be telegraphed to Edmonton. He and his dogs took two weeks —from December 18 to January 1, 1929—to make the long, strenuous, and lonely trip.

As soon as the wire reached Edmonton, the Board of Health at once got in touch with a local flying company and the story became headlines.

"Wop" May and Vic Horner were partners, operating Commercial Airways of Alberta, Ltd., and at that time they owned a small Avro

MISSION ACCOMPLISHED
Wives and mothers greet Victor Horner (*left*) and "Wop" May on their return to Edmonton

Avion, registered G-CAVB, with an engine of only 75 h.p. and an open cockpit. To attempt a winter flight in such a craft, particularly as they had no skis for winter landings, required courage. But when asked if they would fly to Fort Vermilion, May and Horner immediately consented.

Health authorities procured 600,000 units of diphtheria antitoxin and sufficient toxoid to treat 200 cases. The material was wrapped in woollen rugs, with a charcoal-burner inside to prevent possible freezing, and the parcel was handed to the airmen personally by Dr. M. R. Bow, Deputy Minister of Health for Alberta. It was then stowed carefully aboard the plane, together with a special tracheotomy set loaned by the Royal Alexandra Hospital of Edmonton, and an intubation set loaned by Dr. R. F. Nicholls of the same city.

At 12:45 P.M. on the day after the message was received, the two airmen set off.

After three hours in the air, they were forced to land at McLennan Junction, 265 miles from their starting-point, unable to penetrate further because of dense weather and approaching darkness. Fortunately, the authorities at Edmonton had foreseen this contingency, and had wired ahead to have a landing-ground prepared. The pilots came down safely on the cleared surface of a frozen lake.

Next morning they left at 9:40 A.M.—just as soon as it was light enough to fly—and the 50 miles to Peace River were covered by 10:32 A.M. The airmen stayed only long enough to refuel and snatch a bite to eat, then were on the wing again, heading through the grey half-light on the last lap to Fort Vermilion. They did a marvelous job of flying through tough weather, and covered the final 280 miles just before nightfall at 4:30 P.M.

The life-giving serum was immediately loaded on a sleigh, and a musher and a dog team carried it that night to Little Red River.

The two flyers took their time on the return trip. Exposure to below-zero temperatures in the open cockpit had been extremely exhausting and they both suffered severe, though not dangerous, frostbite.

The flight had become widely known and an admiring crowd welcomed them at Edmonton on Sunday the 6th. Later in the year the city honoured them with a reception at which they were each presented with an illuminated address and a suitably engraved gold watch.

STRANDED ON QUEEN MAUD GULF

The two aircraft of the MacAlpine party a few days after their arrival at Dease Point

The arrival of the serum at Little Red River prevented the spread of the diphtheria; there were a number of additional cases but only the one death. The courage and endurance of a musher and his dog team, and the daring of two young airmen, had saved a settlement.

S.O.S.: MAROONED IN THE ARCTIC

Not long ago on a summer afternoon my wife and my sister dropped in at a tiny tea-shop on Granville Street, Vancouver, to discover a couple they knew at a table, sipping cups of tea before a silver dish containing an appetizing assortment of dainty French pastries. The man was Tommy Thompson, who was then an official in the city office of the British Columbia branch of the Department of Air Transport. As Tommy munched his confections, no one could remotely have guessed that together with seven other unkempt, half-starved men, on the coast of Queen Maud Gulf, he had once squatted on his haunches beside a smoky, moss-fed fire, gnawing the while on nauseous hunks of rancid fish, all prospect of rescue as far away as the moon.

The story of this group of men, known to posterity as the MacAlpine party, affords a perfect illustration of the bush pilot at his best and a fitting conclusion to this chapter of flying in the Canadian north.

Richard Pearce, editor of the *Northern Miner*, was one of the party of marooned men, and it is to his diary that I am indebted for many of the following details. Although much was related and published in the press at the time, many experiences of the lost flyers and the rescuers have never been told.

In the summer of 1929, Dominion Explorers had come to have such confidence in their pilots and planes that they planned a more ambitious flight than any they had undertaken before. The trip was to cover some 20,000 air miles. Starting from Winnipeg, the route was to be up to Hudson Bay, to Baker Lake, thence along the Arctic Coast to Aklavik at the mouth of the Mackenzie River. From there, side trips would be made into the mountainous area of the Yukon. The return flight to Winnipeg was planned by way of Great Bear Lake, Great Slave Lake, across country to The Pas, and home.

The two airplanes left Winnipeg at 10:00 A.M. on August 24, 1929. One machine, flown by G. A. Thompson, was a Fokker, G-CASP, chartered from Western Canada Airways; the passengers were Colonel C. D. H. MacAlpine, president of Dominion Explorers, Richard Pearce of the *Northern Miner*, and A. D. Goodwin, engineer. The other aircraft, CF-AAO, a Fairchild owned by Dominion Explorers, was piloted by Stan McMillan. His passengers were E. A. Boadway, mining engineer, who was also a pilot, and Alex Milne, engineer.

Because of smoke from forest fires, visibility was poor almost from the start. Both planes eventually reached Norway House but only after getting lost several times and making forced landings to learn their whereabouts. Similar difficulties beset them *en route* to Churchill, which was finally reached on the 26th. (Thompson had landed at Jackfish Isles to get more fuel for his tanks, but the barrels at the cache there were empty, so he flew on to Churchill with what he had and made it with only a few drops to spare.)

The schooner *Morso*, under charter by Dominion Explorers, was due to call at Churchill to discharge supplies, but it had not shown up, nor did repeated radio calls from the Churchill station bring any replies from her.

On the evening of the 27th two ship's boats were spotted coming into the harbour from the open water. They proved to be from the *Morso*, with her crew aboard. The schooner had caught fire and the crew had been forced to abandon ship. When dynamite in her hold ignited, she blew up and sank. This was a heavy blow to Dominion Explorers, as the ship was taking fuel and supplies to Baker Lake and other bases.

Misfortune continued. The Fokker aircraft, moored off shore, dragged her anchor in the night and the strong rip-tide carried her five miles out to sea. The S.S. *Arcadia* spotted her, still afloat, but with only her wing showing. When an attempt was made to tow her, the line broke, and the plane plunged down and sank. Through a strange trick of fate, when the *Arcadia* hauled up her anchor, the battered aircraft was found hooked to it! The plane was dragged aboard but was found to be beyond hope of repair.

Although attached to a 500-pound anchor, machine AO almost suffered a like fate. At 1:30 A.M. on the 29th, she was discovered at the mouth of the harbour, and was rescued by the tug *Graham Bell*. To save further trouble, the Fairchild was flown to Tavani later in the day, then went on to Baker Lake to await the other men from Churchill, who had radioed to Winnipeg for a new machine. It was September 6 before Roy Brown of Western Canada Airways reached Churchill with another Fokker, G-CASK, and the next day, Thompson and his passengers flew on to Baker Lake. They were more than ten days behind schedule.

At Baker Lake Thompson took on another passenger—Major Robert Baker, bound for Burnside River. Condensation of damp air in the fuel tanks, creating frost particles in the carburetor, delayed SK's departure for some hours, but both planes made Beverley Lake before dark and the party camped out that night.

The planes set off again at 10 A.M. the morning of September 9, using the sun compass as a navigation aid. Pelly Lake was passed

at 11:30 A.M. and they kept to a course well north, so as not to miss Bathurst Inlet. A large lake, bigger than Pelly, was followed for over an hour, then they ran into violent storms. After much difficulty, they managed to land on a small lake at 2:50 P.M. The temperature began to drop alarmingly, and they noticed "dead" ice along the banks of creeks and rivers, left from no one knows how many winters before.

A fuel check showed that each machine had sufficient fuel left for from two to three hours' flying. As the storm eased, they got into the air once more. Visibility grew worse as they proceeded, but at length they were able to spot a large body of water, with ice-floes showing. Believing they were following the coastline of a bay in Bathurst Inlet, they carried on for 30 minutes. Suddenly a small Eskimo camp was seen. They immediately landed on the water near by, taxied to shore, and pitched camp for the night. Two Eskimo men, a woman, and a baby, were the sole occupants of the tiny camp. The natives could not tell the airmen where they were, but indicated the direction of a Hudson's Bay Company post. The explorers knew they had drifted north from their route, and believed they were on the sea coast somewhere near Dease Point, on Queen Maud Gulf, but were very uncertain about it. Meanwhile, a check-up showed that AO had fuel left for one hour's flight, and SK enough for two hours.

For the next two days, the weather was too bad for flying. It was decided to put 16 gallons of AO's fuel into SK's tank, and to try to make the post which, according to the natives, was in a northerly direction, four hours away by motorboat.

The younger Eskimo, originally named Keninya, but now re-named Joe, agreed to fly with Thompson and Milne in SK to show them the route. (Joe and his wife were great smokers, and during consultation the airmen's tobacco supply dwindled steadily.) On the 12th, the attempt was made to fly to the post, but SK developed carburetor trouble again. After much taxiing about and waste of good fuel, it was decided to transfer all the fuel to AO. Pilot McMillan then took off in AO with Baker, Milne, and Joe, but returned almost at once. The four reported that no land could be seen in the direction they were supposed to fly, and

THE MacALPINE PARTY

Left: The sod house, fuel supply, and entrance tent *Right*: Boadway, McMillan, Goodwin, Milne, MacAlpine, Eskimo "Joe," Thompson, Baker

as visibility was bad and fuel low, they thought it advisable to return while still able to do so. The men now knew that they were stranded, and with no radio communication with the outside world. After a conference, they decided to sit tight where they were for the time being, and Major Baker was put in charge. Rationing went into effect at once, with two scanty meals per day.

The season between summer and winter, when aircraft can use neither floats nor skis, was upon them. At that time of the year, too, ground travel in the north is almost at a standstill, awaiting the general freeze-up. All suggestions made by the white men to try to proceed on foot, or in the natives' canoe, were met with point-blank refusal by the Eskimos, who knew how foolhardy such attempts would be.

With bitter weather near, the eight stranded men got busy. They built a sod shack and erected a small tent at one end, but the first heavy snowfall finished the tent. Plans to use AO's wings as a roof had to be abandoned; the position of the stranded plane, in shallow water, prevented them from getting the wing off. So a canvas tarpaulin was stretched from side to side as a top for the sod walls. (It was to cause endless trouble in snow and rainy weather.) When finished the shack was 12 by 14 feet, and 5 feet high.

The weary group took up quarters in the shack on September 18. Their helplessness worried them, of course, but not half as much as the anxiety which they knew their families would be suffering.

The weather was bad from the time they arrived, so smoke-signals were useless, even if there had been anyone to see them. A make-shift stove, fabricated from an engine cowling, was used at first, but later McMillan traded a pair of field glasses for a tin stove from the Eskimos, and all meals were cooked on it. Fuel was a tough problem—gathering moss and willow twigs was a perpetual job for all hands. Although big piles were gathered, they vanished quickly. Engine oil was mixed with this fuel for cooking purposes, and it worked out fairly well. As their own food dwindled, the party fell back on the Eskimos' supplies, but the fare was hard on the white men's stomachs. Dried salmon and whitefish, months old, and more or less rancid, became the staple dish.

Two additional Eskimo men and a woman showed up on the 20th, then, while the airmen were having supper, all the Eskimos left without saying a word. This was disturbing, until they came back bringing dried fish and some fur clothing.

One of the new arrivals (real name Penuka, new name Charlie) could speak English fairly well, and from him they learned at last where they were—on Queen Maud Gulf. Charlie also told them that the name of the distant post was Cambridge Bay, and that the journey afoot could be made in three days and nights just as soon as the sea was frozen to a depth of four inches or more.

One Eskimo brought back a hunk of caribou which he proudly presented to the whites, but on cutting it up, MacAlpine judged its demise to have occurred at least twelve months before; it did not prove popular. The men were well supplied with ammunition and shot a few ptarmigan and ground squirrels. They also tried ducks, which were of the hell diver species, but they proved unfit to eat.

By September 22 the food rations had de-

TOMMY THOMPSON AND DON GOODWIN
Taken the day before the trek to Cambridge Bay

creased alarmingly. A check showed the following items: 25 pounds of flour, a tin and a half of baking soda, 5 tins of milk, a quarter-pound of jam, a pound of salt, 10 pounds of dried beans, a half-pound of tea, 4½ pounds of sugar, 2 pounds of currants, 3 tins of Fray Bentos meat, 2 pounds of dried apricots, one tin each of tomatoes and sausages, a bottle of prepared coffee, a package of dates, 6 hard-tack biscuits, 28 chocolate bars, and 30 Oxo cubes. This food was divided into three lots: one for the sick list, one for use later on the trail, and one for ordinary use; the last mentioned was scanty indeed.

At this time Baker developed an abscessed tooth. After suffering a good deal of agony, he persuaded Pearce to lance the affected part with a pen-knife, after sterilizing the blade with blazing rum (by general admission a waste of good rum). This treatment was effective, and recovery was rapid.

McMillan found a willow on the 26th, the largest shrub they saw in that area. The trunk was an inch through at the base, and the diameter of the spread of the branches lying almost flat on the ground was over three feet. The men tried to gauge its age by counting the growth rings. As each was of cigarette-paper thinness, they gave up with the guess that the plant was many hundreds of years old.

One day some of the men went down to the stranded AO to drain out some gas for a primus stove. They received quite a shock when they found only a pint or two. It was indeed fortunate that the airmen had turned back on their trip on September 12, otherwise they would probably have perished at sea.

There had been hard frost on most nights up to the 29th, which made the men hope that they could soon be on their way. But the next morning a deluge of rain was beating on the roof, causing ominous bulges in the tarpaulin.

Despite the discomfort, and the fact that the last of the tobacco was now gone, morale was good. The strain was beginning to tell on their physiques, however, and most of the party suffered daily cramps from the unchanging diet of fish. Pearce had promised to give the Eskimos his watch if they could shoot a caribou, but so far he still had his timepiece.

On October 1 the weather was mild, the snow gone, and the ground soft, with only slight frosts at night. This was very disheartening—indeed alarming. It was decided that Baker should be asked to draw up a report on their condition, in case the worst happened. By this time they were all staying in bed until 11:00 A.M. each day, to conserve their energy and to save on food.

The natives had left again to fetch more fish from a distant cache. Day after day passed with no sign of them. Spirits again began to droop, until finally, after eleven days, Charlie and his brother showed up with a good supply of fish, and, best of all, part of a caribou. Pearce surrendered his watch gladly, while everyone indulged in a mild orgy. Most of them suffered a bit afterwards, but all agreed that the feed was worth the aftermath.

By mid-October temperatures were consistently colder, and hopes of a departure were high, for surely the ice would soon be thick enough for them to cross the gulf. A maze of islands lies off shore at this part of the coast, and the stranded men realized that they could never reach the post without Eskimo guides.

On October 17 the temperature dropped sharply, and in the night new groups of Eskimos came into camp; by morning there were sixteen of them. Whites and natives piled into the sod shack, 24 altogether. Not much space was left for the cook, but Pearce managed to feed everybody.

A blizzard that night helped to freeze the waters of the gulf, and two days later, October 20, the natives finally decreed that a start could be made the following day. By the next evening the entire party, with three sleighs and dog teams, was 25 miles away from the sod shack they had called home for the past six weeks.

In the days that followed, the airmen found that mushing with a team did not faintly resemble the movies, where a man rides comfortably on the sleigh much of the time. They had to slug along on foot, sometimes through heavy snow and bitter wind. The unaccustomed effort at first caused the undernourished white men much pain at night, but they endured manfully. Lack of tobacco was a hardship. Food was eaten on the trail, uncooked and cold.

By October 24, they had struggled to a point on the Kent Peninsula opposite which, across Queen Maud Gulf, the Eskimos said Cambridge Bay was located. Time and again they were stopped by huge walls of ice hummocks, and only after back-breaking efforts were they able to surmount them. They were still on the peninsula, three days later, when the Eskimos told them they must wait until some of the party went all the way back to Dease Point to obtain food from caches there. Supplies were running low. What they had left at this time was herring and whitefish—actually the feed for the dogs. The men began to think that the 40 miles across to the post might as well be a thousand. They had now been lost almost seven weeks, and were on the slimmest rations they had yet experienced.

Five days later the supply party arrived back, bringing flour and sugar and—almost too good to be true—a small supply of tobacco. The natives reported that Eskimos living at Dease Point had heard an airplane several days before; and the Dominion Explorers men took fresh heart.

The trek across the sea began on November 2, and although it did not appear to worry the natives very much, it became a nightmare to the white men. Hour after hour they followed, back-tracked, turned left or right, wove endlessly about, picking out a rugged path across the high-piled ice-floes, or over jagged hummocks which held no route at all. Many times they fell into deep holes and had to be pulled out. Sleds were tugged and pulled over the most inaccessible places; even the dogs had to be boosted over. It was a nerve-racking experience. On one occasion, the wife of one of the Eskimos broke through thin ice to her knees. Being wise to the business, she did not struggle but simply lay flat until help came. Then, in spite of the freezing wind, she changed clothes right in the open, and carried on as if nothing had happened.

Towards the end of the day, the party struck really thin ice, and a halt was called at once. All the whites were nearly played out. Goodwin's feet were bothering him, and he suffered acutely from frostbite. The Eskimos built snowhouses in the dark, while the white men stood by, unable to do more than assist in chinking up cracks.

Before dawn on the 3rd they were on the move again, slower now, for the thin ice was extremely unsafe. It was touch and go whether all of them would make the land. The final lap began at 5:00 A.M., on the basis of a scanty breakfast of cold fish. The temperature was down to 27° below zero, and the wind was cutting. Toes, fingers, and noses froze time and again. As the day wore on, the different groups straggled out, and became lost to each other. It was a tough grind for all concerned—even the dogs were dead beat.

The first sled and group of men reached Cambridge Bay post at 4:30 P.M., and by 7:30 P.M. the other groups had staggered in. The staffs of the Hudson's Bay Company and the R.C.M.P. received them with joy and solicitude. They were all quickly fed. Then came warmth and relaxation and kindly ministrations, and before long another meal.

The S.S. *Bay Maud*, a ship Amundsen had used in his Arctic wanderings in bygone years, was at the post, having been purchased by the Hudson's Bay Company; in fact it had wintered there the past three years. A company man, Mr. McKinnon, was an amateur radio fan, and

G-CASQ BREAKS THROUGH THE ICE
When landing at Burnside River, Bathurst Inlet

had a small set aboard the *Bay Maud*, using her high masts for his aerial hook-up. Although he could not send direct to outside points, he was in touch with a ship named the *Fort James*, which was wintering some 200 miles north-east, near the magnetic pole. This craft had a large radio set. Through her co-operation messages were sent outside, and anxious families and company officials then learned for the first time the whereabouts of the missing explorers.

Colonel MacAlpine did not minimize the fact that the Eskimos had been the salvation of the flyers, and he saw to it that they were adequately rewarded—not only with rifles, ammunition, sewing machines, clothing, and tobacco, but also with several hundred dollars, left with the post manager to meet their later needs.

By this time the Dominion Explorers men knew that a wide aerial search had been going on for them but just how extensive and

hazardous it had been they did not at first appreciate.

In any winter rescue expedition in the far north, it is obvious that the searchers run into the same difficulties and dangers that beset the missing party; it may also happen that the urgency of their mission creates additional hazards. So it was that, in the earlier days especially, a search party was fortunate if its own misadventures fell short of tragedy.

The expedition that went to seek Mac-Alpine's party totalled six planes, yet at different times four of them were out of commission during the search, and three during the return journey, as a result of accident. But not a man was injured, and rescued and rescuers eventually all got home safely.

Dominion Explorers had planned that the MacAlpine party would reach Stony Rapids on September 20, there to meet two other of the Company's planes, which were to fly in from a western trip.

This second party, in two Fairchilds, CF-AAN and CF-ACZ, piloted by Charles Sutton and Jimmie Spence respectively, reached Stony Rapids on schedule, and found that the MacAlpine planes had not arrived. The weather was not good, so the delay occasioned no anxiety until the 24th, when a radio message was received from Bathurst Inlet, saying the MacAlpine group had not reached there. Then it was realized that something had happened.

The aerial search was begun that very day. Radio messages were sent south and two Fokker aircraft owned by Western Canada Airways flew up to Stony Rapids, G-CASL,

CRUICKSHANK'S MACHINE EMERGES FROM ITS ICY DIP

CHANGING FROM FLOATS TO SKIS
All five MacAlpine search planes were grounded at Baker Lake on October 14 to be fitted with skis

piloted by H. Hollick-Kenyon, and G-CASO, piloted by Roy Brown.

Guy Blanchet, an official of the Department of the Interior and a well-known Arctic traveller, was one of the party who flew in from the west to Stony Rapids. As he was also Dominion Explorers' representative at the time, he was eminently the right choice as leader of the search party.

The decision was reached to fly to Bathurst Inlet and make that the search headquarters. Since there were no gasoline caches between Stony Rapids and Beverley Lake, it was necessary to fly fuel in and establish such bases, if all four machines were to reach Bathurst. One plane went on to Beverley Lake, while the others ferried the fuel. At this time the weather was bad, snow squalls were frequent, and heavy cloud layers shrouded the higher points of land, making flying dangerous in the extreme. The Barrens around Dubawnt Lake, where the first gas cache was established, is a maze of smaller lakes and watercourses. Dubawnt Lake has innumerable small islands and a deeply indented shoreline, and large areas of it are almost unrecognizable as a lake at all. The danger of becoming lost was very obvious, and the flyers had to exercise great caution in flying from one cache to another. When the first aircraft reached Beverley

Lake on September 26, the flyers could see that the MacAlpine group had been there. It was logical therefore to conclude that the missing men must be down somewhere between this cache and Bathurst Inlet. Bad weather then settled in, and it was decided to go to Baker Lake, 140 miles to the east, there to reorganize the search with the help of radio. When the rescuers arrived at this base on the 28th, they found fuel and supplies so inadequate that they had to request a government tug to ferry supplies in from Churchill. The weather still was bad, and although they made many flights to the west, they could not reach Bathurst owing to the dense cloud formations which lay right on the ground. To add to their difficulties, lakes were beginning to freeze over. As all their craft were float-equipped, the rescuers had to take care not to be caught in zero visibility with a low supply of fuel in a place where landing could not safely be effected. When the weather did clear a bit, their fuel had again dwindled in attempts to get through to Bathurst, and more time was lost while supplies came up by boat from Churchill.

The first encouraging flight was made on October 2, when planes were able to cross the height of land and follow the Back River to Pelly Lake. On this flight they noted that

259

all the lakes and most of the waterways on the Arctic watershed were frozen over.

On October 9, two more aircraft reached Baker Lake to lend their aid. One was a Fokker, G-CARK, owned by Northern Aerial Minerals Exploration and flown by Jimmie Vance and Brian Blasdale; the other a Western Canada Airways Fokker, G-CASQ, piloted by Andy Cruickshank. During the next three days, the newcomers tried to reach Bathurst with their float-equipped machines, carrying ski-undergear in the cabins, but they could not safely penetrate the dense, low, cloud formations.

Then on October 13 the water of Baker Lake froze over. This brought an end to pontoon flying, and all five aircraft were hauled to the beach to be changed from seaplanes to "ski-planes."

On the night of the 17th a wild gale sprang up, lashing the thin ice of the lake into a churning mass. Pounding on the shore, it undercut the beach where the planes were, and damaged the tail assembly of Vance's Fokker beyond immediate repair. Pilots and engineers had a rugged time during the night, fighting the waves to save the other aircraft, but they managed to drag the machines to higher ground. It was a week later before the four serviceable planes once again rose into the air, and even then the clouds were still a formidable barrier. The planes spread out over a wide area and soon became separated, but each went down independently to a safe landing and camped out the night. On the following morning, by coincidence, all the aircraft took to the air at almost the same time, and spotting each other, they again flew west,

going over Pelly Lake, eventually to strike the coast of Bathurst Inlet near Gordon Bay. The Arctic sea at last!

They were making for a base and fuel cache at Burnside River, and, spotting a clear stretch of ice along the shore, Spence went down for a landing. As soon as the skis took the weight of the plane, he felt that the ice was rubbery and insecure, so he gunned the engine and speeded at full clip for the rougher ice near shore. Brown in SO did the same and he too arived safely. Cruickshank in SQ was not so fortunate. Probably weakened by the passing of the others, the ice suddenly gave way and SQ plunged through and settled down into the water until only the wings were flush with the surface. The crew escaped through the front hatch above the pilot's cockpit. Kenyon, in SL, landing on rough ice, experienced no trouble. The airplanes were now sixty miles from the post at Bathurst.

As no attempt could be made yet to salvage Cruickshank's plane, it was decided to continue the search with two planes, holding the third in reserve in case of emergency. October 27 broke fine for a change, and Spence and Brown were soon on the wing with observers aboard. Heading over a mountainous range of hills to the Arctic sea, they flew eastward along the coast for many miles. The country was a mass of black rocks, sticking out of the snow in every direction. To straining eyes their different forms took on every conceivable shape, fooling the searchers time and again into thinking they had spotted a grounded aircraft. They flew over Dease Point, where the marooned party had spent such a lengthy

BREAKING CAMP
At Musk Ox Lake at 30° below zero

CF-ACZ AT AYLMER LAKE

period, but of course by that date the missing men were on their trek to Cambridge Bay, and the shack and the downed planes were obliterated by snow. Visibility became so bad that the searchers flew right over the neck of the Kent Peninsula without glimpsing it. It was a long, uncomfortable flight back to Burnside, after a disappointing trip.

Since the weather was unfit for search during the next seven days, the entire party pitched in to salvage the submerged SQ. The Fokker was slowly raised by block and tackle until the skis were clear, and was not lowered until the water froze again where the ice had been chopped away. The engine was given a complete overhaul, and eventually the machine was put back into good condition. It was probably the most remarkable aircraft salvaging job ever accomplished in the north, and reflected great credit on T. W. Siers and the other engineers of the rescue party.

Lengthy flights were made on November 3 and 4. On the 5th, as the engines were being warmed up, a dog team came into sight, travelling fast. The Eskimo driver was excitedly waving a slip of paper and shouting, "They fine 'em! They fine 'em!" The search was over.

The note was from the radio operator at Bathurst, who had sent it by runner to a camp 30 miles south, where Colonel J. K. Cornwell had established a base with Eskimos in order to inaugurate a widespread ground search as soon as travelling conditions would permit. The Eskimo had made the 30 miles to the planes in 10 hours, travelling through the night over such a rough and difficult route that his speed

would have been remarkable even in daylight.

The search now became a rescue, but, not to weary the reader, the return trip may be given in outline only. The stranded party was first flown from Cambridge Bay to Burnside, and then on November 12 the four planes left together, with the entire party aboard them. The first hop was to be to the site of Fort Reliance on Great Slave Lake. Two planes made it—Kenyon's and Cruickshank's. Jimmie Spence in CZ, coming down at Aylmer Lake, buckled his ski fittings; and Brown in SO, alighting at Musk Ox Lake, crumpled a wing; two planes were out of commission. Back at Fort Reliance, Kenyon's SL developed a cracked cylinder. So Cruickshank delivered his passengers at Fort Resolution and back-tracked into the Barrens to search for the two missing planes. The marooned parties were not lost, nor in danger of starvation, but it was a trying two weeks that they put in before Cruickshank reached them. He flew them all in to Fort Reliance, taking along also a cylinder from Brown's engine to fit on to Kenyon's.

At last, on November 30, Cruickshank and Kenyon took off from Fort Reliance on a direct hop to Stony Rapids, with full passenger loads aboard. The weather was good and the trip was made without incident. Thus the entire party was flown out in relays, reaching Cranberry Portage on the Hudson Bay Railway, and then The Pas. The final group reached the railhead on December 4.

Jimmie Vance and Brian Blasdale, who had been stranded at Baker Lake since October

17 when their plane was storm-damaged, were communicated with by radio, but rather than expose any flyer to further risks in bad weather, they mushed out to Churchill with a party who were going there with dogs and sleighs.

So the four months' experience came to a safe end. A sentiment quoted from Pearce's diary may well be shared by us all: "For all members of the expedition . . . I will always have the deepest admiration and respect."

PART 7

FOR FAME AND FUN, 1927-

AFTER the first conquest of the Atlantic in 1919 by the non-stop flight of the British airmen, Alcock and Brown, an interval of eight years passed without any further Atlantic exploits. The hazards were still too great to face without sufficient incentive. Then suddenly, in May 1927, Charles Lindbergh made his magnificent non-stop solo flight from New York to Paris. The tremendous publicity given to this event by press and radio stimulated anew the interest in trans-Atlantic flight.

The years that followed were packed with thrills, achievement, and tragedy—a period such as will not be known again. The passing of time, and, still more, the events of five years of total war and several years of disturbed peace, have pushed these deeds far back in memory. The cloak of glory has become threadbare; the heroes are forgotten, and their exploits are confused with others'.

Here we find ourselves on the territorial and national fringe of Canadian aviation; an area where Canada, and what was then her sister Dominion of Newfoundland, were the objects rather than the subjects of history; they provided the stage, rather than the actors. Some of the world's most famous flights, a majority of them made by nationals of Europe and of the United States, passed over the maritime islands and peninsulas of eastern North America, and often started there. What better evidence could there be that peacetime flying promotes international co-operation, and that Canada is strategically situated on the air highways of the world?

Limited as we are to the story of aviation as it involved flights made over what is today Canadian territory, or flights from Canada by Canadians, the history of the Atlantic flights of the late twenties and of the thirties is outside our scope. And yet it would be a serious omission if we did not, at the very least, glimpse these heroic—and sometimes suicidal—attempts as they were seen through Canadian and Newfoundland eyes.

THE NEWFOUNDLAND STORY

The year of Lindbergh's flight, 1927, saw two trans-Atlantic flying-craft in Newfoundland skies. De Pinedo, flying an Italian-built, twin-engined, twin-hulled, Savoia-Marchetti flying-boat, named the *Santa Maria*, swept up the Atlantic Coast from a circumnavigation of the South Atlantic Ocean, and dropped out of the sky to land at Trepassey Bay on May 19. A brief stop-over, and he was gone, home to Italy *via* the Azores. In August, the Americans Brock and Schlee landed at Harbour Grace, refuelled, and were off on the morning of the 27th for the fifth Atlantic crossing, completing the next day the first direct flight from Newfoundland to English soil.

The thrill of 1928 was a glimpse of Eckener's huge Graf Zeppelin passing over Trinity Bay on her way home to Germany. On June 17 the air enthusiasts of Trepassey Bay rolled out to see pilot William Stultz and his passenger, Amelia Earhart, disappear into the grey dawn. By this successful flight Miss Earhart became the first woman to cross the Atlantic by air.

THE *SANTA MARIA*
At Trepassey Bay, Newfoundland, May 19, 1927

On October 17 a handful of spectators from near-by St. John's silently watched the Englishman, MacDonald, take off with a cheery, "So long." His little De Havilland Gypsy Moth was last seen at a fair height 700 miles outbound.

A year and five days later, the American, Diteman, took his dainty Klemm low-winged monoplane up into the air at Harbour Grace and, like MacDonald, vanished without a trace.

But Harbour Grace residents had a very different experience on the night of June 24, 1930, as their ears strained for the sound of the engines of Kingsford-Smith's *Southern Cross*. The famous Australian and his crew—Van Dyk, Saul, and Stannage—had set off from Portmarnock, Ireland, the day before, bound for New York. At last came the reassuring roar, but it disappeared again. All night the *Southern Cross* circled above the swirling fog, until the mists broke in the morning, and she came down. A day later Harbour Grace cheered her take-off for Long Island.

On October 9 a Canadian pilot from Montreal, J. Erroll Boyd, with an American navigator, Harry P. Connor, took off over the Atlantic from Newfoundland in the *Maple Leaf*, a plane that had already been piloted across the ocean as the *Columbia* by Chamberlin in the third non-stop crossing. The two airmen had planned a non-stop flight to Croydon, but a diminishing oil supply obliged them to descend to a sandy beach on the Scilly Isles, just short of Land's End, Cornwall, after a flight of some 2,200 miles in 23 hours and 44 minutes. The following day they flew on to London without further difficulties, and thus Boyd became the first Canadian airman to pilot a plane across the Atlantic.

Between June 23 and July 1, 1931, two American airmen, Wiley Post and Harold Gatty, made a sensational aerial dash round the world. Their aircraft, the *Winnie Mae*, never faltered throughout the long grind of 8 days and 14 hours in which it encircled the globe. The route led from New York to Newfoundland; thence across the Atlantic to Chester, England. Berlin was the next stop, then Moscow, and so on across the U.S.S.R. and the Bering Sea to Nome, Alaska. After a stop at Fairbanks, and a long flight over the Yukon wilderness, the *Winnie Mae* reached Edmonton on June 30. Leaving the next morning, Post and Gatty made a brief stop at Cleveland and arrived in New York the same day.

The endurance of the two airmen was phenomenal. They allowed themselves the briefest time for rest and meals, and stayed only one full night on the ground; that was at Edmonton, where they were held by the weather. In all, they were in the air for 207 hours.

That year, 1931, saw the Danes, Hoiriis and Hillig, take a trans-Atlantic "pleasure trip" from New York to Krefeld, Germany, *via* Harbour Grace; and two Hungarians, Endres and Magyar, make a non-stop flight from Harbour Grace to Budapest in their native land.

COMMANDER MacDONALD
At Harbour Grace, Newfoundland

MacDONALD TAKES OFF
The Gypsy Moth leaves Harbour Grace, October 17, 1928

URBAN DITEMAN'S LOW-WING MONOPLANE
At Lester's Field, Newfoundland

THE *SOUTHERN CROSS*
At Harbour Grace, June 25, 1930

THE *MAPLE LEAF*
Erroll Boyd and Harry Connor with the Bellanca (still bearing the name *Columbia*). This photograph was taken at Charlottetown, P.E.I., where the flyers were guests of the Jenkins family, shown in the picture with them

THE *WINNIE MAE*
The Post and Gatty aircraft getting a long "drink" at Harbour Grace, June 23, 1931

Lou Reichers, an American, failed to qualify when he landed in the Atlantic only 47 miles from the Irish coast, where he was picked up by a passing ship on May 13, 1932. That was the year that the giant German machine, the DO-X, a twelve-engined flying-boat, refuelled at Dildo Arm, near Trinity Bay, on her way from New York to Germany; and Amelia Earhart Putnam rose from Harbour Grace on May 21 to become the first woman pilot to make the non-stop solo flight. And in July, on the 5th, Mattern and Griffin got off to a flying start on their attempt to break Post's and Gatty's round-the-world record. It was learned in Newfoundland the next day that after breaking all records for speed to Berlin they wrecked their Lockheed in a forced landing in Poland, and had to give up the attempt. In August, Lee and Bochkon, also Americans, set off for Oslo, and were lost in a series of violent storms.

The next year, 1933, began with the spectacular sight overhead of General Italo Balbo's fleet of 24 flying-boats winging their way on the Labrador-Shediac lap of their epic flight to the Chicago World's Fair from Italy.

In 1934 took place the flight of Hoiriis and the Adamowicz brothers from New York to their native Poland by way of Harbour Grace, where the brothers had cracked up in 1933. A quiet year in Newfoundland skies followed, but 1936 saw two record flights. Merrill and Richman averaged 210 m.p.h. from Harbour Grace to Wales; and Jimmy Mollison, the Englishman who had already flown the difficult east-west flight twice, reversed directions, and in his 700 h.p. Bellanca, *Dorothy*, sped swiftly from Harbour Grace to London on October

29–30, making the long, lonely solo hop in the record time of 13 hours and 15 minutes.

The story of Newfoundland's connections with trans-Atlantic flights is necessarily brief and fragmentary. In many cases planes would come one day, refuel, and be gone in the morning, or even overnight. This is largely true, too, of the Canadian scene. But two factors brought the Canadian mainland further into the picture. The first was the fact that flights from and to the United States across the North Atlantic frequently involved flying over fairly large sections of Canada; the second, that the province of Quebec lay under the great circle route from North America to Europe *via*

AMELIA EARHART PUTNAM
With her Lockheed Vega monoplane at Harbour Grace, May 21, 1932

THE *SIR JOHN CARLING*

Terrence Tully (*left*) and James Medcalf with their Stinson-Detroiter monoplane before the take-off at London, Ont., August 29, 1927

Greenland and Iceland. This resulted, early in World War II, in the establishment at Goose Bay, Labrador, of airport facilities that grew to rival those of the great Gander airport in Newfoundland. Today Newfoundland, with Labrador, is Canada's tenth great province; the separation of territories in this narrative has, of course, been necessitated by the fact that in the thirties this union had not been achieved.

Appropriately enough, the Canadian story begins with a Canadian-sponsored trans-Atlantic attempt.

TULLY AND MEDCALF, 1927

This flight was sponsored by the Carling Brewery Company of London, Ont., although the idea originated with Arthur Carty, a reporter on the London *Advertiser*. The Carling directors announced a prize of $25,000 to be awarded to any Canadian or British pilot making a flight from London, Ont., to London, England.

Of the scores of applicants, only one offered to supply his own plane—Phil Wood of Windsor, Ont., brother of Gar Wood, the famous mortorboat racer. As Wood was an American, his offer could not be accepted.

The Carling Breweries then announced that they would supply an airplane, in addition to paying the prize money. Mr. Carty was put in charge of all arrangements, and a high-ranking officer of the R.C.A.F. was asked to co-operate in the final selection.

Thirty of the most likely applicants were brought to London, Ont., and interviewed. Captain Roy Maxwell was the first choice, but he was in charge of the Ontario Provincial Air Service, and the provincial government would not release him from this important post.

Next in line was Terrence Bernard Tully, and he accepted on the condition—which was granted—that his friend, James Victor Medcalf, be navigator. Both pilots were Irish-born, and married. Each had a lengthy and illustrious war record, and had been employed for several years by the Ontario Provincial Air Service.

A Stinson-Detroiter was purchased, and after

being christened the *Sir John Carling* at the Detroit plant, was flown to the Ontario city. (Sir John, the founder of Carling Breweries, had been a prominent member of the Legislative Assembly of Canada and later of the federal Parliament.)

The Carling Company spent $2,500 preparing a field for landings and take-offs, situated quite close to the city. Nothing that could possibly be anticipated was left to chance in organizing the flight. A recent invention—the new earth-inductor compass—was one of the instruments included.

Interest ran high throughout London. The townsfolk had been informed that aerial bombs would be fired from the grounds of the brewery the night before the take-off. The boom of explosions and the siren of the plant were heard on the night of August 28, 1927. By daylight the next morning, over ten thousand people were on hand to see the airmen make a perfect take-off.

Unfortunately, they ran into an impenetrable wall of fog near Kingston, and had to return to London at noon. The original plans to fly non-stop from London to London were modified, and plans now called for a stop at Harbour Grace, Nfld. But when the flyers set off on September 1, they again ran into fog, and for five days were weather-bound at Caribou, Me. Finally they arrived at Harbour Grace.

An airplane named the *Old Glory*, sponsored by American interests, set off on September 6 from New York in an attempt to fly non-stop to Rome. It was piloted by Lloyd Bertrand and James Hill, with Philip Payne, editor of the New York *Mirror*, as passenger. By the time Tully and Medcalf were ready to try their luck, there was anxiety over the whereabouts of the *Old Glory*.

Fully refuelled and in good condition, the *Sir John Carling* left Harbour Grace at 9:45 A.M. on the 7th, and quickly rising into the eastern sky, dwindled to a dot and soon was lost to sight. From that moment, the great spaces of the North Atlantic swallowed her up, and what became of her we can only guess. The *Old Glory* vanished just as completely. Although Hearst, the American publisher, offered a reward of $25,000 for the rescue of her crew, they were never found.

The Canadian representatives of Lloyd's of London had assumed a risk of $10,000 on both Tully and Medcalf. Although the arrangement had been only verbal, Lloyd's paid up without any question, after deducting the amount of $1500 from each payment to cover the cost of the premiums they had not received. This $17,000, plus the $25,000 intended prize

THE HOUR BEFORE THE DAWN

The *Sir John Carling* awaits the dawn of September 7, 1927, at Harbour Grace money, was used to establish a $40,000 trust fund for the benefit of the widows and three children of the airmen.

SCHILLER AND WOOD, 1927

When the American, Phil Wood, was refused entry in the London-to-London flight, he at once began to make plans for a rival flight. He purchased a Stinson-Detroiter monoplane, and engaged C. A. ("Duke") Schiller, a pilot in the Ontario Provincial Air Service, to fly it. The flight was to be from Windsor, Ont., to Windsor, England, and the plane was named the *Royal Windsor*. Everything was ready by the time the *Sir John Carling* was to leave, and a noisy and friendly rivalry sprang up between the two Ontario cities.

Schiller and Wood left Windsor on September 1, the same day the *Sir John Carling* left London. They, too, ran into bad weather, and had to land at St. Johns, Que. They flew to Old Orchard, Me., on the 6th, and the next day arrived at Harbour Grace, some hours after the departure of Tully and Medcalf.

The *Royal Windsor* could not proceed until a small leak in one of the tanks had been repaired. By that time, the *Sir John Carling* was long overdue, and as two aircraft had vanished over the Atlantic in the past week, friends prevailed upon Schiller and Wood to give up the attempt. They then took off on a lengthy search for the London plane, during which they flew hundreds of miles out over the Atlantic. With poor visibility, and tremendous seas running, the quest was hopeless from the start, and the pair returned to Harbour Grace sad at heart, without finding a trace of their missing friends. When all hope was abandoned for the *Sir John Carling*, Schiller and Wood flew back to Windsor.

Thirteen years later, in 1940, "Duke's" ambition to fly the Atlantic was fulfilled when he ferried bombers from America to England. He lost his life on March 14, 1943, in a forced night landing at Bermuda, while ferrying a bomber to South Africa.

MINCHIN, 1927

Thirteen years after his barnstorming flights at Winnipeg, F. F. Minchin, who had served with distinction in the R.F.C. and R.A.F. in World War I, was ominously headlined in

THE *ROYAL WINDSOR*
At Harbour Grace, September 1927

newspapers the world over. On September 2, 1927, he set out from England in an attempt to fly the tough east-west route across the North Atlantic, accompanied by another pilot, Captain Leslie P. Hamilton, and by a woman passenger, the Princess Lowenstein-Werntheim, whose financial support had sponsored the endeavour. The *Saint Raphael* and its three brave occupants vanished without a trace.

Before 1927 ended, a total of 17 lives had been lost in flight attempts over the Pacific and Atlantic oceans and the Canadian government actually considered passing legislation which would ban such flying to or from Canadian points. Fortunately, the act was never passed; it would certainly have had to be rescinded before many years had gone by.

THE *Bremen*, 1928

The year 1928 brought the dramatic attempt of the *Bremen* to cross east to west. Because of the continuous west winds, this route—"the hard way"—was most unpopular. Five persons had perished in attempting it, before three men set off on April 12, from Dublin, Ireland, in the *Bremen*, an all-metal Junkers monoplane. They were Major Fitzmaurice, an Irishman, and two Germans, Baron Huenefeld and Captain Koehl. After 37 hours of incredibly rough flying, they sighted a lighthouse. Fuel was running low, so they came down as best they could. The undergear was wiped off and the propeller was shattered, but they were lucky to suffer nothing worse. They found they were on lonely Greenly Island, in the Strait of Belle Isle, between New-

THE *BREMEN*
Undergoing repairs at Greenly Island, April 1928

foundland and Labrador. News of their landing was sent by radio to the mainland, and soon help came by air. The story of their rescue by the Canadian airmen "Duke" Schiller and Romeo Vachon has already been told in Chapter 28. The Junkers was later shipped to the United States, reconditioned, and put on permanent display in the Ford Aeronautical Museum at Dearborn, Mich.

HASSELL AND CRAMER, 1928

That same year, two American airmen, B. R. J. ("Fish") Hassell and Parker D. Cramer, attempted a flight to Europe *via* Canada, Greenland, and Iceland. They began their flight at Rockford, Ill., on August 16, and the first lap took them to Cochrane, Ont. They took off from the Canadian town on August 18, on a tremendously risky flight which carried them over the Quebec wilderness to Ungava Bay and across Hudson and Davis straits to Greenland. After this 2,400-mile flight, their fuel was getting low. They searched hurriedly for a suitable spot to land near the coast, but were obliged to fly inland and made a precarious landing on the edge of the Greenland ice-cap. They abandoned their aircraft, now completely out of fuel, and set off to walk to a point on the coast where they knew members of a United States exploration expedition were camped. They had expected to be able

to reach their destination in a little over a day, but only after nearly two exhausting weeks of the roughest kind of travel did they finally make it. Although the airmen failed in their attempt to reach Europe, they did set a record by being the first to fly from Canada to Greenland.

Sixteen years later, during World War II, an American military plane, while on a routine flight, spotted the abandoned Stinson upside-down on the ice-pack. The discovery of the wrecked aircraft caused some disquietude until a photograph of it was identified by its original pilot, now Lieutenant Colonel Hassell, Commanding Officer of the U.S. Army Air Force base at Goose Bay, Labrador.

GAST, CRAMER, AND WOOD, 1929

Although numerous flights were made across other oceans during 1929, there were few attempts over the Atlantic, and of those, only two touched on Canada.

The first was another effort to fly to Europe *via* the Greenland–Iceland route. Aboard were Robert H. Gast, pilot, Parker Cramer, co-pilot and navigator, who had made the attempt the previous year with Hassell, and the aviation editor of the Chicago *Tribune*, Robert Wood. The Sikorsky amphibian was owned by the *Tribune*.

The airmen left Chicago on July 3, touched

down at Sault Ste Marie, Ont., for the night, and on the following day flew across northern Quebec to a safe landing on Hudson Strait. They found the sea covered with ice-floes. A northerly gale sprang up and drove great masses of ice-blocks into the cove at Port Burwell where their ship was moored. The sides of the amphibian were soon crushed by the weight and the anchor ropes severed. The machine sank in deep water, despite herculean efforts by her crew and a number of willing Eskimos. The crew had to return to the United States by ship and train.

COMMANDER BOOTH AND CREW, 1930

Eight non-stop flights had been made over the North Atlantic by airships owned either by Britain or Germany, before the flight of the R-100 took place, but none had touched Canadian territory. For some years the British government had considered the establishment of airship routes connecting the far-flung parts of the Commonwealth. Montreal was chosen as one such point, and two British experts came over to select a suitable site for a mooring-mast. The place finally chosen was St. Hubert, seven miles from the city, and between November of 1927 and the spring of 1929 the work of clearing the area, constructing highways, and erecting the "mast" was carried on.

The "mast" to which a dirigible was moored was a much vaster piece of construction than is suggested by the term. It consisted of a 207-foot steel tower surmounted by a "head." Electrically driven mooring machinery went into the tower head, as well as large electrical elevators for transporting people and supplies. The diameter of the passenger platform in the tower head was 38 feet. Power pumps forced fuel to the top, where it could be fed to the tanks of the airship. A big silicol-process hydrogen plant consisting of two units, each unit capable of producing 20,000 cubic feet, was erected near by. Special radio and meteorological staffs, with the necessary buildings and equipment, were established.

The Montreal tower was financed jointly by Canada and Britain and was placed in command of Lieutenant Commander A. B. Pressy of the Royal Canadian Navy. With four experienced ratings, he was sent to England to undergo eighteen months of training at Cardington Royal Airship Works. All in all, the preparations for the first visit of a mammoth airship to Canada were carefully planned, very complete, and most expensive. It proved to be the greatest single air event in terms of capturing the interest of the public, that ever occurred in Canada.

The R-100 left Cardington, England, at 3:30 A.M. on July 29, 1930, and reached Montreal on the night of the 31st. She had to circle about until dawn, then was safely hooked to the mast. She had taken 78 hours and 52 minutes to make the journey, the fastest time to that date for such a flight by an airship.

THE *GREATER ROCKFORD*
At Cochrane, Ontario, August 1928; Hassell (*left*) and Cramer

ON THE ICE-CAP

The remains of the *Greater Rockford* photographed from a U.S. Army Air Force airplane in 1944

The dirigible had encountered a violent storm over the St. Lawrence near Three Rivers, Que., which had damaged the fabric of the immense port stabilizer fin. The repairing of this was entrusted to Canadian Vickers of Montreal who performed an efficient rush job the day following the arrival. Then, with V.I.P.'s of the government, air force, and military as passengers, Commander Booth took his vast ship off on a leisurely flight over the St. Lawrence and Ottawa rivers and Lake Ontario.

The ship reached Ottawa after dark, but her huge bulk showed clearly against the summer sky and practically every citizen was out of doors to watch her pass. A two-way radio system between the airship and Station CNRP was established, and thus residents were able to listen in on the conversations between ship and ground. As she slowly circled above the Parliament Buildings, six powerful searchlights played upon her silver sides, making an impressive picture.

From Ottawa the R-100 flew at a high altitude over eastern Ontario during the night. By dawn she was following the Canadian shore-line of Lake Ontario and by 6:00 A.M. she was over Niagara Falls. Turning back to head for Toronto at greatly reduced speed, she circled above the city at 9:00 A.M., disrupting morning traffic on the down-town streets until she passed out of sight to the east. Streets and roof-tops were black with people.

In leisurely manner the ship returned to Montreal, arriving late in the afternoon, where the trained crew again performed the intricate job of mooring her.

At 9:30 P.M. on August 13, the R-100 was released from the mast, rose slowly to clear all obstructions, then circled the air field and headed off into the east for home. The full crew complement was 42 men, but on the trip back she carried 55 persons. One of her engines went out of commission during the return flight, which slowed her down a bit, but she arrived safely at Cardington on August 16, bringing to a happy ending the one and only Atlantic airship flight to Canada.

It was confidently expected by various governments that a period of great expansion in airship travel was beginning, but this hope slowly subsided when disaster overtook some of the world's largest airships. Only two months after the visit of the R-100 to Canada, her sister ship, the R-101, crashed and burned in France while on a flight to India, carrying 46 persons to their death. In April 1933, the United States airship, the *Akron*, crashed into the Atlantic at night, 20 miles off the New Jersey coast, with the loss of 73 men, and with only 4 survivors. Then in February 1935, another United States airship, the *Macon*, collapsed in the air over the Pacific near California, but in this instance the loss of life fortunately was limited to 2 of the 83 persons aboard.

The finale of the great airships came with the loss of the immense, 812–foot, German ship, the *Von Hindenburg*. On May 6, 1937, she reached Lakehurst, N.J., after a long trip from Germany. With only minutes separating her from a safe mooring, flames suddenly sprang from the rear of the hydrogen-filled craft, and in a few frightful moments she was a blazing mass. Of the 97 persons aboard, 36 perished.

With the *Von Hindenburg's* passing, airship development ceased, and the ships still remaining were dismantled.

It had been hoped that the inflammable gas used in the majority of the ships could be replaced by non-inflammable helium, but the United States government refused to supply any helium to the Germans. Then, too, the huge unwieldy bulk of the airships made them an easy prey to the elements, while their leisurely rate of travel and the difficulty of mooring them showed that they were less practical than airplanes. Several of the major

274

airship disasters were officially proved to have been caused by faulty construction or downright bad handling, and even today, many aviation experts believe that the big airships may stage a comeback.

Whatever the future holds, it is only fair to mention briefly the very successful career of one airship in the thirties. The *Graf Zeppelin* alone made 111 South or North Atlantic crossings between September 1928, and December 1935. In that period she carried 12,000 passengers (or, if the crew is included, 32,962 persons) besides 35 tons of mail and 50 tons of freight, all without a single fatality.

I am indebted for these details to W. B. Burchall, retired secretary of the Air Industries and Transport Association of Canada. His interest, however, is not confined to statistics. Since the days when he was associated with the Sopwith Aviation Co. and the Hawker flight of 1919, Mr. Burchall had been keenly interested in the international rivalry in establishing air services over the North Atlantic. When it was announced that the *Von Hindenberg* would commence scheduled flights between Germany and New York he realized that this would make possible for the first time regular travel by air from Canadian points to Europe. He made the necessary reservations and left Winnipeg on May 9, 1936. Travelling *via* Northwest Airlines and American Airlines

to New York and Lakehurst, N.J., he left aboard the *Von Hindenburg* on that airship's inaugural west-to-east flight to Frankfurt, thence *via* Deutsche Luft Hansa and Imperial Airways to Croydon, England, where he arrived at 10:45 A.M., on May 15. (A two-day stop-over was made at New York and one night was spent in Brussels.)

This was the first journey by any Canadian using scheduled air services from Canada to London, England, and pre-dated by three years the flying-boat services inaugurated by Imperial Airways and Pan American Airways in 1939.

The huge mooring-mast at St. Hubert, Montreal, remained standing for many years, but as the area developed into a busy airport, the tower became a hazard to flying, and it was taken down in 1938. With its disappearance Canada's only link with the airship era vanished.

Von Gronau and Crew, 1930

Before 1930, only two airplanes, those of the United States Army Air Service round-the-world flight, had negotiated the Iceland–Greenland–Canada route from Europe to America. The third successful flight was made by the *Dornier Wal*, manned by Von Gronau and his crew; it was the first flying-boat to conquer this route. The *Wal* (Whale) with its four-man crew consisting of Commander Wolf-

THE R-100
At St. Hubert Airport, Montreal, August 1930

THE *GRONLAND-WAL*

At Prince Rupert, B.C., August 1932

gang Von Gronau, Eduard Zimmer, co-pilot, Franz Hack, engineer, and Fritz Albrecht, wireless operator, set out from List on the island of Sylt, on August 18, 1930, New York bound, *via* the Faeroe Islands, Iceland, and Greenland. Early on the morning of the 23rd they hit Cartwright harbour, Labrador, "right on the nose," 6 hours and 40 minutes out of Greenland.

Halifax, N.S., was the next destination planned, but after the *Wal* left Cartwright on the 24th the weather closed in. Von Gronau was obliged to take refuge in a small cove near the fishing village of Queensport, on the coast of Nova Scotia. The next morning the airmen were able to fly only a few miles further to the tiny hamlet of Marie Joseph before they were brought down by dense fog. The mists lifted later in the day, and they flew on to reach Halifax by 3:30 P.M. On the 26th they covered the 600 air miles to New York, landing just off shore from the Battery. They had accomplished a flight estimated at 4,670 miles in 47 hours' flying time. After a typically tumultuous welcome in the American metropolis, the airmen and their flying-boat returned to Germany aboard ship.

CRAMER AND PAQUETTE, 1931

In midsummer, 1931, an official flight was sponsored by the Transamerican Airlines Corporation, with the hope of starting a trans-Atlantic air-mail route to Europe *via* Greenland and Iceland. They engaged as pilot Parker Cramer, who had tried twice before, but without success, and as radio operator a French-Canadian, Oliver Paquette, whom Cramer had

met at Cochrane, Ont., during his 1928 flight to Greenland with "Fish" Hassell.

The airmen left Detroit July 28, stopping at Abitibi Lake, near Cochrane, Ont., so that Paquette could bid good-bye to his wife and parents, who resided there. On the 29th they flew to Rupert House on James Bay, and went on to Great Whale River on the 30th. They reached Wakeham Bay on Hudson Strait on the 31st, and on August 1 they landed at Pangnirtung on the east coast of Baffin Island. The surprised residents of Holsteinborg, Greenland, greeted them on August 2, and there the airmen stayed a little over two days, preparing their craft for the flight over the Greenland ice-cap. Their five-hour flight to Angmagssalik on August 5 was the first direct flight over Greenland. Several long over-water hops now lay before them. On August 7 they arrived at Reykjavik, Iceland; after refuelling, they hopped off again and reached the Faeroe Islands, at Thorshavn. On this lap, engine trouble forced them down on the open sea, where Cramer made a repair and took off successfully in very heavy seas—a fine achievement. The airmen reached Lerwick in the Shetland Islands on August 9, but when they left for Copenhagen after refuelling, they flew into storms of terrific intensity. At noon the fliers reported by radio that they were in sight of the Norwegian coast. Silence followed. The hours ran into days, and hope for Cramer and Paquette was finally abandoned when the British trawler *Lord Trent* came across their battered aircraft and salvaged the pitiful wreckage. Six months later, a Dutch trawler picked up a soggy bundle floating in the sea off the Orkneys. It contained, in a watertight case, personal papers which had been cast to the mercy of the sea by the gallant airmen before their death.

Von Gronau and Crew, 1931

Accompanied by the same three crewmen who were with him on his previous Iceland–Greenland North Atlantic flight, Von Gronau in a new flying-boat, the *Gronland-Wal* (Greenland Whale), flew from Germany to Iceland as in 1930. From Iceland he crossed to Scoresby Sound, about half way up the east coast of Greenland, reaching that isolated spot on August 13.

The next day Von Gronau made flying history. Just nine days earlier, Cramer and Paquette had been the first to fly across the ice-cap. Now the Germans set off to conquer it in the opposite direction, and successfully completed the hazardous flight when they landed at Sukkertoppen on the west coast of Greenland 10 hours later. On this flight the airmen charted a hitherto unknown range of mountains, some of the peaks of which rise to over 12,000 feet. The Danish government later named the range Gronau Nunatakker in honour of the German pilot.

Resuming the journey, the *Gronland-Wal* encountered engine trouble and severe storms, which caused Von Gronau to stop at various points including Godthaab in Greenland, and Povungnituk and Port Harrison on Hudson Bay. He landed at Longlac, Ont., on August 31.

The next day the airmen left Canada and flew to Chicago in six hours, then on to New York. Their intention had been to return to Germany by back-tracking over the Greenland route, but because of engine difficulties they decided otherwise, and all returned home aboard ship.

Bert Hinkler, 1931

For having accomplished "the most daring flying feat of the year" (1931) Bert Hinkler was awarded the Seagrave Memorial Trophy by the British Aeronautical Society in February 1932. The Australian airman had flown from Toronto, Ont., to London, England, by way of South America, the South Atlantic, and Africa.

This nonchalant young man was habitually so casual about his long-distance flights that few people knew of his plans beforehand. In 1931 he had arrived at Toronto and placed an order with De Havilland Aircraft for a Puss Moth monoplane, to be built to special specifications for long-distance flying. It was fitted with a Gypsy III engine, and when completed was given Canadian registration CF-APK. On October 20, Hinkler left Toronto without any publicity, making a direct hop to New York. He made a leisurely stop there;

CRAMER AND PAQUETTE
Parker Cramer (*right*) and Oliver Paquette, with the diesel-engined Bellanca seaplane in which they flew over the Greenland ice-cap

BERT HINKLER
With his De Havilland Puss Moth at Toronto, 1931

On October 27, people in Kingston, Jamaica, saw a trim monoplane come in for a landing, and were astonished when they learned that Hinkler and his Puss Moth had flown non-stop the 1,472 miles from New York. Forty-one days later, Hinkler landed at Henworth Airport, London, having dropped in at Port of Spain, Trinidad, and at Natal, in Brazil, hopped the South Atlantic to Bathurst, British West Africa, and so to England over Morocco, Spain, and France.

VON GRONAU AND CREW, 1932

In 1932 Von Gronau, who had twice flown the Iceland–Greenland route from Germany to America, made a third trip; he arrived at Cartwright on the Labrador coast July 25, and went on the next day in a non-stop hop to Montreal. Thence he flew to Ottawa, leaving Canada *via* Windsor, Ont., on the 30th.

It was now divulged that the flight was to be much longer than just over the Atlantic; Von Gronau planned to fly on right round the world. With this intent, he and his crew went on to Chicago, Ill., then to Minneapolis, Minn., and by August 9 they were again flying over Canadian territory *en route* to Lac du Bonnet, near Winnipeg, Man. On August 13, they were greeted by personnel of the R.C.A.F. at the Cormorant Lake base, and by the 16th the German airmen had reached Lac la Biche, northeast of Edmonton, Alta. Their next hop, on August 18, carried them in a direct flight to

indeed he did not hurry to leave any stopping-place on this trip. How easy-going were his flying habits can be judged by the fact that the maps he used were torn from ordinary atlases.

THE *HEART'S CONTENT*

Jimmy Mollison in his De Havilland Puss Moth at Pennfield Ridge, N.B. Note the huge fuel tanks in front of and behind the pilot

world flight made in a flying-boat, and took four months.

MOLLISON, 1932; LINDBERGHS, 1933

In 1932, Jimmy Mollison made his magnificent east-west solo flight: 2,700 miles into the teeth of a headwind and continually bad weather from Portmarnock, Ireland, to his landing in a farmer's field at Pennfield Ridge, N.B., a few miles from Saint John. It was the first solo trans-Atlantic flight from Europe, and the first flight made direct from Europe to Canadian soil.

The following year the equally famous Lindbergh with his wife, Anne, touched down at Halifax, N.S., St. John's, Nfld., and Cartwright and Hopedale, Labrador, during a flight they were making to study bases and weather conditions in the North Atlantic in order to determine the routes most suitable for Pan American Airways. Two years earlier, on a round-the-world flight, they had flown up the east coast of Hudson Bay to Baker Lake in the Northwest Territories, and across the "top" of Canada to the Yukon and Alaska.

THE HUTCHINSON FAMILY AND CREW

In August 1932 George B. Hutchinson of Philadelphia, Pa., decided to fly to Europe, and to take along his wife and two daughters in a newly purchased twin-engined Sikorsky amphibian, the *City of Richmond*. Hutchinson himself was a fully licensed pilot, and his crew consisted of Peter Redpath, a Canadian navi-

JIMMY MOLLISON
Taken shortly after his record-making solo flight

Prince Rupert on the Pacific Coast. They continued their long journey on the 21st, and Canada saw them no more. Their flight home carried them *via* Alaska to Japan, thence to India and the Mediterranean. It was the first

ANNE AND CHARLES
LINDBERGH AND FRIENDS

At Baker Lake, N.W.T., 1931

THE *ROBERT BRUCE*

John Grierson's De Havilland Fox Moth receiving a final check-up in the aircraft plant at Rochester, Kent

gator, Joseph Ruff, engineer, Gerald Altifissh, radio operator, and Norman W. Alley, cameraman. Hutchinson planned a route to Europe *via* Canada, Greenland, and Iceland.

After arrival at Saint John, N.B., the wheels were removed from the machine, so that it was no longer amphibious. On August 24 the "Flying Hutchinsons" flew to Port Menier on Anticosti Island in the St. Lawrence River, going on to Hopedale on the Labrador coast on the 30th. Up to this time the Danish government had firmly refused to allow the plane to touch at Greenland, because they believed the risks involved were too great with Mrs. Hutchinson and the two small girls aboard. Nothing daunted, Hutchinson flew the 626 miles to Godthaab on September 2, immediately running into trouble with the authorities there for illegal entry. However, money paid in fines smoothed things over and he was allowed to proceed to Julianehaab on September 7. On the 11th the plane set off for Angmagssalik, but was soon engulfed in dense fog. A leaking gasoline pipe added to the predicament, and a landing was made on the sea. Waves were high, and great masses of ice-blocks floated all about the plane. The situation of the Hutchinsons and their crew was extremely dangerous, but they could not take off until the weather cleared and the seas subsided. For 24 hours they taxied in the direction they knew the shore to be, drenched to the skin, and in danger at every moment of being rammed and sunk by the pitching floes. At length they reached a small rocky island and made their

way safely ashore, but there they were stranded —the heavy surf soon battered the Sikorsky to pieces against the rocks. The castaways were eventually rescued by the crew of an Aberdeen

JOHN GRIERSON AT ROCKCLIFFE

John Grierson being congratulated by Wing Commander Earl Godfrey upon arrival at the Rockcliffe R.C.A.F. Station

280

THE *TRAIL OF THE CARIBOU*
At Wasaga Beach, Georgian Bay, Ont., August 8, 1934

trawler, the *Lord Talbot*. They were extremely lucky to have survived the ordeal.

JOHN GRIERSON, 1934

On July 20, 1934, John Grierson left the river Thames, at Richmond, England, *en route* to Canada *via* the Greenland ice-cap. Hopping to Londonderry, Ireland, and then to Iceland,

AYLING AND REID
James R. Ayling (*left*) and Leonard G. Reid before taking off at Wasaga Beach, Ont.

he made a safe landing at Reykjavik on the 23rd, although he damaged one float of his De Havilland Fox Moth seaplane, the *Robert Bruce*. It was not until August 21 that he got away on the 480-mile hop to Greenland. Losing his bearings, he landed in a small cove along one of the most inhospitable coastlines of the world. Fortunately, his weak radio calls were heard, and his machine had also been sighted from Angmagssalik. Pastor Rosing and six Eskimos finally located the airman and directed him to their tiny hamlet.

Heavy cloud layers over the ice-cap delayed Grierson for a day or so but at last he rose into the air on the 25th, and after 450 miles with nothing but the great ice-cap below, set the *Robert Bruce* down to a landing on the harbour at Godthaab.

En route from Greenland the airman made stops at the Canadian points of Lake Harbour on Hudson Strait, Povungnituk on Hudson Bay, and Fort George and Eastmain on James Bay. He reached Ottawa on August 30, landing on the river just off shore from the R.C.A.F. Station, where his machine was given every care.

Wheels, obtained from the De Havilland factory at Toronto, were fitted to the Fox Moth, and Grierson left Ottawa on September 10, reaching New York after a brief stop-over at Albany. At the Curtiss Field, several mechanics looked casually into the cockpit of the *Robert Bruce*, and one remarked, "Say, she looks as if she had enough equipment to fly an ocean." Grierson, standing near by, spoke up laconically, "She has." There was a puzzled silence for a moment, then one fellow asked, "What ocean?" "The Atlantic," replied John.

One finally voiced their common scepticism as they turned away—"Oh, yeah?"

REID AND AYLING, 1934

A notable flight was made in 1934 by pilot Leonard G. Reid, born in Montreal, and James R. Ayling, of Berhampur, India, in a twin-engined De Havilland Dragon biplane that had been purchased by the Mollisons for an Atlantic flight attempt which they later abandoned. The machine had been stored with De Havilland Aircraft at Toronto, and after giving it a new name, *Trail of the Caribou*, the two airmen flew it to Wasaga Beach, on Georgian Bay near Collingwood, Ont. From this smooth wide beach they took off on August 8, with the intention of flying non-stop to Baghdad, Iraq, in an effort to establish a new world record for distance flight.

On the way across the Atlantic the main throttle control jammed almost wide open, and

MRS. BERYL MARKHAM
Waving to the crowds in London, England, following her trans-Atlantic flight

rather than take a chance on freeing it and perhaps having it become stuck in a closed position, the airmen flew on in this manner, using up much more fuel than they had intended. After 30 hours and 55 minutes in the air, the flyers decided to land in England, and came down at Heston, Middlesex, just after noon on the 9th—the first airmen to fly the west-to-east route over the Atlantic on a direct non-stop flight from the Canadian mainland to England. Not until World War II was this again achieved.

LIGHT AND WILSON, 1934

In 1934, an interesting air cruise round the world was made by Dr. Richard Upjohn Light and Robert F. Wilson, a radio technician. The journey began at New Haven, Conn., on August 20, 1934, and ended at College Park, New York, on January 24, 1935.

The two American airmen touched at Canadian points *en route* to Greenland, first at Sydney, N.S., then at Deer Lake, Nfld., and finally at Cartwright, Labrador. The famous Grenfell Mission is situated at Cartwright, and the authorities there were naturally interested in taking their colleague, Dr. Light, on a tour of the hospital.

The only portion of the trip which Light and Wilson did not fly was the journey across the Pacific; this they made aboard a C.P.R. liner, the *Empress of Canada*. Assembling their machine again at Vancouver, B.C., they flew down the Pacific Coast to Mexico; across to the Gulf of Mexico; over to Cuba; and then up the east coast of the United States, making many stops as they went along.

The world read little about this trip, at the time, but it was a splendid achievement. The risks were great and the flyers had several close calls, but good judgment and careful planning carried them safely through.

SOLBERG AND OSCANYAN, 1935

Back in 1932, Carl Solberg had endeavoured to fly to Oslo, Norway, from New York, but had come to grief in the surf off the shores of Newfoundland. Nothing daunted, he had another shot at it in 1935, and this second attempt proved successful. On July 18 he and his companion, Paul Oscanyan, arrived at St. Hubert airport, Montreal, from New York. On the 19th they went on to Havre St. Pierre,

THE *MESSENGER*
At Bauliene Cove, Cape Breton, September 4, 1936

Que., flying to Cartwright, Labrador, on the 21st. They waited for good weather before setting off to Greenland, and made the hop on the 28th without mishap. With some short and some lengthy flights, they covered the distances from Greenland to Iceland and to the Faeroe Islands, reaching Oslo on August 19.

BERYL MARKHAM, 1936

In the darkness of the early hours of September 4, 1936, a young woman flew alone, bound for New York from England in a Percival Vega Gull monoplane. The courageous pilot was Mrs. Beryl Markham, an English-born resident of Kenya, British East Africa, whose flight was sponsored by Lord Carberry, owner of the plane.

The dawn was long in coming, for the airwoman flew west with the night, and when daylight did come she was still a long way from land. In the afternoon, at last, she sighted Newfoundland. Headwinds had cut down her speed, and her fuel gauges showed that the gasoline in the tanks was nearing a low level. She obviously could not reach New York nonstop as planned, so, as the Nova Scotia coast came into view, Mrs. Markham decided to make a landing. She had flown over 2,700 miles and was exceedingly tired. She landed at Bauliene Cove, Cape Breton Island, not far from the tiny village of Little Lorraine. Luck

had been with her right up to the moment the wheels of the *Messenger* touched down, but then it deserted her. The area proved to be a bog, and as the craft landed, the wheels sank deep in the mire, and the trim craft whipped to a stop with its nose buried in the mud.

Mrs. Markham was royally welcomed in Cape Breton, and members of the Cape Breton Flying Club undertook the arduous task of salvaging the undamaged *Messenger* from her undignified position. Later the aircraft followed the pilot back to Kenya, securely crated in the hold of a ship, to complete her life as it had begun, doing commercial work over the African bush.

So ended the first non-stop solo flight to America from England by a woman, a record which stands to this day.

KOKKINAKI AND GORDIENKO, 1939

The last important trans-Atlantic flight to end in Canada before the outbreak of World War II was also one of the most significant. In the early years of the Russian Revolution little had been heard of Soviet aviation by the outside world. One of the earliest indications that this isolation was to end came on April 28, 1939, when two Russians. General Vladimir Kokkinaki and Major Mikhail Gordienko, set out in a twin-engined monoplane, the *Moscow*, in an attempt to fly non-stop to New York.

THE *MOSCOW*
General Kokkinaki in conversation with a Canadian doctor
who examined the airmen for injuries at Miscou Island

From the Tshelkovo airdrome near Moscow
they climbed slowly into the western sky, for
their heavily loaded plane weighed 12½ tons.
They flew over Iceland and Greenland to the
Labrador coast, and for much of the way
they remained at an altitude of 18,000 feet.
They met cyclonic winds and outside temperatures of 20° to 30° below zero. To add to their
discomfort, the automatic pilot went out of
order not long after leaving Moscow, putting
the strain of flying entirely upon Kokkinaki for
the rest of the trip. The plane was equipped
with a radio compass, and they were able
to follow an exact course.

After the flyers reached Canada, bad weather
closed in on them, and they were obliged to
climb to almost 30,000 feet to keep above it.
As they approached the southern shore of the
St. Lawrence River, radio messages from New
York informed them that the country for miles
around that city was enveloped in dense fog.
Because of the strong headwinds the plane had
encountered, the fuel supply was dwindling

fast, and the flyers finally decided to try to get
down as soon as they could make a landfall.
Low clouds and ground mists made it difficult
for them to see, but eventually they chose a
promising spot and Kokkinaki belly-landed the
aircraft with the wheels up. The airmen received a bad shaking, and the machine,
although not too severely damaged, could not
be flown to New York.

Kokkinaki and Gordienko had landed on tiny
Miscou Island, at the tip of New Brunswick,
on the Bay of Chaleur. They had been 22 hours
and 56 minutes in the air, and had flown 5,000
miles. They were flown to New York by commercial plane and met with much acclaim.
The damaged aircraft was dismantled and
placed on barges, which took it to Halifax,
where it was placed aboard a Russian ship and
returned to its homeland.

The outbreak of World War II brought to
an end all Atlantic flights which did not have
governmental or military sanction.* Nevertheless, flying over both the North and South
Atlantic oceans grew to tremendous proportions
during the war years, because of the passage of
thousands of aircraft on missions to and from
the various theatres of war. With the cessation
of hostilities, commercial aviation took over,
and tremendous strides were made immediately. Carriage of mail and passengers
between Canada and many points beyond the
ocean became commonplace.

Accounts of pioneer Atlantic flights tend to
become a little stereotyped, but it should never
be overlooked that in those early days every
flight meant danger. Unexpected and severe
conditions had to be met with courage and
perseverance. The interminable hours of flying
in the dark, often against tremendous winds or
through dense fog; the terrific weariness after
hours of strain; the ever-present possibility of
engine failure: all these hazards the early trans-
Atlantic pilots met on every flight, and tried
to conquer. All the pilots who entered the
battle against the Atlantic elements took their
life in their hands, and fully earned their place
in the history of world aviation.

*Additional flights of a commercial nature across the
Atlantic, which took place up to the outbreak of World
War II and shortly after, will be found in Chapter 33.

THAT so few Canadians attempted the Atlantic crossings in the "death or glory" days was probably due to the relative lack of money to spend on what was not only a dangerous but an expensive venture. Perhaps, too, Canadians were influenced by proximity to the pioneer days when the grim task of wrestling a living from the newly broken soil left little time or inclination for less practical things. Certainly, the course of Canadian aviation is seldom starred with flights for fame. From the end of the twenties, and through the thirties, there were scarcely a dozen flights that were made for no other purpose than to establish or to break records.

Of these, the first might be termed as much a flight for fortune as for fame.

FROM CANADA TO CALIFORNIA

In September 1929, a race from Windsor, Ont., to Los Angeles, Calif., was announced as a part of the ceremonies to mark the opening of a new air field at Windsor. The air field was a donation of 300 acres of land and $10,000 from two wealthy distillers, the brothers Hiram and Harrington Walker, of Windsor.

The prize-money was first announced as $5,000, but later donations brought it up to $10,000. The rules required that the pilots be Canadian, and that the engines should not exceed 800 cubic inches in displacement. Each plane also had to have a passenger. Eight obligatory stops were to be made, at Chicago, Iowa City, Omaha, Cheyenne, Rock Springs, Salt Lake City, Las Vegas, and Los Angeles, in that order. There were five entries:

William H. E. Drury, London Flying Club, flew a Waco Ten biplane. Milo E. Oliphant of Ann Arbor, Mich., was his passenger, and owner of the machine.

Lieutenant Kenneth Whyte from Grimsby, Ont., of the Hamilton Aero Club, flew a De Havilland Moth, G-CAUE, carrying Harry R. Campbell, a Grimsby hotel-owner.

Charles V. Towns of the Border Cities Aero Club, with a famous American airman, Eddie Stinson, as passenger, flew a Stinson-Detroiter.

S. T. Stanton, of Windsor, flew a Martin biplane, G-CAVE, owned by his passenger, E. V. Hemple, of the same city.

Captain G. S. Abbott, of the Department of National Defence, Ottawa, flew a De Havilland Moth owned by the Border Cities Aero Club, with Frank Mallard as passenger.

The contestants were to find that long-distance, cross-country air racing is a gruelling task for men and machines. Of the five entries, only two finished.

Abbott, in a forced landing at Amherstburg, Ont., was out of the race almost before it had started. Towns was making good time until a cracked cylinder ended his flight at Earlham, Iowa. Stanton was finally forced out of the running at Las Vegas. At 2:45 P.M. on the 12th, after an approximate flying time of 21 hours, Drury and Oliphant landed at Los Angeles to the cheers of 25,000 spectators. They got $5,000 with an additional $1,000 for the best perform-

THE WINNERS

The successful contestants in the International Air Race *Left to right*: Milo Oliphant, Lieutenant Whyte, Harry Campbell, Bill Drury

ance. Thirty-four minutes later Whyte and Campbell arrived, to receive a second prize of $2,500 and $500 lap money. (What happened to the other $1,000 of the original $10,000 I have never been able to find out.)

On August 10, 1929, air history was made when the American airman, Tex Rankin, flew his small biplane non-stop from the Lulu Island air field near Vancouver, B.C., to a landing at Agua Caliente, Mexico, in an elapsed time of eleven hours.

Of the many flights made in the 1930's to lower established marks, one of the earliest was by Roscoe Turner, an American, who flew the 1,350 miles from Vancouver to the Mexican border on July 17, 1930, in 9 hours, 14 minutes. Then came noteworthy flights by two other American flyers—Jimmy Wedell and Frank Hawks—both crack airmen already famous for speed and distance flights in their own country.

At the time Hawks entered Canadian skies, he was piloting a fast, specially built racing monoplane, manufactured by the Travelair Company in the United States. He was sponsored by the Texas Company, and the machine, *Texaco 13*, received widespread publicity, being popularly referred to as the "Travelair Mystery S."

In this monoplane Hawks established many flight records in the United States, and some in Canada. In July 1931 he flew to eastern Canada from New York. In one day he flew from Ottawa to Montreal in 48 minutes, from Montreal to Toronto in 1 hour and 48 minutes, from Toronto to Ottawa in 1 hour and 10 minutes, and back to Montreal from Ottawa in 1 hour and 10 minutes. To do all this in the one day was good going for that period. His average speed over the entire route was 184 miles per hour. Shortly afterwards Hawks left Montreal, hopping to New York at a 200 m.p.h. clip which landed him there in 1 hour and 45 minutes.

Then came still longer hops between Canada and Mexico. Major Jimmy Doolittle established a new record by flying from Ottawa to Mexico City in 1931. This endeavour was termed "The Three Flag—Three Capital Flight," since the capital cities of the three nations were involved. Flying a Pratt and Whitney Wasp-engined, Laird biplane, Major Doolittle left Ottawa at 4:40 A.M., October 20, 1931. After touching briefly at Washington, D.C., Birmingham, and Corpus Christi, he landed at Aalhuena Field, Mexico City, at 3:15 M.S.T. His total time for the 2,500 miles from Canada to Mexico was 12 hours and 36 minutes, of which 11 hours and 45 minutes were spent in the air.

A year later, Jimmy Wedell undertook the same flight. He hopped from Ottawa on October 23, 1932, dropping down at Washington, D.C., for a very short stay. Eleven hours and 53 minutes after leaving Ottawa he landed at Mexico City.

On December 1, 1931, flying his *Texaco 13*, Hawks set off from Vancouver, B.C., while Jimmy Wedell, flying a specially built Wedell-Williams racer, took off from Agua Caliente, Mexico. Both airmen were endeavouring to set a speed record between the two points. Escap-

TEXACO 13

286

JIMMY WEDELL AND HIS
RACER

ing exhaust fumes forced Hawks to a landing near Grenada, Calif., and put him out of the contest. He was lucky to escape with his life, as he was completely overcome after making his forced landing.

Wedell shot away from Agua Caliente, Mexico, at a terrific clip. Five hours and 22 minutes later he was clocked as he sped past the Vancouver airport. A strong tail wind had helped to push him along, and he reached the British Columbia city much sooner than he had anticipated. He had gone about 100 miles farther north along the coast before he realized his mistake and turned back. His flight from Mexico to Vancouver was officially timed at 6 hours and 27 minutes, being tabulated to the time he landed. He had left Mexico at 6:00 A.M., touching down at Vancouver at 12:27 P.M.

Thus began the interest in the so-called "Three Flag Flights," others of which were made as the years went by.

In 1932, Hawks flew his *Texaco 13* on a round trip from Mexico to Canada and back. He left Agua Caliente at 4:11 A.M. on January 23, landing at Oakland, Calif., to refuel 2 hours and 23 minutes out of Mexico, then going on to land at Portland, Ore., after another 2-hour, 52-minute hop. His time from the Oregon city to Vancouver, B.C., was 1 hour and 16 minutes. When he reached the Sea Island airport he merely touched his wheels on the main runway at 11:04 A.M., then was off again, these few seconds constituting the total length of his

visit to Canada on that occasion. He was back at Portland by 12:24 P.M. to refuel there. Then in one direct hop he flew on to Mexico, landing at Agua Caliente at 5:55 P.M. Hawks had made a "Three Flag Flight" from Mexico to Canada and back in one day!

During 1933, Hawks was still flying under the sponsorship of the Texas Company, but by this time he had another aircraft, an all-metal, low-wing, Northrop Gamma monoplane, powered with one of the first twin-row Wright radial engines, which had a total of 14 cylinders and was rated at 700 h.p. In its day, it was a fast machine. At an altitude of 9,000 feet it could make 225 m.p.h. Among the improvements on the craft suggested by Hawks were a streamlined wing stub, where the wings were attached to the fuselage; set-off ailerons; and built-in wing flaps to retard speed in landing. This Gamma, named the *Texaco Sky Chief*, was a single-seater, built specially for speedy, long-distance flying. Its non-stop radius was 2,500 miles.

Hawks and his machine arrived in Canada in 1933, landing at Regina, Sask., preparatory to making an attempt to fly non-stop to the United States Atlantic seaboard. At 4:33 A.M. on August 4, the airman hopped off from Regina, heading eastward into the dawning skies, and by 12:31 P.M., Regina time, he had reached and landed at Bridgeport, Conn., after a flight which had carried him approximately 1,700 miles. He made a special flight to Bridgeport in order to pay a visit to the recuperating English flyers, Amy and Jimmy Mollison, who had been hospitalized there after the crash landing which climaxed their non-stop flight from Britain to Bridgeport in July.

On August 22, Hawks and his famous *Texaco Sky Chief* showed up at Vancouver, B.C., where the pilot stated the he had come with the intention of establishing a non-stop flight across Canada. At 5:10 P.M. on the 25th, the American airman left the Vancouver airport, where a lengthy run of 1,600 feet was required to lift his heavily loaded Northrop clear of the ground. He carried a small sack of official air mail, which bore a special *cachet* and included letters from Mayor R. H. Gale of Vancouver to the Mayor of the City of Quebec and to the Lord Mayor of London. The overseas letter was to be delivered to the C.P.R. liner, the *Empress of Australia*, which was due to sail from Quebec at 2:00 P.M. on August 26.

Hawks left the Pacific Coast behind and climbed steadily eastward, flying by way of

FRANK HAWKS

the Fraser Valley, and over the smoke-laden air of the coastal range where many forest fires were burning. He passed at a great height over the Okanagan country, and over the main range of the Rockies going by way of the Crowsnest Pass, a route which had first been flown the "hard way" by Captain Hoy in a Curtiss Jenny back in 1919.

Night overtook Hawks over the Canadian prairies. He kept sending out radio signals at regular intervals, and his course was followed intently by many amateur short-wave radio fans.

As dawn came Hawks was in the vicinity of Cochrane, Ont., then all trace of him vanished. For a time it seemed that misfortune might have overtaken him, but finally word flashed over the news teletypes that he had landed at Kingston, Ont. He had been forced off his course by strong winds, and had then encountered fog. Spotting the airport at Kingston, he went down to find out where he was. He was soon in the air again and reached Quebec City at 1:30 P.M., eastern time. He had been 17 hours and 10 minutes flying from the Pacific Coast. A speedy taxi carried Hawks to the dock where he had time to hand the letter for the Lord Mayor of London to the captain of the *Empress of Australia* just before the liner sailed, slightly behind schedule.

This cross-Canada flight stood for a long time as the only accomplishment of its kind in Canada.

The last notable record-breaking flight of the thirties came in 1935, when an American pilot, Earl Ortman, took off on the "Three Flag Route" in an effort to better the existing mark between Vancouver and Mexico. He worked out a different procedure. Taking off from Seattle, Wash., at 10:30 A.M. on July 3, Ortman arrived over the Sea Island airport at 11:06 A.M. After circling the field he swung off again to the south without landing, in order not to waste time reporting to Customs officials at his first stop in the United States. After stopping for fuel at Sacramento, Calif., Ortman reached Agua Caliente at 4:34 P.M., establishing a new record of 5 hours and 28 minutes from Canada to Mexico. His actual flying time was 5 hours and 10 minutes.

Both Wedell and Hawks lost their lives in air accidents. Wedell died in a crash at Patter-

ON COMPANY BUSINESS

In this De Havilland Puss Moth, G. J. Mickleborough (*left*) of De Havilland Aircraft, and Geoff S. O'Brian, Company pilot, made a trip from Toronto to Vancouver and back. The journey occupied several weeks and totalled 6,050 air miles and 57 flying hours.

son, La., June 24, 1934, while Hawks lost his life on August 23, 1938, at Buffalo, N.Y.

Flying for fame or fortune chalked up no further records and won no more races in Canada before World War II. But flying for fun was going on from coast to coast all through the thirties. While bush pilots were penetrating the remote parts of Canada and mail pilots were pioneering the air-mail routes of the nation, amateur flying and air hobbies were gradually becoming an established feature of Canadian aviation.

THE FLYING CLUBS

BY the middle twenties, as the demand for pilots became more pressing, it was realized that a source of supply other than the pool of former war pilots must be found to fill the need. In 1927, the Department of National Defence decided on a programme of assistance in the formation of flying clubs which would help to stimulate aviation and provide training facilities for prospective pilots.

Conditions for the organization and entry of members to clubs were covered by Order-in-Council. The government offered to provide two light aircraft of two-seater type to any community, on condition that the community engage a licensed flying instructor and an air engineer and provide a licensed air field with housing and maintenance for the machines. Thereafter, for every additional aircraft that the club might buy the government would donate one. Furthermore, the government promised to pay $100 for each pupil who qualified for the private pilot's certificate.

Tuition cost about $250 per pupil, and to begin with the instructors were pilots who had served in World War I.

Enthusiasm throughout the country surpassed all expectations. In 1928, the first year, 16 clubs began active operations, 2,400 members joined, 8,124 hours of flying time were recorded, and 209 pupils flew solo. Of that number, 111 had earned their private pilot's certificates, and 28 their commercial licence, before the close of the year. The Toronto Flying Club was the most active, the Winnipeg Flying Club second, and the Montreal Light Aeroplane Club a close third.

Membership more than doubled in 1929, increasing to 5,233. Flying hours doubled, reaching 16,612. Four hundred and forty-two pupils took off on their first solo flight, 183 won their private pilot's licence, and 58 their commercial licence. Toronto again led with a total flying time of 1,514 hours and 35 minutes, and Winnipeg again was second, with 1,464 hours and 8 minutes, but Ottawa's club nosed out Montreal that year, to come third with 1,290 hours and 20 minutes.

Unfortunately but inevitably, accidents occured. During 1929, eleven people lost their lives, and seven were injured. Five of the accidents resulted from stunting.

Coincident with the club movement, the number of good air fields increased, and as the years passed the clubs became a major part of the nation's aviation activities. With the start of World War II, the R.C.A.F. obtained many fine pilots who had first soloed in DH Moths

SIR JAMES MacBRIEN

as members of one of the clubs, and today club graduates may be found at the controls of military and commercial aircraft, not only in Canada but in other parts of the world.

On November 1, 1929, the Canadian Flying Clubs Association was formed to co-ordinate the numerous clubs under a central organization. Sir James MacBrien became the first president. The Association was given an annual grant by the Department of National Defence with the proviso that it employ a permanent secretary with qualifications and duties acceptable to the Department. George M. Ross received the appointment and was installed as executive secretary, with headquarters at Ottawa. In 1931, the Association took over the publication of *Canadian Aviation* as its official organ. George Ross was its editor until his resignation in 1939, when the magazine was taken over by the Maclean-Hunter Publishing Company.

As a further stimulus to Canadian aviation, the Canadian Flying Clubs Association organized the trans-Canada annual air pageants. The first of 26 great flying meets was held on July 1, 1931, at Hamilton, Ont. From that date until September 12, twenty aircraft, together with a picked group of pilots from the Royal Canadian Air Force flying Siskin fighters, covered Canada, going as far west as Vancouver, B.C., then back to Hamilton, on through the Maritimes, westward again to Cleveland, Ohio, ending the tour with a show at London, Ont.

The first autogyro ever seen by most Canadians performed in the 1931 assembly. It was a Pitcairn, flown by Godfrey Dean, who took it over the entire trip, including a flight over the Rocky Mountains. He once looped the autogyro, not, however, when flying with the pageant. Godfrey Dean lost his life the following year in the crash of a Junkers at Kagian-agani Lake, Ontario.

THE WEBSTER MEMORIAL TROPHY

In July 1931, the only entrant from Canada in the King's Cup race held in England was John Webster of Shediac, N.B. He and his Curtiss-Reid Rambler came in thirteenth out of 40 entrants, over a course of 980 miles flown in exceptionally bad weather. The welcome and praise that John received on his return to Montreal were richly deserved.

During that summer, John was selected to take part in aerobatic flying displays in connection with the trans-Canada air pageant.

JOHN WEBSTER

THE WEBSTER TROPHY

arranged by the Royal Canadian Flying Clubs; and the winners of the different zone trials compete in the finals at some central point. Marks are awarded for precision flying and for cross-country flights involving intricate navigation. In addition to the individual bronze medallion, the champion receives a silver tray, awarded by the R.C.F.C. Association; this organization also rewards each winner of a zone trial with an engraved gold wrist-watch. The contests put aircraftsmanship to a severe trial, and the winning of the finals is recognized as a great achievement.

1932, 1933, 1934. EDWARD C. COX. Edward Cox had planned a career as architect, but by the time he had qualified for this profession it was 1918 and he was 18 years old. He therefore joined the R.A.F. and was taking the flying course at Deseronto when the war ended. For the next ten years he worked as an architect.

While residing in Detroit in 1928 he joined the Border Cities Aero Club at Walkerville, Ont., and qualified for the private pilot's and the commercial licences. Two years later he went to Montreal and soon became an outstanding member of the local Light Aeroplane Club. For three years running he won their Lytell Trophy for "the most competent pilot in the club," which made him the logical choice as the club's representative in the first Webster Trophy competition held at Hamilton in 1932. He won first place, and he repeated this success for the next two years.

In 1939, he became a professional airman by joining the R.C.A.F. and served as Squadron Leader during the war.

1935, 1936, 1938. GORDON R. MCGREGOR. Gordon McGregor was born in Montreal and received his early schooling there. He continued his education at St. Andrew's College, near Toronto, and graduated in engineering at McGill University, Montreal. In 1923 he entered the employ of the Bell Telephone Company and rose through various positions until in 1938 he became central district manager at Montreal.

McGregor's first flying lessons began in 1933 as a pupil with the Flying Club of Kingston, under the instruction of Captain H. B. Free. He progressed quickly, obtaining both his

While on a practice flight at the Montreal airport, his machine stalled during a steep climb and crashed, and the young airman died from his injuries.

In remembrance, the Webster family donated the Webster Trophy, to be awarded annually to the best amateur pilot in Canada. The beautiful memorial is a bronze, two feet in height, of a four-winged figure poised on a globe, the work of the famous Canadian sculptor, Dr. Tait McKenzie. The trophy remains on permanent view at the National Art Gallery at Ottawa, where one may see inscribed on the pedestal the names of the annual winners. Each winner receives a suitable medallion to mark his achievement.

The Webster Trophy is awarded for an entirely different achievement from that recognized by the McKee Trophy. Professional pilots cannot compete for it, and it is given strictly for flying ability. Local elimination tests are

private and commercial licences by 1934. While still connected with this club he won the national competitions for the Webster Trophy in 1935, 1936, and 1938.

In January 1939, with the demand for pilots increasing, he enlisted with the No. 115 Auxiliary Squadron and was commissioned as a Flying Officer. By the next year he was in active combat, and earned the D.F.C. in September. He commanded No. 1 and No. 2 Fighter Squadrons, and in December 1941, was made Director of Air Staff at the overseas headquarters of the R.C.A.F.

In May 1942, McGregor was sent back to Canada and was appointed Commanding Officer of "X" Wing, R.C.A.F., in Alaska. Subsequently, he headed No. 14 Fighter Squadron in the Aleutians, and commanded the R.C.A.F. Station at Patricia Bay, B.C. He received the O.B.E. in 1944.

When peace came, Gordon R. McGregor turned to civilian aviation. He joined the staff of Trans-Canada Air Lines, where his knowledge and executive ability quickly carried him through different administrative positions until finally he became president.

1937. B. J. BOURCHIER. A native of Ontario, B. J. Bourchier graduated in mechanical engineering from the University of Toronto, then spent a year at the Institute of Technology at Flint, Mich. In 1930 he joined the London Flying Club, and was selected by the R.C.A.F. from numerous applicants as one of the young pilots to receive provisional pilot officers' training at Camp Borden. During the summers of 1930, 1931, and 1932, Bourchier took advanced Air Force instruction. When at last the plan was discontinued, he had approximately 40 hours' dual and 43 hours' solo flying to his credit. Thereafter he did little flying until 1936. By this time he had been commissioned as a Flying Officer of No. 110 City of Toronto R.C.A.F. Auxiliary Squadron, which commenced flying in March 1936. The next year he won the Webster Trophy as representative of the Toronto Club.

Bourchier enlisted with the R.C.A.F. on the very day the war began in Europe—September 3, 1939. After four months as a pilot on coastal artillery co-operation work on the Atlantic Coast, he was made chief instructor with No. 11 S.F.T.S. at Yorkton, Sask. In January 1943 he was sent overseas for a three months' course at the Empire Flying School in England. In June he was shot down and spent the rest of the war years as a prisoner. When he returned home he went into business in eastern Canada.

With the outbreak of World War II the

| E. C. COX | G. R. McGREGOR | B. J. BOURCHIER |

BADGE OF THE FLYING SEVEN

Webster Trophy contests were suspended as young Canadians by the thousands enlisted in the R.C.A.F. for a grimmer kind of competition.

CANADA'S FIRST WOMEN PILOTS

With the organization of flying clubs across the country in 1928, women as well as men were to be found at the controls of club machines. In Vancouver a unique club was formed to which no male flyer could be admitted. "The Flying Seven" originally consisted of Margaret Fane, Betsy Flaherty, Alma Gilbert, Rolie Moore, Jean Pike, Elainne Roberge, and Tosca Trascolini. The membership of the club has steadily increased since then, and has done much to further its aims of encouraging flying among women.

But from that August day in 1913 when the famous American airwoman, Alys Bryant, flew at Vancouver, B.C., fifteen years had slipped by before any Canadian woman earned the distinction of becoming an air pilot.

To Eileen Vollick goes the honour of being the first, an honour shared to some degree by Jack V. Elliot's Air Services flying school at Hamilton, Ont., where she learned to fly under the instruction of Leonard J. Tripp. She was a diminutive young lady, only 5 feet 1 inch in height, weighing only 89 pounds, but what she lacked in size was amply compensated by ability and determination. Since there were no airplane clubs in Canada at the time she decided to learn to fly, she had to enrol with one of the commercial flying companies.

She made her first solo flight in February 1928, taking off from the ice on Hamilton Bay on Lake Ontario, in a ski-equipped Curtiss JN4 aircraft, which was registered G-CANY. On March 13, 1928, with a total of 16 hours dual and solo flying to her credit, Miss Vollick passed her flight tests before the government inspector, and became the proud owner of private pilot's certificate No. 77, the first to be issued to a woman pilot in Canada. Not long after she had learned to fly, Eileen married and moved to New York to live.

The second woman to obtain a pilot's licence in Canada was another Eileen, this time of Winnipeg—Eileen S. Magill.

Miss Magill's dual instruction began on May 28, 1928, under the guidance of Michael de Blicquy of the Winnipeg Flying Club, one of the newly formed light airplane clubs. After 11 hours of such flying, spread over a month's time, Miss Magill soloed for 28 minutes in the same De Havilland Moth that she had learned to fly in. On August 10, she passed her tests successfully before the government inspector, Squadron Leader L. Stevenson, and received private pilot's certicate No. 142, dated October 24, 1928.

Before Miss Magill discontinued flying at the close of 1929, her log book showed a total of 57 hours and 33 minutes' flying time. One notable flight made in company with J. Morgan, and another machine flown by club instructors de Blicquy and J. Sully, was a "Goodwill" trip from Winnipeg to Minneapolis and back, made on April 29, 1929. The machines were in the air for a total of 15 hours.

Eileen Magill was married in 1932, becoming Mrs. Cera, and when last heard from was residing in Woodbridge, Ont.

Dorothy Bell and Mrs. H. A. Oaks were the next women in Manitoba to become qualified pilots. As the wife of a famous Canadian airman, H. A. ("Doc") Oaks, first recipient of the McKee Trophy, Mrs. Oaks's desire to become a pilot can well be understood.

In the province of Alberta, Gertrude de la Vergne was the first woman to earn a pilot's certificate. Her dual flying began in September 1928, with Wilfred "Bill" Rutledge, pilot and instructor at the Calgary Aero Club. After 14 hours training, Miss de la Vergne soloed in one of the Club's Moths, and ten days later passed her flying tests. Her private pilot's licence was No. 157, dated November 21, 1928.

When Miss de la Vergne was aloft on her first solo, her instructor and others on the field were so interested in watching her that they did not notice that a discarded cigarette had set fire to the grass of their prairie airdrome.

Gertrude de la Vergne became Mrs. C. R. Tanner, resident in Calgary.

A close second to Miss de la Vergne in receiving a private pilot's certificate in Alberta was Louise Burka, who earned hers under instructors Rutledge and Jock Palmer.

Daphne Paterson, daughter of Dr. and Mrs. A. P. Paterson of Saint John, N.B., was an active member of both the Saint John Flying Club and the Montreal Light Aeroplane Club. She began her dual instruction at Montreal on February 2, 1929, under H. "Tony" Spooner, and was allowed to go solo after only 7 hours and 10 minutes' flying. On May 29, 1929 she passed her flying tests before Inspector Stuart Graham, and received private pilot's certificate No. 327; then she went on and qualified for her commercial pilot's certificate, No. 658, dated March 15, 1930.

Miss Paterson became very well known as a pilot in eastern Canada. She won the De Havilland Trophy in Class "A" in the Maritime zone competitions in 1931, and next year ranked second to Edward Cox, winner of the Webster Trophy. Finally, after some years of flying, the ambitious young woman qualified as a licensed transport pilot, the only woman to hold such a licence in Canada prior to World War II. By the time the flying clubs suspended operations in the war years, she had flown a total of 410 hours.

During the war, Miss Paterson married Squadron Leader A. J. Shelfoon, R.C.A.F.

Nellie Irene Carson was the first Canadian girl to solo a plane in the province of Saskatchewan; this she did as a member of the Saskatoon Aero Club after 15 hours and 8 minutes of dual training. She passed the tests for a certificate on September 30 before the government inspector, H. C. Ingram of the Civil Aviation Department.

On June 8, 1930, Miss Carson established an altitude record for women when she took her Moth up to 16,000 feet, as high as it would climb, completely out of sight and hearing of spectators on the ground. Later in the month she engaged in a bit of barnstorming around the countryside, taking up passengers at $3 a flip, and dropping in where an exhibition or picnic was being held.

During this time, she was also the very able secretary of Dr. H. C. Broughton, medical

EILEEN VOLLICK   EILEEN MAGILL   NELLIE CARSON

superintendent, at the Saskatoon Sanatorium, a post she held until January 1, 1942, when she resigned to enlist in the R.C.A.F. (W.D.). At that time she had 50 hours' flying time to her credit.

During the war she was a Corporal in the Administrative Branch of the Women's Division, and when last heard from in 1942 had been recommended for an officer's course at Toronto.

Another Saskatoon girl, Grace Hutchinson, was the second woman to solo in Saskatchewan.

Although her log book showed only 22 hours' solo to the end of 1940, Jeanne Gilbert was the most travelled airwoman in Canada, prior to World War II.

She was born at Kamloops, B.C., of French-Canadian parents. In later years she lived in Winnipeg, where she married the well-known pilot Walter Gilbert. When he was sent to the Pacific Coast by Western Canada Airways, she went with him, and enrolled with the Aero Club of British Columbia at Vancouver in September 1928. Her training was twice interrupted, first by the death of her instructor, Percy Hainstock, then by the transfer of her husband to Stewart Lake. She finally passed her tests—in very bad weather, too—in December 1929, and became the first woman to receive a private pilot's licence in British Columbia.

In the years that followed Mrs. Gilbert flew more often as a passenger than as a pilot. In 1928, at Winnipeg, she had the honour of a short flight in Admiral Richard Byrd's tri-motored Ford plane, with the famous pilots Floyd Bennett and Bernt Balchen at the controls. But her other passenger flights were with her husband, and because commercial flying in the north was at that time in the pioneer stages, every flight was an adventure. The Gilberts transported Hudson's Bay Company officials on their inspection trips throughout the Prairie Provinces, and to many isolated areas "down north," going even as far as Coppermine on the Arctic Ocean.

In 1940, Jeanne Gilbert took a refresher course of five hours' (each) dual and solo flying. In recent years she has resided in Montreal.

The first woman to fly in Prince Edward Island was Louise M. Jenkins of Charlottetown. She began as a pupil with the Curtiss-Reid Flying School, at Cartierville, Que., her instructors being "Gath" Edwards, "Tony" Spooner, and J. Phinney. After 10 hours and 15 minutes of dual, she soloed on December 17, 1931, proving she was an apt pupil.

On February 2, 1932, during the coldest snap of that winter, she went aloft in a Rambler, CF-ACI, and passed all her tests before the inspector, Stuart Graham, except the figure-eights, which she completed on February 24.

As soon as she had earned her licence, Mrs. Jenkins purchased a machine from the De

GERTRUDE DE LA VERGNE   LOUISE JENKINS   DAPHNE PATERSON

Havilland Company at Toronto. She had set her mind on having distinctive registration letters for her craft and persisted until, with the helpful intervention by Prime Minister R. B. Bennett, her wish was granted, and her new Puss Moth appeared with the proud markings, CF-PEI.

On February 23 Mrs. Jenkins made a non-stop hop from Montreal to Charlottetown in 4 hours and 40 minutes. She made a fine landing on the area adjacent to her home. Her husband, Colonel J. S. Jenkins, later granted permission to commercial companies to use the area as an air field, and it was in use until the new Charlottetown airport was built.

Mrs. Jenkins had intended making a cross-Canada flight and had completed all her arrangements by March, but was prevailed upon by friends to give up the project. From that date until July 1933, she flew to many cities in the Maritimes and performed at the air pageants. Family responsibilities finally compelled her to abandon flying. She sold her beloved plane and for a long time could scarcely bear to hear an aircraft in the sky. When her flying career came to a close, after nineteen months, Mrs. Jenkins had over 210 hours in the air, 150 of them solo.

One of the highlights of the 1935 show at St. Hubert's airport, Montreal, sponsored by the Canadian Flying Clubs Association, was an altitude flight by Edith Freeman, who climbed to a height of 16,300 feet in one of the Montreal Flying Club machines. The young pilot had only six hours' solo flying to her credit when she made this flight, which established a women's altitude record for Canada.

Many Canadian women have qualified as private pilots and some have earned the coveted transport pilot's certificate, yet the piloting of commercial planes remains in general a man's job. During World War II Canadian girls and women found a place—not indeed in the air—but in almost every other department of the R.C.A.F. Over 10,000 of them were in the Women's Division, and gave excellent service.

One woman's name must be included, though she never piloted a plane.
Elizabeth (Elsie) Gregory MacGill was born in Vancouver, B.C., a daughter of the late

JEANNE GILBERT

James Henry MacGill, who practised law in that city for over forty years, and of the late Judge Helen Gregory MacGill, who was the first woman jurist in the province and Judge of the Juvenile Court in Vancouver for twenty-two years. Elsie seems to have inherited her mother's pioneering spirit, for she became the first woman to graduate in electrical engineering from the University of Toronto, where she obtained her B.A.Sc. in 1927, and also the first woman to obtain the master's degree in aeronautical engineering at the University of Michigan, where she obtained her M.S.E. in 1929.

An attack of acute myelitis then interrupted Miss MacGill's formal studies but, during the long and trying time that she had to spend in a wheelchair, she did work on aeronautical design and wrote articles on aviation. Once sufficiently recovered to walk with the aid of a cane, Elsie went to the Massachusetts Institute of Technology to continue her postgraduate work. While in Boston, she accepted a job in Canada as assistant aeronautical en-

ELSIE MacGILL

gineer with Fairchild Aircraft, Ltd., at Montreal. There she did stress analysis on the prototype of the first all-metal aircraft designed and built in Canada; and design, aerodynamic, and stress calculations on other commercial prototypes. She represented the Company when model tests of a new design were undertaken at the wind tunnel of the National Research Council of Canada, and, when the prototype aircraft were put through their flying trials at Longueuil to determine flying characteristics before certification, she went with the pilot as observer on all the test flights.

The year 1938 marked the achievement of another "first," when Miss MacGill was elected to corporate membership in the Engineering Institute of Canada.

In the same year she left Fairchild Aircraft, Ltd., to become the chief aeronautical engineer of Canada Car and Foundry Co. Ltd., Montreal. Her first task was to assume responsibility for carrying through the calculations necessary to obtain a Certificate of Airworthiness, Acrobatic Category, for a prototype fighter aircraft designed and built in the Company's plants. In 1939, that task completed, her department undertook the design of a special duty primary

trainer aircraft, the Maple Leaf Trainer, for use in Mexico. The prototype was built and test flown at Fort William, Ont., and Elsie flew as observer on all the test flights, including the first. Within only 8 months from the time the design was begun the aircraft received the Certificate of Airworthiness, Acrobatic Category—a record-breaking performance.

During World War II, Miss MacGill, as chief aeronautical engineer, was in charge of all engineering work in connection with the production, by Canada Car and Foundry Co. Ltd., of Hawker Hurricane fighter aircraft for the British government—the famous Hurricanes of the Battle of Britain. She was responsible also for the development of a "winterized" version of the Hurricane. It was the first time that a high-speed aircraft had been fitted with skis (instead of wheels), with propellor de-icing, and with de-icers at the leading edges of the main planes and tail surfaces. The test flights for this novel conversion were carried out under Miss MacGill's direction, and the conversion was a success. The contract for the Hurricanes once completed, the Company entered on the production of Curtis-Wright Helldivers for the United States Navy, with Elsie again responsible for the work.

In 1941 another honour was conferred upon her: she was awarded the Gzowski Medal by the Engineering Institute of Canada for her paper entitled "Factors Affecting the Mass Production of Aeroplanes."

In 1943 she set up her own business in Toronto, opening a consulting office in aeronautical engineering, and in that year, too, she married an aircraft associate, E. J. Soulsby of Victory Aircraft Ltd.

In 1946 Miss MacGill became the first woman to serve as Technical Adviser to the International Civil Aviation Organization (a United Nations body) when she attended the sessions of the Airworthiness Section and helped to draft the international airworthiness regulations for the design and production of commercial aircraft. She has since attended other sessions of the organization. In 1947 she was chosen to be chairman of the Stress Analysis Committee, one of the five regular committees—a unique honour for Canada and for herself, for she was the first, and to date the only woman to have chaired such a committee.

In March 1953 she received a further distinction when the American Society of Women Engineers made her an honorary member, named her "Woman Engineer of the Year," and presented her with their medal, awarded annually, "in recognition of her meritorious contribution to aeronautical engineering." It was the first time that the award had gone outside the United States.

But her contribution had not gone unnoticed in her own country—earlier in the same month, her photograph appeared with those of 49 men in the Gevaert Gallery of Canadian Executives at Montreal, Que., "a tribute to the 50 Canadians responsible for this country's great industrial development during the past 15 years." And in her own community she is known as a citizen who by no means confines her interests to purely technical and professional matters, but is actively engaged in promoting the welfare of society as a whole.

Between Two Wars: Gliding

The attempts at gliding in Canada came to a halt during World War I, but after 1918 a few adventurous spirits, mostly in western Canada, took up the sport again.

In 1922, Norman Bruce, a fifteen-year-old boy of Medicine Hat, Alta., began to build a biplane glider of the "hang-on" type. (The pilot hangs suspended from the armpits while in flight, and depends on his own underpinning when landing.) Bruce, however, added two wheels to his glider. It was finished in two years, but lacking proper material to treat the wings, he used boat-varnish on them. Bruce had never seen an airplane at close range, nor received any help in his construction.

He flew his glider very successfully as a kite, and on at least one occasion he went with it, running fast down a slope and hanging on to the glider like grim death. He was thrilled when it lifted him clear for a few feet.

When this glider was smashed by the wind, Norman began another. It received a friendly inspection by an officer from the High River Air Station near by, who congratulated the boy but warned his father that the glider should never be flown. Norman was then taken to see the aircraft at High River. This led him to scrap his second glider, and build a third of a much more sucessful design. It had a wing span of 22 feet and was equipped with a wicker chair seat. Although it lacked a controllable rudder, Bruce made three good flights in it before it was smashed.

When Norman became old enough, he took a course in aeronautical engineering in the United States. When he returned in 1930, he organized the Cloud Rangers Gliding Club, and built his fourth glider, which was of much better type and construction than its predecessors. His two brothers became pilots at this period, but most regrettably one of them lost his life when he stalled the glider during a towed take-off and dived straight down from 200 feet. Public opinion in that district then turned against gliding, and the club was disbanded.

But Norman could not give up the work, so he formed a women's group—the Skylarks Gliding Club. Two Medicine Hat women soon distinguished themselves at gliding, Mrs. Ellick and Patricia Terril.

In 1934, in conjunction with James Fretwell

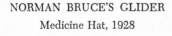
NORMAN BRUCE'S GLIDER
Medicine Hat, 1928

THE WEBERS' GLIDER
North Battleford, 1928

of the Stettler Gliding Club and Paul Pelletier of Calgary, Norman formed the Canadian Gliding Boosters. With a knocked-down glider in a trailer attached to their car, the three set out on a lengthy tour of the Prairie Provinces to give glider exhibitions. From May 25 to July 1, 1935, they were seen in action at 22 different towns and cities between Calgary, Alta., and Winnipeg, Man., their tour covering over 1,082 miles by road. A grand total of 410 glider flights was made during that period, often under very trying conditions, and thousands of Canadians who had become quite used to airplanes were given their first sight of a glider. Collections were made at each point to help defray costs, but people were not very generous—the total earnings of the entire tour were only $214.

One summer night, while the men were asleep in a small tent, a tornado-like wind sprang up, whisked up the glider like a feather, and smashed it to matchsticks. The three enthusiasts were obliged to drive home sadly to Medicine Hat, with the remains of their little craft packed in a small box.

Other Canadian enthusiasts who became prominent in gliding were Fred and Charles Weber of North Battleford, Sask. After reading of the exploits of German glider pilots, the brothers began construction of a primary-type glider in 1927. (A primary type is one specially built for strength and downhill gliding. A sailplane, on the other hand, is built for soaring, and is strong but light.)

The Webers had to meet and overcome many difficulties in construction. Everything that went into the job had to be made by hand, but by 1928 the work was completed and actual tests began. The bush-sprinkled prairie was their air field, a 1926 Chevrolet touring car their towing apparatus. Later a higher-powered Hudson truck was pressed into service. A height of about 150 to 200 feet could be gained before casting off the tow-rope. On occasion during early morning hours, the Webers used an eight-mile highway, where there were no poles or wires. With a 2,000 foot tow-line, the glider could be carried up to 500 feet, then cast loose, and flights of 15 to 30 minutes were often made.

In 1926 Brandon, Man., had a glider fan in Dick Noonan. In his early teens Noonan constructed a biplane type of hang-on glider and made attempts to fly it from a hill near the city, but with no success. Later, he and his friend Fred White built two different machines, both of which had closed-in fuselages. Brief towed flights were made with them before they were damaged beyond repair.

Beginning in 1929, a wave of enthusiasm for gliding swept over the American continent. The MacDonald Tobacco Company presented a Falcon glider to the McGill Glider Club at Montreal. Not to be outdone, the tobacco firm of Tuckett Limited, makers of Buckingham cigarettes, donated a primary glider to the Toronto Flying Club, which was ceremoniously christened *Buckingham Booster* by the wife of the mayor of Toronto. It is too bad that this rivalry among Canadian firms did not continue, for many clubs were forming across Canada, some doing very well, others just managing to scrape along.

During the late 1930's, Lethbridge, Alta., produced an ambitious group of gliders. One woman member, Evelyn Fletcher, manipulated an H-17 sailplane to a height of approximately 5,000 feet during a 50-minute flight, and established a soaring flight record for a woman in Canada. Miss Fletcher made her start with a motor-towed take-off from flat country. Once released from the tow, at about 600 feet in height above the ground, the airwoman made use of rising curents, known as thermals, in order to gain the altitude she reached. Her 50-minute trip also covered a distance of 10 miles, although the sailplane had circled many times in covering that distance.

When the activities of World War II brought civil gliding to a close in the late thirties, it was only a case of suppression for the time being, and just as soon as the war was over gliding enthusiasts again turned their thoughts towards the sky.

Of all those who indulged in the art during the twenties and thirties the name of Norman Bruce is high on the list. He had left Canada to reside in England, where he remained until 1947, but during that period he became an expert at gliding. Upon his return to Calgary, his log book showed a total of 882 gliding and soaring flights, with a logged flying time of 50 hours, accomplished on 14 different types of craft since the day in 1922 when his feet almost dragged along the ground on his first momentous flight. Recently Mr. Bruce has built himself a modern, all-metal, Schweiger sailplane, equipped with blind flying instruments and two-way radio, which is a long way from his first "antique" craft with its wobbly wings and its kitchen-chair seat.

SOARING AND GLIDING IN CANADA AFTER WORLD WAR II

Throughout the ages, men have gazed upwards at the soaring of the birds, and yearned for wings to fly in such graceful, seemingly effortless motion. The modern sailplane, in free flight, without the aid of mechanical motive power, comes nearest to satisfying this longing. In the course of many years the sport of gliding has developed remarkably. Associations have been formed, standards set, and tests drawn up in keeping with rules of the Fédération Aéronautique Internationale, and certificates of varying importance are granted for expert soaring.

Shortly after the outbreak of World War II, civilian gliding and soaring in Canada came to a halt. Military gliders, of course, came into prominence during the war. Most of these were large machines in which a full load of armed soldiers could be carried. They were probably used most effectively by the Germans in the capture of Crete.

After the war, the scattered glider clubs throughout the country were united in a single national organization known as the Soaring Association of Canada, with headquarters at Ottawa. The first president was J. A. Simpson. In 1949, the association published its first year book, edited by Douglas A. Shenstone.

International certificates of proficiency have been available to enthusiasts since the war,

EVELYN FLETCHER
In the H-17 sailplane

SOARING ASSOCIATION BADGES

but none had been issued in Canada prior to the end of 1939, although two Albertans had applied for them by that time: Evelyn Fletcher of Coronation and Arthur Larsen of Lethbridge. To the end of 1949 some 200 certificates had been issued, and many more have been issued since. It was in 1949 that the first national meet was held at Kingston, Ont. So many gliding and soaring fans turned up that it was decided to hold annual meets. In 1950, the meet was held at St. Eugène, Ont., in 1951 at Waterloo, Ont., and in 1952 again at St. Eugène. In 1950 the first national gliding champion was named, Albie Pow. The title was to be awarded annually for the best flight or flights of the year. Pow made two flights of 70 miles each, and a duration flight of over 6 hours' soaring. These were not world records, to be sure, but a good start. Indeed, there is no reason why the pilots of Canadian sailplanes should not equal or pass the best records made, because many parts of Canada are ideal for gliding.

In 1952, nineteen gliding and soaring clubs were affiliated under the Soaring Association of Canada, together with four private owners. At the same time six additional clubs were being formed. Altogether, approximately 500 persons were engaged in soaring and gliding. This is not a large total for the whole of Canada, but most of the people interested in this branch of flying are not by any means wealthy. It is not an expensive sport, but it costs some money. In 1952, no manufacturer in Canada was turning out a glider or sailplane for sale to the public. During World War II the Sparrow glider was designed under the supervision of W. Czerwinski and W. J. Jakimiuk, and constructed by members of the De Havilland Aircraft of Canada who had formed their own glider club. The craft was completed and successfully flown by T. B. S. Tarczynski in 1942, but has never been produced on a commercial basis.

The "Loudon" glider is the result of a project in design set for students in Aeronautical Engineering at the University of Toronto. The supervising designers were B. Shenstone, W. Czerwinski, and B. Etkin, and the project was checked by T. R. Loudon. The finished design was built with the help of the students in the university workshops and when completed was christened "The Loudon" as suggested by the students. This craft showed great promise in the air; but in 1952 it was not possible to obtain such a craft, although blueprints of its construction could be secured from the Toronto Gliding Club. The price of a good new glider delivered in Canada in 1952 varied a great deal, running from about $1,500 to $5,000, depending on its type and the country from which it came. A tandem two-place trainer in kit form, bought from Slingsby Sailplanes Ltd. of Kirbymoorside, Yorkshire, England, cost £345, or £587 complete. A similar type from Wilhelm Eicke & Company of Bremen, Germany, cost $2,360 f.o.b. Canada. High-performance sailplanes might run into higher brackets, a good German type anything up to $5,000, and a similar English plane around £1,000; both, of course, two-seaters. These prices do not include shipping costs or sales tax. Clubs might have been able to afford such prices if there were sufficient members, but for the individual a good sailplane came high.

In 1952, the Soaring Association of Canada was trying to establish a fund which would make loans to certified clubs at a low rate of interest to enable them to procure gliders and equipment, and certainly contributions to this fund should be an effective encouragement to gliding in Canada.

Through the Soaring Association of Canada, badges in keeping with F.A.I. rules are awarded to pilots able to earn them. The "A" badge has a single bird upon it, wings outstretched against a blue background. The "B" has two birds, and the "C" three. After them comes the Silver "C" which has three birds in flight, surrounded by a wreath; and a similar design in gold is a top award. The first three awards are for elementary gliding, but the Silver "C" and the Golden "C" are difficult to earn. Both are awarded only for soaring flight. The former requires a stipulated gain of 3,300 feet, a duration of 5 hours, and a distance flight of at least 32 miles. The Golden "C" demands a soaring flight of at least 10,000 feet above the release point, and a distance from take-off to landing in a direct line of at least 187 miles. Only a few Golden "C" soaring badges have so far been issued anywhere in the world, but Canada has the climatic conditions needed to gain such records, and she certainly has the human material. All that is required is money to back the enthusiasts, and the records will come rolling in.

The first Silver "C" ever issued to a Canadian went to Ovila ("Shorty") Boudreault of the Gatineau Gliding Club of Ottawa, who obtained it in 1949.

The free flight of sailplaning is the nearest approach that man can make to the soaring flight of the great birds. To be able to rise to thousands of feet in the air without mechanical power, and to remain there, wheeling in great circles close to the vast, rolling masses of snow-white cumulus clouds, is to pay a visit to another world—a wonderfully exhilarating experience which far more Canadians should be able to enjoy than present facilities permit.

INSIGNIA OF THE PARACHUTE CLUB OF CANADA

THE PARACHUTE CLUB OF CANADA

Since the end of World War II, the sport of parachuting and sky diving has been sponsored by the governments of several European nations, and because of this support, it has grown to immense proportions in Russia, France, Czechoslovakia, Poland, Bulgaria, Holland, and elsewhere. Interested and eligible men and women are extended free instruction and equipment and airfields have been set aside purely for use in connection with this thrilling sport. It is estimated that in the countries behind the Iron Curtain there are at least two million *civilian* parachutists. By the time the first World Parachuting Competitions were held in 1951, at Tivat, Yugoslavia, the sport was receiving considerable support in Europe, but in Canada and the United States not one single sports parachutist then existed.

Interest in the United States was sparked in 1955 by the efforts of parachute enthusiast Jacques Istel. In Canada, by 1958, three clubs —with an average total of 300 jumps per year— had become operative. The parent body of the sport in Canada is the Parachute Club of Canada, which is closely affiliated with the Royal Canadian Flying Clubs Association, and, as this is written, there are fourteen parachute and sky diving clubs in active operation throughout the country with a total yearly jump average of over 2,000.

To date, there have been five World Parachuting Competitions, all held in Europe, all but the first having been won by parachutists from Iron Curtain countries. The first, held in 1951, was won by France. Because of lack of financial support, Canada has had to take a back seat in the contests, but in 1958 a single representative, Floyd Martineau, a member of

FLOYD MARTINEAU

the St. Catharines Parachute Club, represented Canada. The competitions are now held every other year, and in 1960 Canada had four members taking part at the contest, held at Sofia, Bulgaria. They were Glenn Masterson, Daryl

Henry, Mike Thouard, and Al Coxall. They made a good showing, with points placing them in seventh position among eleven contesting countries; but once again Canada could not compete in the full sense, with two men short of the required team total of six.

The World Parachuting Competitions for 1962 are to be held in the western hemisphere for the first time, at Orange, Massachusetts. It can be expected that a strong team of Canadian parachutists will at last be able to show what they can do. Whatever the outcome, it will not be for want of trying, and it is to be hoped that the Canadian government, through its Department of Transport, will assist in every possible way to enable a fully trained Canadian team to participate, in an effort to bring to Canada the coveted Parachuting and Sky Diving Trophies.

SOMETIMES I look at the youngsters today playing with their accurately modelled plastic toy airplanes, or see the older boys absorbed in the delicate work of constructing out of balsa wood flying models in which they will later install miniature gasoline engines; and I feel a little envious. In my boyhood there were no airplane kits with detailed plans and instructions. My own room as a teen-ager in Nottingham in 1910–1912 was filled with models constructed for the most part from pictures of full-sized aircraft—most of them very poor illustrations of the pioneer machines.

All envious feelings aside, I know that it was the lads who had a consuming passion for everything that flew, who later had a finger in making (or writing about) aviation history. And so, I suspect, it will be in the future, no matter how completely model-making and flying instruction become standardized. Perhaps the day may come when high octane aviation gas will become standard diet for teething young airmen. But the mainspring of advances in aviation has always been, and will continue to be, the interest, enthusiasm, and imagination of youth. No youngster truly interested in flight is satisfied until he has been up. And it is never enough merely to be a passenger. The big moment comes when he himself is at the controls. Today the legal limit for solo flight, in gliders or planes, is 17.

OUR FIRST BOY PILOTS

The first boy to earn the honour of becoming Canada's youngest pilot was Kirkpatrick Maclure Sclanders. Born in Saskatoon, Sask., on January 2, 1916, Pat had grown up in Saint John, N.B. Long before he reached his teens he was passionately devoted to airplanes. When he was fifteen he won a contest sponsored in the United States by pointing out the greatest number of technical errors in an article on aviation.

In 1931, when Pat was fifteen and a half, he became a member of the Saint John Flying Club, commencing his dual flying lessons on August 18 under Clifford S. Kent in a De Havilland Moth, CF-CBG. He soon showed himself so capable in the air that the instructor had no qualms in allowing him to solo on October 18, just two months after he first began to fly. This flight made Pat the youngest pilot to solo in Canada. By the end of the year he had totalled 11 hours and 5 minutes' flying time. Although he was too young to receive a private pilot's certificate, he was not debarred from further solo flying. During 1932 he took an active part in the air pageants at Charlottetown, P.E.I., and at Saint John, N.B., where he entertained the large crowds with an original act. The Moth was first flown by the instructor, who landed in front of the grandstand, climbed out of the cockpit, and left the machine with the engine running. As soon as the pilot was out of sight, Pat Sclanders, dressed in a Boy Scout uniform that made him look even younger than sixteen, wandered out from the crowd and pretended to examine the plane. In a few minutes Pat climbed into the pilot's seat. Then suddenly the engine broke into a roar, and with a few crazy swerves, the Moth sped down the field and took to the air. For a time its erratic manœuvres held the spectators spellbound, for they believed the youngster had tampered with the controls and taken off accidentally. Eventually the announcer ended the tension by explaining that the young lad aboard the high-flying Moth was in reality a trained pilot. Pat returned to the field to genuine and spontaneous applause.

January 2, 1933, was the great day of Pat's seventeenth birthday, and before the month was out he received private pilot's certificate No. 1175. Now he could at last take up passengers, and on February 19 he took up his first from the Saint John air field in a ski-equipped Moth.

By the end of the year he had 31 hours and 25 minutes' flying time to his credit, and at the close of 1934 his total was 50 hours and 30 minutes.

For the next two years, young Sclanders had to give up flying and take on a job, but at the end of 1936 he went to England and enlisted in the Royal Air Force. After training, he received a commission as Flight Lieutenant. Then an unfortunate illness intervened which caused him to be invalided out of the service, and he returned home to Saint John sad at heart. After his recovery he edited a periodical in the United States connected with the oil industry.

Then came the war. Unable because of his previous ill-health to gain entry into the R.C.A.F., Pat immediately joined a group of United States pilots who planned to fly for the Finnish army. He arrived in Finland just as that particular phase of the war came to a close. He went to Paris and enlisted with the French Air Force, but was obliged to flee to England after the German invasion. For three weeks he and several other pilots dodged the Nazis afoot, until they could make their way to a French port and board a vessel going to England. By that time the demand for trained pilots was urgent, his application to re-muster into the R.A.F. was quickly accepted, and almost overnight he once again found himself wearing the Air Force blue.

Pat was soon in the thick of the Battle of Britain. On one sortie with ten other aircraft, Pat and his comrades ran into a force of 110

PAT SCLANDERS

German machines. They attacked at once, and a desperate battle took place over the English Channel. From this encounter, Pat Sclanders never returned. He is one of the "few" immortalized by the words of Prime Minister Winston Churchill as the saviours of Britain.

The second young Canadian to earn his "flying spurs" early in his teens was Wesley B. Hodgson, who was born in Regina, Sask., October 22, 1916, and was raised in that prairie city.

By means of a paper route and numerous odd jobs, Hodgson made enough money to take instruction from the Regina Flying Club, under the expert tuition of Roland J. Groome. His first dual flight was in 1932, when he was fifteen. Throughout 1932 and 1933 he slugged away, earning money to spend on flying lessons.

In June 1933, Wesley made his first solo in one of the Moth aircraft belonging to the Club, and in August he passed his flying tests before the critical examination of the government inspector, Squadron Leader A. T. Cowley, who congratulated him on his fine airmanship. Wes was obliged to wait a couple of months for his seventeenth birthday before he became eligible for his private pilot's certificate. The Civil Aviation Branch of the Department of

WES HODGSON

National Defence gave it to him as a birthday present. It was dated October 22, 1933, and was numbered 1350.

In April, Wesley purchased a Heath-type monoplane, which had been built by members in the workshops of the Regina Flying Club. In this craft he continued to fly, adding to his time in the air until August 1935, when he passed his tests for a commercial licence; for this certificate he again waited two months—this time to reach his nineteenth birthday on October 22.

During Hodgson's prairie flying career, he won the Webster Memorial Trophy zone contest for western Canada in 1935, and thus became the representative of the West in the finals at Montreal. There he lost by a narrow margin; his total being brought down by the low marks he obtained on the cross-country flight during the competition.

On the outbreak of war in 1939, Hodgson enlisted in the R.C.A.F. He served overseas for a year and ten months. Then the authorities sent him back to Canada, where, with the rank of Squadron Leader, he served in various capacities until the end of the war.

The Model Aircraft League of Canada

When the Canadian Flying Clubs Association was formed in 1929 as an amalgamation of the numerous light airplane clubs, it instituted an organization which became a kind of fairy godfather to many Canadian boys. All over the country, countless boys who were "crazy about flying" were inventing and constructing model airplanes. Among them were many whose keenness, cleverness, and devotion to hard work deserved official encouragement. The Association recognized that such encouragement to deserving youths would in later years be repaid by the development of Canadian aviation. So the Model Aircraft League of Canada came into existence.

The M.A.L.C. was affiliated with the Society of Model Aeronautical Engineers of Great Britain, and local clubs were soon being formed throughout Canada under General Sir James MacBrien, president, and Captain C. C. Hirst, secretary.

There was no scarcity of entries for the contests that the League sponsored, at first local, then provincial, and finally federal. The first of the federal competitions was held at Ottawa on July 4 and 5, 1930. Commercial and industrial firms contributed prizes, and a special award for the winner of the greatest number of points was a trip to England and entry into similar competition there.

There were two classes, junior up to the age of 16, and senior up to the age of 21.

The results of the first federal flying model meet were as follows:

Senior Class

Indoor Flying Stick Model

1st. ROSS FARQUHARSON, Vancouver, B.C., with a flight of 5 minutes. Awarded the Dale and Company Trophy.

Indoor Fuselage Model

1st. ROSS FARQUHARSON, Vancouver, B.C., with a flight of 2 minutes and 21 seconds. Awarded the Imperial Oil Trophy.

Outdoor Flying Stick Model

1st. WALTER ALLDER, Vancouver, B.C., with a flight of 18 minutes and 49½ seconds. (World record.) Awarded the Goodyear Trophy.

BADGE OF THE MODEL AIRCRAFT LEAGUE OF CANADA

ROSS FARQUHARSON

Outdoor Fuselage Model

1st. RICHARD HISCOCKS, Toronto, Ont., with a flight of 5 minutes and 45 seconds. Awarded the Aero Engines of Canada Trophy.

Non-Flying Scale Models

1st. VICTOR HILL, Vancouver, B.C., with a De Havilland Moth. Awarded the Southam Trophy.
2nd. BRANSON ST. JOHN, Winnipeg, Man., with a tri-motored Fokker. Awarded the Simpson Cup.

Junior Class

Indoor Flying Stick Model

1st. J. A. CHAMBERLAIN, Toronto, Ont., with a flight of 3 minutes and 49⅜ seconds. Awarded the T. Eaton Trophy.

Indoor Fuselage Model

1st. GEORGE RYCKMAN, Hamilton, Ont., with a flight of 3 minutes and 17⅜ seconds. Awarded the De Havilland Trophy.

Outdoor Flying Stick Model

1st. JACK PURVIS, Toronto, Ont., with a flight of 9 minutes and 26 seconds. Awarded the Fairchild Trophy.

Outdoor Fuselage Model

1st. LEIGH BEGG, Vancouver, B.C., with a flight of 2 minutes and 12 seconds. Awarded the Pratt and Whitney Trophy.

Non-Flying Scale Models

1st. G. RICHARDSON, Montreal, Que., with a Boeing Fighter. Awarded the Dunlop Trophy.
2nd. DONALD RANKIN, Toronto, Ont., with an Avro Avian. Awarded the International Paints Cup.

In 1931, the Earl of Bessborough donated an additional award, the Governor General's Trophy, to go to the winner of the highest number of points in the annual federal contests.

The Wakefield Trophy, awarded by Lord Wakefield of Hythe to the lad gaining the highest points, was won by Ross Farquharson of Vancouver, B.C., with two firsts, one second, and one third. This meant that Ross, then sixteen years of age, left for England five days after the Ottawa meet to compete in the British Empire model aircraft championships held in London.

Ross's models, however, did not win any of the contests, chiefly because they were too light for the boisterous wind that was blowing at the time. They were precision built, of very light construction. Lifting and control surfaces were paper-covered, on a finely built balsa wood framework. (Incidentally, all the models at that date were powered by elastic; the tiny gasoline motors for model aircraft had not then come into use.) So the Canadian lad returned home without trophies, but with a wealth of information and recollections of a wonderful trip.

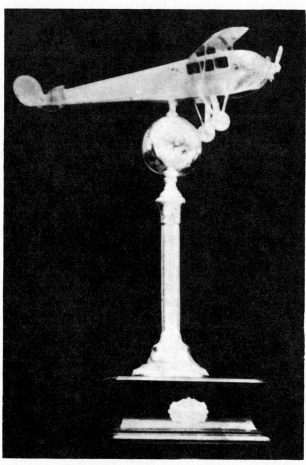

THE WAKEFIELD TROPHY

This young champion was born in Toronto, Ont., lived for some time in Hamilton, Ont., and in 1929 moved to Vancouver, B.C., when his parents went there to reside. He won many other model flying contests before he gave up the hobby in 1932. He had hoped to be able to attend university after leaving high school, but the depression lowered the family income, and a university career seemed impossible. Then one day Ross's mother saw an advertisement of the Fisher automobile body manufacturers, describing their international model coach building contest. Encouraged by his family, Ross began construction of one of the intricate coach models, but unfortunately he was unable to complete it in time to enter the 1932 competitions. He was tempted to give way to despair, for the model had cost a considerable sum to build, but friends came to his assistance and his family urged him on. Again he put his heart and soul into construction of another model, which he was able to enter in the 1933 contest. This was judged the best provincial entry and gained a $100 first prize.

Spurred by this achievement, Ross completed his third model coach after hundreds of hours of patient work and entered it in the 1934 contest. Success at last! It was a wonderful day for the Farquharson family when a telegram arrived from the Fisher Body Company in the United States, stating that Ross's entry had been judged first on the continent, and that he had won a $5,000 scholarship!

So Ross went to an American university and for the next few years studied engineering, specializing in aeronautics and industrial management. In 1937 he obtained a position with Boeing Aircraft in Vancouver. He was there until the war broke out, when he enlisted in the R.C.A.F. In the service he rose to the rank of Chief Engineer Officer, and was engaged on flight test work much of the time. After the war he continued his career as an industrial engineer.

THE ROYAL CANADIAN AIR CADETS*

Among the thousands of men and women in Canada connected with aviation, individually or in groups, no greater enthusiasm can be found than that of the teen-age lads who form

the 234 squadrons of the Royal Canadian Air Cadets. In 1952, 17,000 young Canadians were actively connected with the League's diversified activities, and over 60,000 boys over eighteen had been connected with the organization since the formation of the League was authorized by an Order-in-Council dated November 10, 1940. The Canadian charter was granted on April 9, 1941.

While full credit cannot be given here to all the people who originally fostered the formation of the League in Canada, the nation should be everlastingly grateful to "Nic" Carter of Vancouver; Foster, Illsley, and Merrill of Montreal; and MacLaren of Winnipeg, for the unstinted efforts they gave during the early years of the League.

The organization came into being during World War II, its chief objective then being to train boys to enable them to become members of the R.C.A.F. when they became of age. By September 1944 there were 379 active Air Cadet Squadrons throughout Canada, with over 29,000 cadets enrolled. To take care of this vast "young Air Force," 650 civilian instructors were required, together with over 2,000 other public-spirited citizens from every walk of life who realized they could serve their country no better than by aiding in the work of the Air Cadet League.

The result was that during the war over 3,000 air cadets enrolled in the R.C.A.F. upon reaching their eighteenth birthday, and more than a score of them were decorated for gallantry. The number recruited into the R.C.A.F. by 1952 totalled over 8,000.

The League did not slow down its activities at the close of the war. The main objective now, however, is not to train a steady flow of young men for war duties, but to fit them to become better citizens. Boys have energy to burn, and they need something to occupy their

*When the Air Cadet League of Canada was organized, boys forming the various squadrons were known as Canadian Air Cadets. However, the designation "Royal" was authorized by Act of Parliament (Royal Canadian Air Force Act) on June 26, 1946, and from that time onward, members have been known as Royal Canadian Air Cadets.

minds. No better opportunity is offered the average Canadian boy than to enrol in the Royal Canadian Air Cadets, to mingle with other lads of his own age in healthy sports, exercises, and drill, and to receive the finest guidance the country can offer towards building him mentally and physically into a good citizen.

The advantages of being an air cadet are many, and all that is required for entry is that a boy be medically fit, and between the ages of fourteen and eighteen.

Since it is an air cadet organization, the training and instruction received follows along Air Force lines. Drill there is, to be sure, but not too much of it, and discipline, the latter being a very necessary addition to the training in any group or organization where a number of boys are brought together.

Of deep interest to cadets is instruction in aero engines, airframes, and the theory of flight and aerial navigation. Wireless, meteorology, and other allied crafts connected with aviation are all taught by competent instructors, and in a form which boys can quickly understand.

Summer vacations are spent at various Royal Canadian Air Force camps, where, among other things, cadets are taken for flights in Air Force machines. There is great rivalry in the sports and swimming events. Friendships spring up between cadets from widely separated parts of the country, fostering an atmosphere of goodwill which will be of great value in the years ahead.

Annual contests in rifle shooting and an international drill competition are held. Any boy who wishes to do so many enter any activities.

During 1948, 1949, and 1950, teams drawn from Royal Canadian Air Cadet Squadrons defeated teams of United States Air Cadets in the international drill competitions. In 1951, a Scottish team won the honour, and in 1952 the American boys defeated all comers.

Should cadets desire to follow a career in the R.C.A.F. upon coming of age, scholarships are available which entitle successful competitors to enter the Canadian Service College. The value of the scholarships is $600 each, the money coming from donations by public-spirited citizens. The amount covers the cost of an entrant's first term, and if he is successful in his studies, arrangements are made for him to enrol in the R.C.A.F. during the summer, so that financial difficulties may not interrupt the course of further and final studies. In addition, scholarships for university courses in aeronautical engineering are available.

When a cadet reaches the age of seventeen, with just one more year to run before the age limit musters him out, he is given the opportunity to become a full-fledged pilot. He must be physically fit, of course, and he must pass a written qualification test in aviation subjects before he can be accepted as an applicant. The course entails a total of 60 hours of ground-school tuition, together with 17 hours' flying. Through the generous help and co-operation of the Royal Canadian Flying Clubs and local sponsoring committees, the majority of cadets, once having gained their wings under Royal Canadian Air Cadet supervision, are assisted towards the ultimate goal of thirty hours' flying which is necessary to enable a pilot to obtain a private licence.

From the formation of the League to the end of 1951, 1,190 cadets were taught to fly, and the yearly quota has been raised from 225 to 250.

During the summer months, fifty Royal Canadian Air Cadets are selected annually for exchange visits which take them as a group either to British or United States points. The selections are made purely on the merits of the lads concerned. They must have a first-class record in their squadron, and be able to represent Canada worthily when taking part in such exchange visits. The choice rests squarely on the abilities of the lads themselves. Once selected, the cadets are flown overseas or to the United States. On arrival they are entertained, taken by air, bus, or rail to visit points of interest.

What of the future?

When the youths of today are adults tomorrow, some may ask, is there anything left for them to accomplish? The pioneer period is past: even the sound barrier has been broken. But ask the youngster himself. Saturated in science fiction and space-annihilation comics, he smiles with the slightly superior smile of the very young; and for answer looks up at the stars.

PART 8

WINGS AT WORK, 1927-

33. The air-mail story

AT this point we return to the adult world, where the hard facts of life were making more and more practical demands of the airman as the twenties moved into the thirties and flying became an established branch of commercial enterprise. Business and government initiative had shown by the middle twenties that passengers, parcels, and mail could be carried by air far more speedily than by any other means, particularly over certain routes. By the thirties both passenger and air-mail service moved out of the experimental stage. Particularly was this true of air mail. In every part of Canada, sooner or later, air transport companies were organized capable of carrying out government mail contracts on schedule, and familiar with flying conditions in their locality.

Official air-mail delivery in Canada began in 1927. Any air mail carried prior to that year had been semi-official only, that is, the postal authorities had merely given permission for mail to be carried and for special stamps, stickers, etc., to be used on the envelopes. Some of these flights were special—to mark a record or celebrate an event; others, which became regular, were business arrangements with commercial flying companies. In these cases the government did not initiate the venture, assume responsibility, or foot the cost.

In 1927, however, the government decided, in view of the success of these operations and the regularity with which they had been carried out, to set aside the sum of $75,000 for the carriage of mail by air. The first flights, largely experimental, were made both ways between Montreal and Rimouski, on the lower St. Lawrence, to expedite outgoing and incoming Atlantic mail.

The service was inaugurated on September 9, 1927, the aircraft used being a Vanessa seaplane, built by Canadian Vickers, which was undergoing experimental tests at the time. It was designed somewhat along the lines of the famous American Bellanca seaplane of that date. Originally, it was fitted with an English Lynx engine, but as this motor would not stand up to cooling in the enclosed, built-in nose section, it was replaced by a 200 h.p. radial, air-cooled Wright Whirlwind.

The Department of National Defence arranged for the first pick-up of air mail from Rimouski, using the Vickers machine. The pilot who had been in charge of conducting the flying tests, J. H. Tudhope, was assigned the job of taking on the mail at Rimouski and flying it to Montreal. With him as engineer was Gerald LaGrave.

Tudhope and LaGrave flew the machine from Montreal on September 8, arriving at Rimouski during a heavy wind and thunderstorm. In spite of the high seas running, Tudhope managed to bring the craft down in the lee of the Rimouski wharf, but the flotation gear took a very severe pounding, and a strut fitting was damaged, although unnoticed at the time.

At 8:00 A.M. on the 9th, the pilot boat received 500 pounds of selected first-class mail from the incoming *Empress of France*, and the bags were transfered to the waiting aircraft.

313

THE VICKERS VANESSA SEAPLANE

Since there was still a heavy swell on the river, the Vanessa was towed by the pilot boat to calmer seas, in the lee of Barnaby Island about three miles from Rimouski. While the seaplane was taxiing at speed for the take-off, the pounding of the choppy waters against the floats caused the weakened strut fitting to give way. The fuselage immediately tipped to one side, the port wing sliced into the sea and was smashed, and the propeller ripped through the forward section of the port float, cutting it in half.

The pilot boat quickly came to the rescue, transferred the precious mail to its sturdy deck, and made fast a line to the crippled seaplane. By the time the pathetic parade had reached the Rimouski wharf, the rough seas had battered the Vanessa completely under. She was a total wreck; only the engine was worth salvaging. The mail taken off the *Empress* arrived at Montreal quite a time after the liner instead of ahead of her.

However, mishaps such as this led to improvements and corrections. It was an unfortunate ending to the first attempt, but another lesson had been learned.

The first flight to connect with an outward-bound vessel was carried out on September 12, and was quite successful. A Canadian Airways HS2L flying-boat, piloted by H. S. Quigley, was chartered for the trip. Five hundred pounds of mail were taken aboard at Montreal, and Quigley made the 330 miles to Rimouski in two hops, stopping once at Three Rivers to refuel. A short time after Quigley reached Rimouski, the liner *Doric* halted briefly off Father Point to drop the pilot and to take aboard the mail flown from Montreal. The saving in time

for such deliveries to points in Great Britain was from 24 to 96 hours. For incoming mail taken off the ships at Rimouski and flown to Montreal, the time gained in deliveries approximated 48 hours at Montreal and 24 hours to main post offices west of Winnipeg.

The first successful flight with a load of incoming mail was made on September 16 in the same HS2L flying-boat, with H. S. Quigley and Stuart Graham at the controls. The inbound *Empress of Australia* discharged 500 pounds of mail at Rimouski which was flown to Montreal with a stop *en route* at Three Rivers.

Between September 28 and November 11, seven more flights were made between the two points. There was one variation; on October 27, the seaplane piloted by H. M. Pasmore alighted at Quebec City, where a redistribution of mail was made; then it flew on to Ottawa. By November, emergency landing-fields had been prepared along the route and the last two flights were made by aircraft with wheel-equipped landing-gear. J. H. Tudhope was the pilot on these flights and his successful performance at every stage made up for his ill luck on the first flight of the series.

None of the letters carried on any of the ten flights in 1927 received any special franking, but a few were signed by the pilots. The majority of these are not of high value to collectors—an evidence that air mail was gradually becoming commonplace.

The postal authorities had been well pleased with the undertaking as a whole, and in 1928 the federal government awarded a contract to

RIMOUSKI TO OTTAWA

The first official mail plane reached the capital on October 27, 1927

Canadian Transcontinental Airways for delivery of mail over the Ottawa–Montreal–Rimouski route. The record for 1928 was 67,195 pounds of mail carried on 94 flights, totalling about 41,360 miles flown. This service continued from spring to autumn for many years until changes came about with the development of additional routes *via* Newfoundland.

The second air-mail route established under full government contract in 1927 was for winter service exclusively. The contract called for four round trips weekly during the winter months between Leamington, Ont., on the mainland, and Pelee Island, seventeen miles out in Lake Erie. Ice and weather conditions on the lake often kept the island isolated for long periods during the winter months, and it was hoped that the air service would provide a solution.

The firm of London Air Transport, Limited, was awarded the work. A Waco Ten biplane, registered G-CAJE, made the first flight to Pelee on November 30, 1927. The plane was flown by F. I. Banghart, accompanied by a director of the company who was also a pilot, William Drury. When the demand arose, passengers were flown on later flights.

Pelee Island is situated far to the south of all Canadian cities, and its municipal airport is the southernmost in Canada.

The latitude of Pelee Island's most southerly tip is further south than is the northern boundary of the state of California, which is situated along the line of 42°N. latitude. A long way, indeed, from the distant northern "air strips," surveyed by Major Logan on Ellesmere Island in 1922!

CANADIAN TRANSCONTINENTAL AIRWAYS' FAIRCHILD

Used on the Murray Bay–Seven Islands route on Christmas Day, 1927

WELCOME

Large numbers of Magdalen Islanders welcome Pilot Cooper and Engineer Francis in their Fairchild

The third air-mail contract of 1927 was also awarded to Canadian Transcontinental Airways. The Post Office Department required the mails to be flown between Murray Bay and Seven Islands, along the northern shore of the Gulf of St. Lawrence, with flights to Anticosti Island, 45 miles out in the open gulf. This, too, was a winter contract, calling for approximately 1,500 pounds to be flown weekly between the various points along the coast, with one trip per week to Anticosti and back.

The pilot, Romeo Vachon, had gone to New York and had flown a new ski-equipped Fairchild monoplane back to La Malbaie, on the north shore of the St. Lawrence at Murray Bay. He arrived there on December 21, passing over the city of Quebec where mail was dropped by parachute. This particular mail was only semi-official, the covers bearing no stamps. They were, however, officially cancelled at the Quebec post office. This bag of mail was the first recorded to be dropped by parachute in Canada, and individual envelopes are valued at $25 today.

The first and only delivery of air mail to take place in 1927 along the north shore route was on December 25, when Vachon made the run from La Malbaie to Seven Islands, handling 753 pounds of Christmas mail. Deliveries were made at various points *en route*, mostly by parachute. There were many joyful people along the north shore that Christmas Day when the flying postman came along.

The same pilot and machine made the first air-mail flight to Anticosti Island on February 8, 1928, a landing being made on the island at Port Menier.

The fourth air-mail contract of 1927 fur-

CANADA'S AIR-MAIL STAMPS

Dates of issue were:

Number 1	September 21, 1928
Number 2	December 4, 1930
Number 3	June 1, 1935
Number 4	June 15, 1938
Number 5	April 1, 1943
Number 6	September 16, 1946

nished real drama in the lives of a few Canadians. The Magdalen Islands in the Gulf of St. Lawrence had always been practically isolated during the winter. The radio, of course, provided a connection, but no news of a personal nature ever reached the islanders except when a ship occasionally got through the ice. When it was known that the government was going to send a plane in once a week to bring and take out mail, there was much happy excitement.

Most of the round trip from Moncton, N.B., to Grindstone Island, was over the Gulf. Canadian Transcontinental Airways again secured the contract. The inaugural flight actually took place on January 11, 1928; the pilot was J. Cooper.

The landing-gear used on the Fairchild aircraft flown by Cooper was unique in Canada. It was designed to permit the craft to land on water, ice, or snow. The machine was a sea-

plane with standard floats, but underneath each float was a built-in ski. The chassis struts were unsprung, but the skis were ingeniously attached to the underpart of the floats by means of a series of coil springs. They did not remain resilient for long, however, because skis, springs, and floats quickly froze solid in winter weather and remained frozen all winter. Ice- and snow-landings were extremely precarious, but Cooper always contrived to bring his machine down safely. Nine round trips were made between Moncton and the Magdalens during the first winter's operations, with occasional stops *en route* at Charlottetown, P.E.I.

The engineers who looked after the well-being of the seaplane, H. W. Francis and R. Demers, had no hangar facilities during the first season, only a makeshift nose-shed at the Moncton terminus, but they managed to meet their many problems.

An interesting human sidelight on the first flight was the way the pilot was besieged with requests to bring the islanders copies of the mail-order catalogue of a Toronto department store. As these bulky books were too heavy to be given higher than fourth-class mail rating, they could not be sent as air mail, but Cooper procured a few copies on his own account and took them over as gifts. They were eagerly thumbed from cover to cover, and passed from family to family as precious possessions.

Rigid schedules were out of the question because of the extreme weather conditions, but otherwise the plan was an unqualified success, and has been continued every winter since.

The fifth and final air-mail contract of the original group got under way on January 25, 1928, when Western Canada Airways agreed to fly the mail route between Rolling Portage (in the vicinity of Sioux Lookout) and the Red Lake area. On the first flight, 588 pounds of air mail were carried in and 90 pounds were flown out; by the end of the season, 57,566 pounds of mail had been carried and approximately 26,240 miles had been flown.

During 1928, the amount of mail carried over semi-official routes increased considerably, as did the number of such routes, and as time went on a good many official contracts were given to cover the operations. Ten air-mail routes were being operated in 1928 under full government contract with the Canadian

AIR MAIL TO AKLAVIK

Above: Loading mail bags at Fort McMurray. *Below, left to right*: Glyn-Roberts, Bishop Geddes, Colonel Hale and "Wop" May, with the first official air mail to arrive at Aklavik

Post Office Department, and of these, three had actually commenced operations during 1927.

Of the new contracts, the first, which went into effect May 5, was awarded to Canadian Transcontinental Airways, over the original Ottawa–Montreal–Rimouski route. The pilots on the initial flights were C. S. Caldwell, G. M. Dean, J. H. St. Martin, and Romeo Vachon. This service operated during the season of navigation.

The second award was a semi-weekly contract issued to Western Canada Airways Limited to fly the mail between the northern Manitoba points of Lac du Bonnet and Wadhope, *via* Bissett. It was also a summer service.

Third on the list was a yearly contract awarded to Canadian Airways Limited to operate six flights each way per week between Montreal and Toronto. The service commenced May 5. The pilot on the inaugural flight was Harold P. Ayres. By the end of the year some

44,250 pounds of mail had been flown over this route.

The international air service which began on October 1 between Montreal, Que., and Albany, N.Y., was the fourth contract to go into effect. It was awarded to Canadian Colonial Airways, and called for six trips each way per week. By the end of the year, 30,660 pounds of mail had been carried. The original pilots who opened up this route were D. S. Bondurant, W. H. Hughes, Paul Reeder, and O. C. Wallace.

The contract awarded to Western Canada Airways to fly mail between Kississing and The Pas was the fifth contract. This was a summer service operating between the two northern points once each week.

The sixth contract for the year went to the same company to operate a weekly round trip between Sioux Lookout and Jackson Manion, in the Red Lake area.

It was on September 21, 1928, that the first

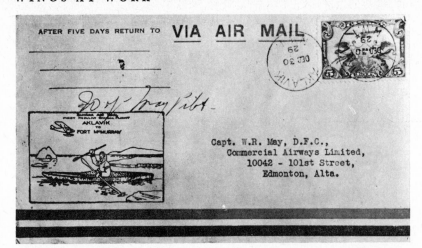

AFTER FIVE DAYS RETURN TO   VIA AIR MAIL

Capt. W.R. May, D.F.C.,
Commercial Airways Limited,
10042 - 101st Street,
Edmonton, Alta.

AN OFFICIAL FIRST FLIGHT
ENVELOPE

Addressed to, and signed by, the pilot

governmental issue of a regular air-mail stamp became effective. The design was for a 5-cent stamp to be used on all flown letters weighing not more than one ounce.

Two special authorizations for the carriage of air mail were also granted by the government, one in the Yukon in 1927, the other in the Northwest Territories in 1929, but neither was under contract. The first allowed the Tredwell Yukon Company Limited to fly mail between White Horse and Dawson; the other was issued to Western Canada Airways Limited, granting permission to carry mail between Fort McMurray and Fort Simpson. Both companies were already flying passengers and freight and the addition of mail brought them increased revenue.

After 1928, the carrying of air mail under government contract began to be extended widely across Canada. In 1929, 2,162 single trips were successfully accomplished by aircraft flying government mail, and 430,636 pounds of mail were carried.

Although carrying the mail by air was now a routine proceeding, it remained a colourful enterprise. The most advertised air-mail flight of the year was the one which took place towards the end of 1929, carrying mail from Fort McMurray under full government contract to points north all the way to Aklavik, on the Arctic coast. The work was undertaken by Commercial Airways Limited, of Edmonton, and the two aircraft which accomplished the job were flown by W. R. ("Wop") May and I. Glyn-Roberts. The mail reached Aklavik on December 30, and was the first fully "official" mail to reach the Canadian Arctic coast. Twenty-six distinctive *cachet* designs

were used on the envelopes addressed to the various destinations *en route,* which were Fort Chipewyan, Fort Fitzgerald, Fort Smith, Fort Resolution, Hay River, Fort Providence, Fort Simpson, Wrigley, Fort Norman, Fort Good Hope, Arctic Red River, Fort McPherson, and Aklavik. First flight collectors from all over the continent, and many points beyond, sent in mail, and by the time the last bag was packed the Edmonton post office had received 120,000 letters, totalling four tons in weight! The Commercial Airways Company had to acquire three new planes—two Bellanca cabin monoplanes, and a five-passenger cabin Lockheed Vega monoplane, all ski-equipped.

Even with this fleet, relays had to be made to transport the great bulk from Edmonton to Fort McMurray. The flight from Fort McMurray over the 1,676-mile route to the north began on December 10, 1929. Seventeen days were required to fly the load to its various destinations, and most of it had to be flown out again to be returned to the collectors who had sent it.

At Fort Simpson, where the airmen spent Christmas Day, the mercury dropped to 60° below zero. It was 44° below when the airmen took off the next morning. The engineers, Sims and Vanderlinden, had kept the engines in running condition throughout the night, right out in the open in the sub-zero temperature. They used blowtorches and plumbers' fire-pots, with flashlights for illumination.

By the time the two airmen had delivered the last bag to Aklavik, a total of 6,000 miles had been flown.

The result for philatelists was ironical. Too many letters had been carried for any one of them to be a rarity! Despite the risk, work, and expense involved in flying them, single specimens today are worth only 50 cents, and only $5 if signed by both the pilots.

From December 10 to December 29, 1928, an experimental daylight service was operated by Western Canada Airways between Winnipeg, Calgary, and Edmonton to test the practicability of rapid air-mail transportation and to gauge public response to the innovation. The results proved conclusively that night flying was essential, so the government proceeded to install rotating beacons, radio, and other aids to navigation.

In 1930, the Prairie Air Mail began to function, when Western Canada Airways became the Western Division of Canadian Airways Limited and received the contract. Ten veteran pilots were assigned to the job: Winnipeg to Regina, W. J. Buchanan and G. A. Thompson; Regina to Calgary, F. R. Brown, A. H. Farrington, A. E. Jarvis, and H. Hollick-Kenyon; Regina to Edmonton, M. E. Ashton, Paul Calder, W. N. Cummings, C. M. G. Farrell, and D. R. MacLaren.

The route between Winnipeg and Calgary was a night run. Every ten miles along the way beacon lights were established, while at seven strategic points powerful searchlights were placed, the beams from them being visible 90 miles away in clear weather. Revolving searchlights marked the main air fields at Winnipeg, Brandon, Regina, Moose Jaw, Medicine Hat, Calgary, Saskatoon, North Battleford, and Edmonton. A score of emergency fields were also marked out. These were the first scheduled night flights in Canada.

The trips made between Regina and Ed-

POSTMASTER TO PILOT
The Edmonton postmaster handing the registered mail for Regina to Paul Calder, March 3, 1930

monton, *via* Saskatoon and North Battleford, were scheduled for daylight flying.

The final link in the prairie air mail came when an extension route was established between Calgary and Lethbridge. The inaugural flight took place on January 15, 1931, when Hollick-Kenyon flew the first trip in. (Kenyon Field, Lethbridge's present-day airport, was named in honour of him.)

These prairie air-mail routes functioned until 1932, when the entire project was discontinued, chiefly because of the cut in postal estimates as a result of the Depression. Nevertheless, the use of air mail continued to increase, slowly but steadily, on other routes, of which in 1935 there were 26 under government control. As poundage increased mileage flown decreased. Whereas in 1930, 1,608,687 miles were actually flown in transporting 474,199 pounds of mail, in 1935 only 734,737 miles were flown in transporting 625,040 pounds.

As the years passed, people wondered how long it would be before regular flights would begin across the Atlantic with passengers and mail.

Much was written and planned about transocean flights, their possibilities and safety, but

LETHBRIDGE JANUARY 15, 1931
The first official air mail arriving

Above: THE *CALEDONIA*
Arriving at Botwood, Newfoundland,
July 6, 1939

Below: THE *CLIPPER III*
Also at Botwood on July 5, 1939

FIRST TRANSATLANTIC
AIR MAIL

THE SHORT-MAYO
COMPOSITE

The *Mercury* rides on the back of
the *Maia* at Foynes, Eire

design lagged and the years slipped by. The decade of the thirties was well on the wane before the first experimental flights with fully commercial ends in view, were begun between Britain and Canada. At that time it was considered that only multi-engined flying-boats could safely be used, particularly as it became clear that Atlantic air journeys would have to be made by as direct a route as possible in order to speed up the schedules. So it came about that the British company, Imperial Airways, instituted its initial Atlantic flights using two huge four-engined Short flying-boats, the *Caledonia* and the *Cambria*. The former arrived at Montreal after the inaugural crossing on July 8, 1937. Experimental flights continued through 1938 and into 1939, until the English company was fully satisfied that trans-ocean flights were feasible and could be commercially successful. During the same period, Pan American Airways were also conducting similar Atlantic flight experiments, using large four-engined Sikorsky Clipper flying-boats.

In June 1939 the *Caledonia* flew from Foynes, Eire, carrying her first load of mail from east to west, and on her return trip from Shediac, N.B., and Botwood, Nfld., on June 24, 1939, she carried the first west-to-east regular mail flown by a British aircraft.

By mutual understanding the two big companies began regular air-mail service on July 5–6, 1939. The Imperial Airways' *Caledonia* completed the east-to-west hop of 1,990 miles from Foynes, arriving at Botwood after being in the air 15 hours and 9 minutes. At the same time Pan American's *Clipper III*, having left Botwood on the 5th, was winging her way across in the opposite direction and reached Foynes in 12 hours and 40 minutes. The two huge craft passed each other almost in mid-Atlantic at 11:00 P.M. on the night of July 5. In touch by radio, each knew where the other was, but as they were 67 miles apart they did not glimpse one another.

The *Caledonia* was under the command of Captain A. S. Wilcockson, with a crew of four. She was flown to Montreal on July 7, going on to Port Washington, N.Y., on the 8th, to complete the flight schedule. *Clipper III* was under the charge of Captain H. E. Gray and a crew of six.

These two flights were the true beginning of scheduled Atlantic air mail by heavier-than-

air craft, and a short time later passengers were carried as well.

In 1938 another major undertaking was conducted by Short Brothers, when a unique method of launching a heavily loaded seaplane was put into practice in an effort to reduce the cost of Atlantic transport by air. For months the firm had been experimenting with what came to be known as a "pick-a-back" launching system. Major Mayo of Short Brothers was the inventor, and the design was described as the Short-Mayo Composite. It comprised two aircraft, the lower component part being the *Maia*, a large four-engined Short flying-boat. The upper component, the *Mercury*, a seaplane, also with four engines, could be released in mid-air. The idea was to provide a means of launching a small fast aircraft which was so heavily loaded with fuel and cargo that it could not take off by itself in the normal way.

After exhaustive trials had been conducted, the long-planned flight of the *Mercury* across the Atlantic took place. Under command of Donald C. T. Bennett, pilot, and Albert Coster, co-pilot and radio operator, the seaplane was released from the *Maia* at Foynes, Eire, to make the crossing on July 20–21, 1938. She was flown non-stop direct to Montreal, carrying chiefly newsreels of the King's visit to Paris, together with consignments of English newspapers, which went on sale in Montreal and New York the day after publication. No mail was carried on this trip. The *Mercury* was flown on to Port Washington, N.Y., on July 21, as the final gesture in completing the Atlantic flight.

As there was no *Maia* available on this side of the Atlantic to aid in launching the *Mercury*, the seaplane had to be flown back to England by shorter stages and without freight, *via* Botwood, Nfld., the Azores, and Portugal. The return journey began on July 25, 1938, from New York, the *Mercury* leaving Botwood for the Azores on the 26th.

The Atlantic flight in July was the only one made to this continent by the *Mercury*, but in October the aircraft was used after a similar launching from the *Maia* to establish a world seaplane record for a non-stop flight from Dundee, Scotland, to South Africa, approximately 6,000 miles. On this spectacular hop,

FIRST YUKON AIR MAIL

the plane carried nearly 7,000 pounds—much more weight than on the Atlantic flight; she lifted in fact, three-quarter times her own tare weight. Again, in December 1938, with another assisted take-off from the *Maia*, Bennett and Coster flew her non-stop from Southampton, England, to Alexandria, Egypt, carrying the Christmas mails. This was the first non-stop night commercial flight between England and Egypt carrying such a payload.

After all her flights the *Mercury* was obliged to return home by easier stages as there was only the one *Maia* designed to help her off and of course the latter always remained in England.

The pick-a-back method of launching a plane to fly long distances was undoubtedly a clever idea and at first aroused enthusiasm; but it soon became apparent that it was uneconomical and that passengers would have an aversion to being launched from another aircraft in mid-air. The scheme was soon discarded.

The big four-engined flying-boats continued to make many Atlantic crossings until well into World War II, when the numerous war restrictions brought such flights to an end. During and after the war, four-engined landtype aircraft took up the job, plying the Atlantic and many other routes with greater speed, larger payloads, and more certain schedules.

Meanwhile, in Canada, one of the most important air-mail schedules was instituted in 1937 when United Air Transport, Limited, began mail and scheduled passenger flights between Edmonton, Alta., and White Horse, Y.T. The initial trip took place on July 7, following a route *via* Fort St. John, B.C., Fort Nelson, B.C.,

and Watson Lake, Y.T. The machine was flown by Grant W. G. McConachie, who was the president of the company. In 1938 an additional step was taken to provide a direct service into the territory from Vancouver. The Ginger Coote Airways amalgamated with United Air Transport, and a new company emerged, Yukon Southern Air Transport, Limited, with Grant McConachie as president.

Original schedules provided for picking up mail from the train at Ashcroft, B.C. and flying it to Fort St. John where it connected with the Edmonton–White Horse route. At this time the air mail was only semi-official, not being flown under government contract.

Direct air-mail flights from Vancouver to the Yukon under post office contract began on August 4, 1938, when two aircraft of the company left Vancouver air harbour for Fort St. John. A twin-engined Fleet seaplane was flown by Grant McConachie with Ginger Coote as co-pilot; a second machine, a Norseman seaplane, was flown by Sheldon Luck. Twenty-six thousand air-mail letters which the Post Office Department had received for this first flight went north with the aircraft, which left Vancouver at 7:00 A.M. Brief stops were made along the way to discharge mail bags at Williams Lake, Quesnel, and Prince George; and both craft finally reached Fort St. John at 2:00 P.M. McConachie then flew the remainder of the mail to White Horse over the already established route, arriving at the northern terminus at 7:00 P.M., where the mail posted at Vancouver the night before was immediately distributed. A tri-motored Ford monoplane then flew the mail on to Dawson, landing at 11:30 P.M. at the northern town, where it was still quite light at that time of the year. In the gold rush days of 1898, months of rough going were required to reach that distant point in the Yukon; the 1938 flight made it in 12 hours.

The culmination of all these air-mail enterprises came with the formation of Trans-Canada Air Lines in 1937.

CANADA'S GREAT AIRLINES

Canadians have reason to be proud of the two vast air networks that have been built up since the middle thirties. To tell the full story as it should be told would be to recount again much that has already been told—and more. It is the story of the airmen who returned from World War I, bought a biplane, and set up in the flying business for themselves. It's the story of the fly-by-night bush outfits that organized a service, cracked up a plane or two, and folded their wings. It is the story of air transport companies financed by mining syndicates and far-seeing investors who established small but permanent airlines wherever they would pay. And finally it is the story of the government itself, and the huge C.P.R. interests, buying up and reorganizing a hundred established or precarious air transport lines.

INSIGNIA OF TRANS-CANADA AIR LINES

Trans-Canada Air Lines

Trans-Canada Air Lines, Canada's huge publicly owned air transport system, was born by act of Parliament on April 10, 1937, and grew rapidly to become one of the greatest airlines in the world.

H. J. Symington, K.C., of Montreal was the first president. O. T. Larson of Winnipeg was vice-president, and R. F. George of Winnipeg was operations manager. T.C.A. was originally capitalized at $5,000,000, and 50,000 shares are held by Canadian National Railways. The first fleet of aircraft put into operation consisted of 12 Lockheed 14H ten-passenger machines, and 6 Lockheed Loadstars, each seating fourteen passengers. The original flying personnel included 51 captains, 49 second

TRANS-CANADA AIR LINES

Some of the officials responsible for the development of the system. *Left to right*: Don Saunders; Lew Parmenter, flight engineer; F. I. Banghart, Montreal Airport Manager; W. H. Hobbs, Secretary of T.C.A.; H. J. Symington, President of T.C.A.; Hon. C. D. Howe, Minister of Transport; J. H. Tudhope, pilot; Commander C. P. Edwards, Deputy Minister, Department of Transport; J. D. Hunter, co-pilot; John A. Wilson, Controller of Civil Aviation; G. Wakeman; Don R. MacLaren

TRANS-CONTINENTAL AIR MAIL 1939

officers, and 36 stewardesses, while the entire staff of T.C.A. then totalled approximately 1,200 persons.

Air strips for emergency use were established at 100-mile intervals across Canada from Halifax to Vancouver, Edmonton to Lethbridge, Toronto to Windsor, Toronto to New York, and Vancouver to Seattle. The Department of Transport provided a 24-hour meteorological and radio range service along all the routes; airports and range stations being linked by teletype for collection and dissemination of weather reports for flight control. A teletype circuit of 4,588 miles provided instantaneous communication among all stations and traffic offices.

Following incorporation, T.C.A. took over the service between Vancouver and Seattle which hitherto had been operated by Canadian Airways Limited. T.C.A. also purchased the two Lockheed Electra aircraft which had been used and the service continued without interruption.

During 1937 and into the early months of 1938, experimental and training flights were conducted over the entire route. Hon. C. D. Howe, Minister of Transport, who was head of the organization, made a trip from Montreal to Vancouver as a check-up of the system from end to end. The minister was accompanied by two other officials of the organization, Commander C. P. Edwards, Deputy Minister of Transport (Air), and H. J. Symington, K.C., Director of T.C.A. and of the Canadian National Railways. The pilots of the Lockheed Electra were J. H. Tudhope and J. D. Hunter; Lew Parmenter was the engineer. The party left the Montreal airport at one minute after midnight on July 30, 1937, but were hardly away when a terrific thunderstorm broke and they had to return. Off again at 1:18 A.M., their stops were Gilles, Ont., Sioux Lookout, Ont., Winnipeg, Man. (at 10:00 A.M.), Regina, Sask. (12:35), and Lethbridge, Alta. (3:04 P.M.). Finally, after a quick trip over the Rockies at twelve thousand feet, they made a good landing at Sea Island airport, Vancouver, at 6:29 P.M.

Experimental flights between Montreal and Vancouver started December 1, 1938; coast-to-coast air-mail and express flights were inaugurated on March 1, 1939, and passenger service on April 1, 1939. The route was extended to Moncton, N.B., on February 15, 1940, and to Halifax, N.S., on April 1, 1941. Passenger, air-mail and express service between Toronto and New York began on May 10, 1941.

In the western section of T.C.A. alone, between Winnipeg, Vancouver, and Seattle, the planes carried 7,806 pounds of express, 367,734 pounds of mail, and 2,086 passengers in the first year of operation.

Experimental air-mail operations became effective over the entire system on March 1, 1938, and the passenger schedules from coast to coast followed a short time after.

Full-scale overnight trans-continental air mail was begun on March 1, 1939. To commemorate the event, 40 pictorial designs were used for *cachets* to mark the various envelopes flown between different destinations. The volume of mail which gathered for this flight was beyond expectations, but it was all carried.

The returns for the system during its first year of operation to March 1, 1940, were remarkable. Approximately 85,000 hours were flown and over 19,600,000 pounds of freight and express were carried, exclusive of some 1,860,000 pounds of air mail, and 105,121 passengers.

As traffic increased, T.C.A. began to use the larger four-engined North Star aircraft, built in Canada by Canadair, Limited, of St. Laurent, Que. Numerous schedules have been added to all the routes, and mail, express, and passenger traffic have swelled far beyond original expectations.

In 1938, the McKee Trophy, highest aviation award in Canada, was awarded to P. S. Johnson as representing T.C.A., particular emphasis being placed on the work of the maintenance crews in determining the award.

T.C.A. was in full operation during World War II and contributed tremendously towards the final success. The flying personnel, maintenance crews, and administrative and office staffs can well be proud of their showing at that strenuous time. The full story of T.C.A. cannot be told here, but when it is written, it will be a glowing record of Canadian achievement in aviation. Gordon R. McGregor, Webster Trophy winner and celebrated military airman, is, as previously mentioned, president.

INSIGNIA OF CANADIAN PACIFIC AIRLINES
This Canada Goose is painted on all C.P.A. airliners

Canadian Pacific Airlines

The story of Canadian Pacific Airlines did not begin until after 1939, and hence might be considered beyond the scope of this book. However, some account is needed to complete the story of many of the pioneer flying companies whose activities have been described in preceding pages.

In 1939 the Canadian Pacific Railway, which had had a financial interest in Canadian Airways since 1930, made a survey of the privately owned airlines in Canada, some of which were having a difficult time. During 1940–1941, the Canadian Pacific Railway gradually acquired controlling interest in the majority of the feeder lines in western and northwestern Canada and these were integrated as Canadian Pacific Airlines. The companies absorbed included Yukon Southern Air Transport; Mackenzie Air Services; Prairie Airways; Starratt Airways; Wings Limited; Canadian Airways Limited; Quebec Airways Limited; Dominion Skyways Limited, and Arrow Airways Limited. The majority of the personnel of the various firms became employees of C.P.A. and many of Canada's top-ranking pilots and air technicians thus came into the company's service.

World War II was already in progress when the change took place, and C.P.A. inherited from many of the private flying companies the responsibility for operating Air Observer and Elementary Flying Training Schools, under the British Commonwealth Air Training Plan. Tens of thousands of recruits drawn from Canada and allied countries throughout the world first passed through these schools before going on to higher training in service schools under the direct supervision of the R.C.A.F.

Another important enterprise entrusted to C.P.A. during the war was the establishment and maintenance of five huge overhaul and repair plants in connection with military aircraft and engines of many types.

C.P.A. played an important part in the Air Transport Auxiliary which was organized in 1940 to ferry aircraft, built in the United States and Canada, across the Atlantic. As soon as the call went out for experienced flyers, a steady stream of pilots of various nationalities began arriving at headquarters in Montreal. Some of them were flyers of international fame. A later call for radio operators also found ready response. The majority of these were Canadians, drawn from the Radio Branch of the Department of Transport.

On November 10, 1940, the first transport flight, consisting of seven Hudson aircraft, took off from Newfoundland, landing in Britain 10 hours later. This was the forerunner of thousands of similar flights, in which tens of thousands of aircraft were delivered overseas to take their part in the fray.

When the Battle of Britain was at its height in 1940, the operations of the Air Transport Auxiliary were greatly enlarged. Under C. H. "Punch" Dickins, then on the staff of C.P.A., extensive new airport facilities were provided, especially at Gander, Nfld., where the volume of planes congregating for Atlantic take-offs had grown in proportion to the general increase in aircraft production.

The job of administering the Atlantic aircraft was taken over later in the war by the R.A.F. Ferry Command, which continued the immense enterprise until the end of the war.

The tradition of enterprise and service set by the men who established the pioneer air companies is splendidly maintained by Canadian Pacific Airlines today in technical research and far-flung extension of services.

Although C.P.A. does not compete along routes served by T.C.A., it has many connections along the way, and flights on scheduled and non-scheduled routes reach out to distant points in Canada and overseas. The Canadian north country is well served by C.P.A. aircraft, and regular schedules are now in full effect to Alaska, Mexico, the Orient, and Australia. The latter run began in mid-July 1949, when C.P.A.'s Canadair Four, the *Empress of Sydney*, inaugurated the service, flying from Vancouver to Sydney *via* San Francisco, Hawaii, and Fiji.

The Orient service was established by C.P.A. three months after the commencement of the Australian route, and follows a course around the North Pacific to Tokyo, and on to Hong Kong. The Korean War brought increased demands, and Canadair-built four-engined aircraft owned by C.P.A. steadily operated on the Korean air lift. Personnel and vital supplies were flown on the westward trips, and wounded men and returning personnel on the homeward journeys.

The vast operations of Canadian Pacific Airlines are directed by Grant W. G. McConachie, its president, a famous veteran of commercial flying in the Canadian northwest.

In Canada today, air mail and air travel have become commonplace. All letters up to one ounce in weight mailed in Canada for delivery in Canada are carried by air in all cases where scheduled airlines operate. Business men and holiday travellers automatically include air transportation when they plan their routes or tours.

The history of commercial Canadian aviation has been generally—though not entirely—free of serious accidents involving fatalities. Both of Canada's huge airlines have established enviable records for safety and reliability. That this is so is not a matter of chance; rather it is a heritage from those magnificent pioneers of commercial flying—the bush pilots. Today, too, it is a consequence of the heritage of superb training of flying personnel and engineering staffs, and traditions have already been established that point to a great and certain future.

34. Wings on many missions

THE golden age of bush flying is, perhaps, already fading into the past, to make way for new ages whose nature we may only guess. And yet through the thirties and into the forties—and even today—the number of actual flights into the roughest and most isolated areas has kept increasing, long past the point where it would be possible to tell their story in any kind of detail. Here, as in many other instances, we may only select a few stories that must stand for all the varied experiences and achievements by air in Canada's hinterland from the end of the twenties to the outbreak of World War II.

IN AID OF THE LAW

It is only since 1937 that the Royal Canadian Mounted Police have had their own planes and pilots. Before that date the Force was assisted in remote and uninhabited areas either by the R.C.A.F. or by planes chartered from commercial firms.

During the 1920's a good deal of such flying was done along the Pacific Coast in an effort to stop illegal hunting and fishing. In the north, planes assisted in the investigation of all kinds of law-breaking, from fur-stealing and wife-beating to murder. The most sensational manhunt in which a plane played a part was the tracking down of "the mad trapper of Rat River," a white man named Albert Johnson, who led the police a 29-day chase over Arctic wilds in the midwinter of 1932.

It began when A. N. Eames, Commanding Officer of the R.C.M.P. at Herschel Island,

received complaints from Indians that someone—they suggested Johnson—was robbing their traps along the Rat River. When Constable King was sent from Aklavik to investigate, Johnson in threatening terms refused to open the door of his cabin. King returned a few days later with another constable. This time the trapper answered with shots, one of which wounded King in the chest.

On January 10 several policemen marched on the cabin, but in the 15-hour gun battle that followed, the trapper could not be dislodged. When a still larger patrol returned a few days later, the cabin was empty.

Then began a four weeks' long pursuit over some 600 miles of rugged Arctic terrain, in the course of which Johnson, from ambush, shot and killed one of the Mounties. He was zigzagging craftily towards the mountains, sometimes removing his snowshoes and letting his tracks mingle with those left by a caribou herd, but the police, with the greatest difficulty, managed to keep on his trail. Gradually they travelled so far from their base that they were threatened with a shortage of food and ammunition. It was at this stage that an airplane was called into use, both to bring up supplies and to help to spot the fugitive.

"Wop" May of Western Canada Airways was given the job, and he took off from Fort McMurray on February 3, in a ski-equipped Bellanca monoplane, CF-AKI, with a supply of dynamite, tear-gas bombs, and ammunition. May flew the 1,000 miles to Aklavik, where he took on the much-needed supplies for the men, and another constable; then he took off for the distant trail. Afterwards, the airman made several round trips between Aklavik and the slowly advancing patrol. Johnson hid when the plane passed over him but May was able to report his tracks to the posse. One day the posse suddenly came upon Johnson in the open, as they rounded a bend on the Eagle River. The hunted man immediately dropped and opened fire on his pursuers, seriously wounding one of them; then the rifles of the party sent a withering fire back, and Johnson was killed. May flew the wounded policeman back to Aklavik, then returned and brought out the entire posse in a series of flights.

The dead trapper was discovered to have over $2,000 in his possession, as well as a hoard of gold teeth, which made the police suspect that he had murdered numerous persons; but no proof was ever found.

This and other experiences proved that the R.C.M.P. should have a flying patrol of their own. Much of the credit for formation of the Aviation Section of the Force is due to the foresight of Major General James H. Mac-Brien, who became Commissioner of the R.C.M.P. in 1931.

Sixty-four years after the Mounted Police force was first established, the Aviation Section of the R.C.M.P. at Toronto acquired four De Havilland twin-engined Dragonflies. For the first year the four planes were used almost entirely on preventive patrols working in co-operation with the Marine Section of the R.C.M.P. based at Moncton, N.B. The air patrols covered the Gulf of St. Lawrence, the

THE *CLAIRE*

The Tredwell Yukon aircraft on the Liard where first authentic word of the missing Burke party was obtained

Bay of Fundy, and coastal waters off Nova Scotia. In 1938 a fifth machine was acquired, a float-equipped Noorduyn Norseman, procured especially for bush flying operations, but it did not go into actual service until the summer of 1939. In the same year one of the Dragonflies was moved from the Atlantic patrol work to Regina to transport personnel of the crime investigation laboratory there.

The first captains of the Aviation Section were all drawn from personnel of the Force. They were Sub-Inspector T. R. Michelson, Staff Sergeant M. P. Fraser, Sergeant R. H. Barker, Sergeant W. Barnes, and Sergeant W. R. Munro.

Since that time, the R.C.M.P. have used aircraft extensively in many phases of their work, especially in the north, but it seems unlikely that planes will ever make canoes and dogs obsolete in the northern wilds.

THE AFFAIR OF THE MAD BALL PLAYER

Here I cannot resist the temptation to digress briefly in order to tell of an incident that involved the police, but far from happening in the wilderness, it occurred over the city of Toronto, in 1935. In September of that year the Brooklyn Dodgers baseball team was *en route* back to New York by plane, when Len Koencke, one of their outfielders, was put off the plane at Detroit, charged with intoxication. Going over to the Canadian side of the boundary, Koencke chartered a plane from a Canadian company, giving instructions that he wished to be flown to Buffalo. With him as his guest was Irwin Davis, a well-known "batwing" parachute jumper.

Thousands of feet above Toronto, the ball player suddenly attacked Davis, who was smaller than he, beating, biting, and choking him. As it was apparent that Davis was in danger of being killed, the pilot, Bill Mulqueeney, left the controls, and seizing a handy fire extinguisher, beat the raving Koencke into submission. As soon as he could, the pilot landed on the outskirts of Toronto to report the affair to the police. It was then discovered that Koencke was dead.

The law required that both the pilot and the parachutist be charged with manslaughter, but they were honourably acquitted at the trial.

TRAGEDY ON THE LIARD

Few stories of the north hold more pathos or heroism than that of the disappearance of the Burke party in the Yukon wilderness, the search by air, and the final rescue of two survivors.

At the time when Burke and two companions vanished *en route* from the Liard post to Atlin on October 11, 1930, any flying in that rugged, unsurveyed and unmapped area was beset with tremendous risk. Indeed, it still is.

A prospector, Robert Martin, had chartered a float-equipped Junkers monoplane at Atlin, and had been flown to the Liard post (Lower Post) on October 10 by Paddy Burke, accompanied by Emil Kading, engineer. On the 11th, the men left on a flight which should have carried them back to Atlin *via* Teslin, Gladys, and Surprise lakes, but before they could fly over the mountains they were enveloped in a dense snowstorm, and Burke turned back to the Liard River to avoid getting lost. As the storm increased, he decided to land on the river, and once safely down they spent the night there.

It was still snowing heavily when they took off the following day, and fifteen minutes later Burke was again obliged to seek a landing on the river, as visibility was just about zero. Then misfortune overtook them. Submerged rocks ripped into the bottom of the pontoons, and Burke was obliged to run the craft to the shore. But precipitous banks made it impossible to get the aircraft clear of the water and it was equally impossible to effect immediate repairs.

Their predicament gave no immediate cause for alarm because Burke had left word with his wife at Atlin to notify the owners of the machine, the Air Land Manufacturing Company of Vancouver, B.C., if at any time he was too long overdue. A rescue plane could be expected within a week, and the men did not worry; but to be on the safe side, they went on rations of three cupfuls of beans or rice a day per man, together with a little tea. Their emergency supplies were small and they were ill-equipped for a lengthy stay. Only one of the three eiderdown sleeping bags was of winter weight. They had an axe and a 30-30 rifle (with twelve shells), together with a few cooking utensils. Grub consisted of 51 pounds of beans and rice, 3 tins of bully beef, 1 pound of tea, 3 pounds of sugar, 3 tins of dried vege-

THE RESCUERS
Left: Joe Walsh and Everett Wasson

tables, 2 pounds of butter, 3 pounds of raisins, and 6 chocolate bars—scant fare for three men stranded in the mountainous wilds with winter approaching fast. Two hundred feet of rope and the engine tools completed their entire stock.

From the start, Kading hunted each day but saw no signs of game. After a week the food had diminished alarmingly and a consultation was held. The engineer suggested making a raft to float to the nearest habitation, many miles downstream. The logs would have to be bound together with rope which they judged would not last long in the rock-strewn river. As neither Burke nor Martin could swim, the idea was abandoned. It was then decided to leave a message explaining that they had set off on foot for "Junkers Lake" at the Liard headwaters, where they had made a good

329

THE DISABLED JUNKERS DOWN
ON THE LIARD

Guide Joe Walsh stands by the tail
of the craft which was almost hidden
from the air by a deep covering of
snow

cache of food during the summer. From there
they planned to go on to Wolf Lake, where
there was an Indian encampment. They carved
a message on a tree, and on October 17 they
set off.

The river was still open, and they had to
travel along the bank, ploughing through the
deep snow as best they could. Progress was
painfully slow. Lugging along their sleeping
bags, rifle, axe, the remainder of their food,
and an empty fuel can for boiling water, they
made less than four miles the first day. Day
after day they toiled in this fashion, until, at
the end of a week, Burke played out. Then
they decided to just sit tight and hope for the
best.

On October 24 they consumed the last mor-
sel of food, and from then until November 15
their only food consisted of one small duck
and four tiny pine squirrels!

Kading continued to hunt every day, but
deep snow prevented travel far from camp. No
tracks were seen; the white and lonely forest
seemed to hold nothing but the three gaunt
men. On November 15 a caribou wandered to
within two hundred feet of their camp. With
extraordinary marksmanship, considering his
weakened condition, Kading brought the
animal down with a single shot. Burke was
failing fast; his companions made nourishing
soup for him, but it was too late. On November
20, with only the faltering glow of the camp
fire to cast its weak rays in the blackness of

the night and of the forest—his friends standing
helpless by his side—Paddy Burke died.

With what strength they could muster,
Martin and Kading fashioned a crib of logs,
and placing the pilot's remains inside, with
his sleeping bag as a shroud, they muttered
a brief prayer and covered him up. They hung
his boots on a near-by branch and left a note
in one of them: "Paddy Burke died November
20, 6:30 p.m. cause: sickness from lack of food,
having been 23 days without same. Please
pardon our poor efforts as we are in a sinking
condition. Expect to leave here Saturday,
November 23, for Wolf Lake, following the
Liard River until Caribou Creek. Hope we can
make same. Snow very deep and no snowshoes. Bob Martin, Emil Kading."

Packing along what meat they could carry,
they set out, but they could do only five miles
before they played out completely.

Throughout these weeks of ordeal, no sound
of aircraft engine had come to them through
the winter silence. On November 24, for the
first time, a plane was spotted, flying very
high. Caught on the trail, they were unpre-
pared with a signal fire, and by the time one
was lighted and a thin curl of smoke rose
through the heavy timber, the aircraft had
passed from sight.

Now almost too weak to obtain wood for
their fire, they conserved energy by staying
in their sleeping bags from 2:00 P.M. each day
until 8:00 A.M. the next morning. Martin's toes

had become seriously frostbitten; both men were wearing ordinary leather boots; quite unfit for such conditions.

As they lay in their sleeping bags eighteen hours out of every twenty-four, surrounded by the deep silence and the bitter cold, without hope, and far from their fellow men, their thoughts can scarcely be imagined; yet they bore up under the strain as only fictional heroes are supposed to do. One recalls the noble words with which the explorers Amundsen and Lincoln-Ellsworth maintained their morale when marooned in the Arctic in 1925: "From hopes cut down, across the world of fears, we gaze with eyes too passionate for tears, where faith abides, though hope be put to flight."

When her husband failed to return to Atlin, Mrs. Burke had notified the Vancouver office; and after some delay a machine set off, piloted by R. I. Van der Byl, with W. A. Jorss and T. H. Cressy, engineer, aboard. It managed to get as far as Thutade Lake, north of Hazelton, B.C., but no further.

The Air Land Manufacturing Company then requested the Tredwell Yukon Company of White Horse to take up the search. They also got in touch with a pilot named Dorbrandt of International Airways, who was *en route* from Seattle, Wash., to Cordova, Alaska. Dorbrandt reached Atlin on October 26, and made a number of flights in the vicinity in an endeavour to locate the missing Junkers. Unfortunately, he was obliged to abandon his efforts, as numerous prospectors were awaiting him at various isolated points, whose well-

being depended upon being flown out on schedule.

Arrangements at last were made for Everett L. Wasson, pilot of the Tredwell Yukon Company, to take up the search, and he began a series of hazardous flights which continued right into December.

In the meantime, Sam Clerf, a prospector friend of Burke and Martin, had reached Vancouver by steamer from Skagway, *en route* to San Francisco to see his wife and new-born son. On learning that his friends were missing, he cancelled his southern trip, chartered a plane from Alaska Airways, and set off on a search. With him were Pat Renahan, a Canadian pilot, and Frank Hatcher, engineer.

All went well until they left Butedale, on the British Columbia coast, for Prince Rupert, on November 4. Somewhere on the hop to Prince Rupert in thick fog, they and their machine vanished into the sea. Only one wheel, washed up on a small island, was ever found, despite an intensive search by R.C.A.F. pilots.

When Wasson arrived over the Liard post, winter conditions made it impossible for him to make a landing in his float-equipped machine. He dropped a note to Osborne, manager of the post, asking him to write on the snow if Burke had been there. The reply was soon etched on the white surface: "Yes, Burke left here October 11 for Teslin, Teslin." The last word was printed twice.

Thus destiny does its dirty work. Teslin was incorrect—it should have been Atlin; but Osborne thought it right, as only once before in all Burke's trips had he failed to call at the

FOUND

Left to right: Kading, Martin, and Walsh, with Kading holding the makeshift skis

village of Teslin on return trips to Atlin. So Wasson went off to cover hundreds of square miles searching vainly in the wrong areas.

Finally Wasson had to interrupt the search while he proceeded to Mayo to have skis fitted in place of floats. While there he enlisted the aid of one of Yukon's most experienced guides, Joe Walsh, to fly with him.

With full winter equipment obtained from White Horse, the pair set off on November 12 to make a more systematic search of the country between the Liard post and Atlin. The country in that vicinity is a mass of rugged peaks, hundreds of them well over five thousand and some over seven thousand feet high; and the whole wilderness between Teslin Lake and the Liard post was unmapped. It was winter, too, and the airmen encountered storms almost all the time they were in flight.

The first trace of the Burke party came when Wasson landed on the lower Liard near the mouth of the Frances River, and learned from Indians that a "big white swan" had flown up-river several weeks before.

Flying conditions were vicious, with snowstorms occurring hourly. Frequent landings had to be made, many in desperately dangerous surroundings and circumstances. After one such landing, ten attempts to take off were required before the airmen could make a getaway from the deep snow of an unnamed lake where they had spent the night. Often the two men were obliged to trample laboriously with their snowshoes to make a semblance of a runway. On one take-off, Walsh had to run alongside the open door of the cabin until the plane gathered sufficient speed for the wings to lift. He was dragged off his feet as the machine picked up speed, and had to struggle into the cabin. As he wore long snowshoes, and the wind dragged at them, the feat was perilous.

So the search continued. On November 24, and again on December 4, the two flyers made complete return trips from the Liard post to the headwaters of the river. It was on the first of these that Martin and Kading had glimpsed the machine during their last day's trek. It was on this date, too, that Wasson and Walsh spotted a peculiar shadow on the surface of the snow-covered river. Going down for closer scrutiny, they discovered the stranded Junkers.

Great hummocks of snow and ice covered the river's frozen surface and a landing there was out of the question. They flew back to White Horse for additional equipment, then returned and landed on a small lake sixteen miles from the Burke machine, the nearest point at which they could come down safely. Their strenuous trek to the Liard brought disappointment: only the plane was there, and the message stating that the three men had gone upstream.

Back at White Horse, it was decided that the lost men might be with Indians on the Pelly Reserve. So Wasson and Walsh left White Horse on December 6, and it was while they were searching for a suitable pass to follow over the Pelly Range, that they saw a wisp of smoke drifting up through the treetops of the forest far below. As they glided down for nearer observation, two human figures were seen silhouetted against the snow, whose actions left no doubt that the long quest was over.

Flying over at a low altitude, Walsh opened the cabin door and kicked out a box of food. Then, believing it had dropped in deep snow too far for the weakened men to find, Wasson circled still lower, and as they swept past on another run, Walsh heaved out a second box of food—their own supply. Notes attached to the boxes stated the airmen would land on a lake some ten miles away, and would trek from there.

Once on the lake, Wasson and Walsh lost no time in hitting the trail. Darkness soon overtook them, and they were obliged to "siwash" it for the night. (In the vernacular of the North, this means stopping where you are when night falls, without any camp preparations.)

Travelling light, and without food, they were off before dawn. When daylight came they realized that in their eagerness they had overshot the place. Slowly back-tracking, shouting as they went, they twice heard the crack of a rifle. Minutes later they reached the pathetic camp.

Both boxes of food had been found and salvaged. They learned, too, that the shots Kading had fired to attract their attention were the last two shells he had.

Walsh fabricated a pair of makeshift skis for himself and gave his snowshoes to Kading, while Martin put on a spare pair which had

BACK FROM THE MAGNETIC POLE

Left to right: Stan Knight, Richard Finnie, Walter Gilbert, and Major Burwash, at Gjoa Haven with the Fokker that had been abandoned by the MacAlpine party

been brought along. The journey back to the plane ended on December 10. After great difficulty in getting the engine started, the four of them made it to White Horse the same day.

On December 16 Wasson left once again, accompanied by Walsh and Sergeant Leopold, R.C.M.P., to bring out Burke's remains. The final sad rites on December 19 concluded an aerial rescue-search which, for tenacity and heroism, has never been surpassed in Canada's flying history.

The investigation into the death of Burke, held at White Horse on December 20–22, attributed no blame to anyone, but a ruling was made that all aircraft flying in the Yukon Territories between October 1 and April 1 should be required to carry adequate supplies for all aboard.

The federal and the British Columbia governments handsomely acknowledged the fine work accomplished by the two Tredwell Yukon Company airmen. They awarded $1,500 to Wasson and $500 to Walsh in appreciation of the heroic efforts which saved two men from death.

The Junkers plane had a further history. In February 1931 she was salvaged by a party of five in a Bellanca monoplane. Skis were fitted and she was flown to Atlin. Then Pilot McClusky undertook to fly her in easy stages to Van-

couver but the flight came to an ignominious end in a forced landing on the C.P.R. tracks at Boston Bar, B.C. The damaged machine finished the trip on a railway flat-car.

She was rebuilt, however, and resumed a useful career. A well-known pilot, Charles B. Elliott, made a name for himself and the machine in August 1937, when he transported six and a half tons of freight and twenty men from Stewart, B.C., to the otherwise inaccessible MacKay Lake. The Junkers was by this time owned by Pacific Airways Ltd. and was under charter to Premier Gold Mining Company. The operation took place in tremendously rugged country, yet the airman flew all the men and material in to MacKay Lake in relays over a period of two and a half days, making eleven round trips altogether, covering a total distance of 2,200 air miles. Every time he crossed the high mountainous barrier between Portland Canal, where Stewart is situated, and the inland lake where the supplies were delivered, the airman had to rise to 10,000 feet. It was all in the day's work for Elliott, but a fine flying achievement just the same.

The final chapter in the career of the Junkers came on June 27, 1938, when she crashed at Nation Lake, a "tiny spot of water," 100 miles north of Fort St. John. It was a tragic ending. A passenger, Dan Miner, a prospector,

was instantly killed, and the pilot who had taken Elliott's place succumbed to his injuries the day after the crack-up. The Junkers was a total loss, and parts of her are still on the shores of the little lake up north.

SCIENCE TAKES WINGS

In 1930 the Canadian government sponsored an aerial expedition over Arctic latitudes which had as its objects, first, the defining of magnetic areas, second, an aerial circumnavigation of King William Island, and third, the investigation of reports of the remains of Sir John Franklin's expedition of 1845–1848.

A Fokker seaplane, G-CASM, was chartered from Western Canada Airways, with Walter Gilbert as pilot, and Stan Knight as engineer. Major L. T. Burwash was leader of the expedition. Major Burwash had visited King William Island previously in 1929, and had established the point on Boothia Peninsula where one of the great magnetic influences is centred.

The expedition was to start from Coppermine on the Arctic shore, but before setting out, Gilbert carried another company pilot, W. J. Buchanan, to the spot near Dease Point where the Western Canada Airways plane, G-CASK, had been abandoned by the Mac-Alpine party when they had been marooned there in 1929.

The trio left Coppermine on August 14, and

after a flight of several hours they spotted SK at the mouth of the Dease River and landed near by. After fuel had been poured into her tank and fresh oil put into her engine, the airmen had remarkably little trouble getting SK started, in spite of the fact that she had been out in the open during eleven months of Arctic weather. With Buchanan at the controls of SK, the two aircraft returned to Coppermine late next evening, just ahead of a "rip-snorter" from the northwest.

Buchanan left for the south on August 17, and the echoes of SK's engine had just died away when the motor of SM developed major trouble. A radio message to Winnipeg brought a Dominion Explorers pilot up with new parts but SM's engine still misbehaved. Another radio call was made and up came Buchanan flying old SK. This plane was now in sound condition and it was turned over to the party.

Storms delayed the departure another nine days but finally on September 4 the flight was made *via* Cambridge Bay to King William Island. The airplane came down at the little settlement of Gjoa Haven (now Peterson Bay). Here another passenger was taken aboard— Richard Finnie, who had been making studies of Eskimo life for the Department of the Interior—and the trip to Boothia Peninsula and the circumnavigation of the island began.

September 4 was late for flying a seaplane in that region, but the four men flew off with high hopes. They set their course along the east coast of the island, up Rae Strait. One of the Franklin ships is reported to be sunk close off shore from Blinkey Island at the southern entrance to James Ross Strait, and the airmen had hoped to spot it. Unfortunately, the sea was a mass of large ice-floes, and when the plane arrived in the vicinity of Blinkey, both land and sea lay under a heavy fog. This put examination out of the question, and the airmen proceeded north to Boothia Peninsula, where they circled the magnetic pole area. Flying south again, they picked up the coast of King William Island at Cape Felix, following the coast to Victory Point on the northwest side, where they made a landing. All four men searched diligently for any sign of the burial place of Franklin, or of members of his crew. Only a few relics were found—some short lengths of rope, fragments of naval broadcloth, and a piece of a linen tent—which were later forwarded to Ottawa. The plane flew southward along the west coast, and made a stop-over at Terror Bay, where other relics of the Franklin party had been discovered, but an all-too-brief search there brought nothing further to light, and the airmen flew back to Gjoa Haven along Simpson Strait.

Early on the morning of September 7 the airmen said good-bye to Finnie and the settlement and set off for Coppermine, thence flying out to Edmonton *via* the established route of Great Bear and Great Slave lakes.

The unfortunate delays due to mechanical trouble and bad weather brought the King William Island flight too close to freeze-up time for a lengthy stay, and thus prevented a thorough search for Franklin relics. However, the extensive photographing of the shoreline of King William Island will be of great assistance to future searches, when it is hoped additional records of that tragic expedition may yet be found.

When Major Burwash's party reached Edmonton, reporters wrote up a completely false account of the discovery of the relics, which caused much embarrassment to the party. However, although the trip was disappointing in the search for Franklin remains, its contribution to the geographical knowledge of the Canadian Arctic was of great value. One pleasing result of a personal nature was the honour conferred on Walter Gilbert—he was made a Fellow of the Royal Geographical Society.

An interesting development in the northwest has been the aerial exploration of mountain areas. Planes were used by Father Bernard Hubbard when exploring the almost inaccessible mountain area of the Elias Range in Alaska and the Yukon. The work of "the glacier priest" in that rugged country has been shown in movie travelogues throughout the continent, and his reports and articles are known to all who are conversant with the subject.

One of his discoveries led to the surveying of one of the biggest glaciers in the north, now named the Hubbard Glacier. Mount Hubbard in the Yukon, rising to a height of 14,986 feet, was among the many important physical features discovered.

During 1934–1935 the National Geographical Society made an aerial survey of the Elias Range area, operating from a base far inland on a glacier. From air and ground they mapped some 10,000 square miles of territory. Thus in recent years the entire north country of Canada and Alaska has been mapped by the aerial cameras of the high-flying craft of the R.C.A.F. and the U.S.A.F. Before long, topographical detail will replace all the blank areas in maps of the north—blanks which have meant disaster to more than one airman.

Canadian flyers have not only explored the Arctic but have penetrated the icy regions on the other side of the world. The Antarctic ex-

THE LINCOLN ELLSWORTH ANTARCTIC
EXPEDITION 1935
The four Canadian airmen invited to join. *Left to right*: Bert Treris, Herbert Hollick-Kenyon, Harold Lymburner, Pat Howard

FLYING AMBULANCES

Left: The seaplane of Austin Airways, Toronto. *Right*: The Waco of Speers Airways, Regina

plorations of the American explorer, Lincoln Ellsworth, in 1935–1936 and 1938 brought honour to Canadian flyers. When Ellsworth left Montevideo on his third Antarctic expedition, four Canadian airmen were aboard his ship: Herbert Hollick-Kenyon, Harold "Red" Lymburner, Pat Howard, and Bert Treris. Kenyon was first pilot of the expedition, Lymburner and Howard were reserve pilots, and Treris was engineer.

During 1935–1936 Ellsworth and Hollick-Kenyon flew a Northrop Gamma aircraft, the *Polar Star*, from Dundee Island, on the Atlantic side of Antarctica, to a point near the Bay of Whales, situated on the Ross Sea. Some 2,300 air miles were covered, but it was about two months before the two explorers were located by a search ship and rescued from their snow-bound camp at the Bay of Whales.

Lymburner and Treris again accompanied Ellsworth to the Antarctic in 1938, when successful flights were carried out, although not so extensively as in 1935–1936.

Mercy Flights and Rescues

The first Canadian ambulance plane was a regular Curtiss training plane converted into a stretcher-bearing machine by hinging the upper rear of the fuselage so that it opened upwards, thus permitting a stretcher to be passed in and made fast. This machine was prepared in Toronto for the use of the R.F.C. during the winter of 1917–1918, when road travel was especially difficult. It was registered as C-1451 and carried the Red Cross symbol on both sides.

Years later the Waco Aircraft Company of Troy, Ohio, re-designed their 1934 biplane so that it could carry a patient on a stretcher, a doctor, a nurse, and the pilot.

One of these Wacos had a busy career operating in Ontario. In 1934, a well-known test pilot of the De Havilland Company, Leigh Capreol, resigned his position and joined Charles and Jack Austin in running an air service from Toronto harbour. They purchased two Wacos, one of them designed as an ambulance. This plane, CF-AVN, was called into use the very day after it was received at the border —a Toronto heart specialist was flown to a patient in Muskoka. For nine years thereafter, Capreol and Austin Airways transported countless stretcher cases from outlying points to hospitals in Toronto.

Similar Wacos were later used by other Canadian flying firms, the earliest being Dominion Skyways Limited, Gray Rocks Inn Air Service, and McIntyre-Porcupine Mines.

The first ambulance plane to operate in western Canada was also a converted Waco. It was purchased in 1936 by Speers Airways Ltd. of Regina, and was piloted by Charles Skinner. This plane was a godsend, as, in previous emergencies, patients who were being transported in an ordinary plane had often suffered seriously through exposure and the cramped seating accommodation.

Many of the early searches and rescues were thrilling. In August 1936, Flight Lieutenant S. Coleman and Aircraftman Joseph Fortey of the R.C.A.F. were returning in a Fairchild seaplane from a northern patrol when they received word at Fort McMurray to fly engine parts to Hunger Lake, where another Air Force plane was grounded with engine trouble. They delivered the parts, helped to repair the engine, and left for Fort Reliance on August 17.

A bad storm came up, obscuring their vision and confusing their direction, and by the time they were free of it they could not identify the terrain below. Compasses are unreliable in that area, and the flyers soon realized they were lost. They landed once on a small lake and attached a note to an empty oil drum which they left on the shore, stating that they would fly in what they guessed to be a southerly direction. Then they flew till their fuel gave out, and came down on a small lake. They settled down to wait for rescue.

A six-plane search party was soon on the wing, but hope turned to anxiety as weeks passed and their efforts proved in vain. In this vast area a plane and a tent at the edge of a lake are the merest specks, easily missed in poor weather from a plane, even by the keenest eyes. The note on the oil drum was finally found but the intensive search to the south was unsuccessful. It was a bush pilot, Matt Berry, of Canadian Airways, who finally found them, and he did so simply by following a "hunch" about where he would go if he had been in their place.

The marooned men were in fairly good condition but weak from lack of food. Boiled ground squirrel and berries were all they had had to supplement their 12 pounds of rations. They were flown to Edmonton and a short spell in hospital soon brought them back to normal.

A few months later, Berry was responsible for another, more hazardous rescue. In December word came from Aklavik that some priests and Eskimos were stranded and out of food at Letty Harbour along the Arctic coast. Government sources got in touch with Canadian Airways, and Matt Berry, with Rex Terpening, engineer, was assigned the job of bringing them in. Berry and Terpening left Aklavik on December 11 in a ski-equipped Junkers monoplane.

The double difficulty of flying by contact, that is, by identifying the ground, and by the dim light of the Arctic winter, meant that they had to use a flashlight to keep checking their instruments as they flew over landmarks, for it would have been fatal not to know where they were in the event of a forced landing.

On the way to the rescue they encountered a terrific wind, estimated at times to be 75 m.p.h., which whipped up a ground drift of snow to a height of 1,000 feet. To find Letty Harbour was a feat in itself, and to make a landing there was another. The airmen could see the ground only for twenty-five feet ahead, but with the gale against them, their landing-speed was almost nil, so, although they ran into huge drifts when the Junkers touched down, there was no mishap.

The airmen found five Roman Catholic clerics and five Eskimos, mostly children, gathered in the small mission. Their food supply was giving out because bears had cleaned out a cache of fish which they had counted on. Their three dog teams were in a semi-starved condition.

On the 12th the two airmen snatched a

RESCUING THE R.C.A.F.

The R.C.A.F. Fairchild with the Canadian Airways rescue plane beside it on the shores of the small lake where Flight Lieutenant Coleman had been stranded

break in the weather to fly to a distant cache, and they brought back 1,500 pounds of food. Then the weather closed down with a vengeance, and blizzard after blizzard swept over the land for ten days.

On the 22nd the entire party (including the dogs) boarded the Junkers, and Berry took off into what appeared to be good weather. No sooner were they in the air than down the storm came again. It was impossible to fly, so they landed on a small frozen lake. There they spent the night in a temperature of 45° below zero, some under the canvas engine-cover, some in a small tent, and the others in the cabin of the plane.

By 11:00 A.M. the next day there was sufficient twilight to get into the air, so they pushed off, encountering a dense gloom which made it almost impossible to recognize landmarks. As they sped over the mainland, however, they came into clearer weather and Berry was able to reach Aklavik without further difficulty. One gallon of fuel remained in the tank!

How different rescue flights became within a few years may be visualized from the flight of an R.C.A.F. Dakota aircraft and crew of three, which took place in September 1950. They left Edmonton on the 20th and returned to Churchill, Man., on the 22nd after a flight of 3,650 miles which took them to Eureka Sound on north Ellesmere Island—only 680 miles from the North Pole—to rescue a meteorological employee stationed there who was seriously ill. The fourteen years since Berry's heroic flight had wrought an enormous change.

It was not only Canadians who became lost in the Arctic wilderness. Encouraged by two successful flights from Moscow to California in 1937, the Soviet Union launched a pretentious effort on August 12 of the same year. A mighty four-engined monoplane with a crew of six took off for a flight from Moscow to Mexico *via* the Canadian Arctic.

When the plane was 300 miles past the North Pole, a radio message was received: the flyers were having engine trouble and were going down. That was the last word ever heard from them.

The search for the Soviet flyers was international. The U.S.S.R., through their embassy at Washington, commissioned the famous Arctic explorer, Sir Hubert Wilkins, to organize a search party. Sir Hubert procured a twin-engined Consolidated PBY flying-boat, each motor being of 1,000 h.p. The crew he selected were Canadians: Herbert Hollick-Kenyon and Silas A. Cheesman, pilots, Gerald D. Brown, engineer, and Raymond E. Booth, radio operator. Sir Hubert was navigator.

The daily search flights were made from Coppermine and were continued for 30 days, until the winter closed in. Sir Hubert then went south to get a different plane and returned with a ski-equipped Lockheed metal monoplane, the one in which Merrill had flown —wheel-equipped—to England and back in 1937. This plane had a much greater cruising range.

During the sunless Arctic winter, thousands of miles were flown by moonlight and the search was not abandoned until March. Although unsuccessful in its chief object, this long series of flights added very substantially to the knowledge of Arctic climate and geography, some 34,000 square miles north of the Arctic Circle having been covered.

ON COMPANY BUSINESS

As commercial flying became established, some business firms began to use their own aircraft to transport their representatives and goods, and it was not long before larger companies were buying and operating their own planes. Of these the two most important were the Hudson's Bay Company and Imperial Oil Limited.

In 1939 the Hudson's Bay Company bought a twin-engined metal Beechcraft monoplane designed for wheels, skis, or floats. From its base at Edmonton, CF-BMI became a welcome visitor during the next two years through every fur-trading district in the Yukon and parts of Saskatchewan, Manitoba, Ontario, and Quebec. When the original pilots, Harry Winny and Paul Davoud, left the Company's service to join the R.C.A.F., the engineer, McLaren, became the pilot and Jerry Buchan became the new engineer.

As an example of the work the machine and crew accomplished, one flight made during the summer of 1939 stands out.

In that year, the Governor of the Hudson's Bay Company, Patrick Ashley Cooper, visited Canada from England to make a wide tour of inspection of the Company's activities and holdings. He boarded BMI at Edmonton on June 18, accompanied by P. A. Chester, the Canadian General Manager, and R. H. Chesshire, Assistant Manager of the Fur Trade.

Between that date and its arrival at Winnipeg a week later, the Beechcraft flew 4,685 miles and visited 28 of the Company's posts. Post managers had not been warned of the plane's coming and when they strolled down to the water to look the new machine over, they were rather startled to find that one of the visitors was the Governor of the "Honourable Company" in person. Mr. Cooper was, however, not unaccustomed to Arctic flying; he had made two previous inspections by air in 1932 and 1934, when he was flown by Canadian Airways Limited. His visit to the parent post at Rupert House on James Bay in 1934 was the first any governor had ever made.

Between July 10 and 19, 1939, BMI carried the Fur Trade Commissioner, Ralph Parsons, from North Bay, Ont., on a trip of 3,200 miles which touched at many of the Company's posts along Hudson Bay. From that time until her last trip in 1941, BMI flew thousands of miles on the Company's business. The accompanying map shows the many routes followed between April 1939, and August 1941, many of which were covered more than once.

In August 1941, the trim craft came to an untimely end at Richmond Gulf, just after having completed her maiden trip over Labrador territory. Aboard as passengers were P. A. Chester and R. H. Chesshire of the Hudson's Bay Company, and Dr. Frederick F. Tisdall, chairman of the Committee on Nutrition of the Canadian Medical Association. McLaren and Buchan formed the crew. The route followed was from Montreal to Cartwright, Labrador, along the Atlantic seaboard to Hebron, westward inland to Fort Chimo and south

MISSION TO LETTY HARBOUR

Matt Berry's Junkers at Letty Harbour during his rescue of the Mission, 1936

HUDSON'S BAY COMPANY
INSPECTION, 1939

After flying 4,685 miles in a week. *Left to right*: Harold Winny, Patrick Ashley Cooper, Duncan McLaren, Paul Davoud, P. A. Chester, R. H. Chesshire

to Fort McKenzie in northern Quebec, then to Port Harrison on the east coast of Hudson Bay, and finally to Richmond Gulf. When BMI was landing on the Gulf, a submerged rock damaged one of her floats, immobilizing the craft. A radio call was sent out, and soon a Canadian Airways plane flew from Edmonton and picked up the stranded passengers. McLaren went south with them to procure the needed parts, and arrived back by air early in September. Repairs had been made, and BMI was well anchored off shore, when a vicious gale swept out of the north with winds of 50 miles an hour. Caught out in the open, the seaplane resisted bravely for several hours, but was at last dragged under and engulfed.

To replace the Beechcraft, a ship of the same make was purchased, which was registered as CF-BVM. In later years, the Hudson's Bay Company purchased a twin-engined Canso amphibian registered as CF-FOQ and named the *Polar Bear*.

From the time that flying began in Canada, Imperial Oil Limited has not only been one of the largest suppliers of aviation fuel, but has also operated aircraft of its own. A previous chapter described the first flight into the Northwest Territories in 1921, made by machines owned by Imperial Oil Limited. In 1939, the company's latest aircraft went into service, a Beechcraft single-engined biplane, registered as CF-BJD. Until it was withdrawn from service in 1948, it was flown exclusively by the

late T. M. ("Pat") Reid, the Company's widely known aviation sales manager. He covered many of the routes taken by BMI. He made trips over the more populated areas of southern Canada, the area of the Mackenzie River to Aklavik, the British Columbia coast (which BMI never visited), along the Northwest Staging Route to White Horse, around Hudson Bay, along the shores of the St. Lawrence, to and from Anticosti Island, and northward up the Labrador coast.

At the beginning of this chapter I suggested that the golden age of bush flying is now nearly over. Perhaps, if we substitute for the reader comfortably relaxed under his own roof, the bush pilot himself, we might make the statement in reverse. Material that makes for good reading doesn't necessarily produce happy fly-

WINTER PROTECTION
The Hudson's Bay Company Beechcraft at White Horse during one of its flights

ing. One by one the hazards and hardships of flying in the Canadian hinterland are being reduced. Bases, fuelling depots, and well-built caches now exist all over the northland. Radio transmission and reception have improved enormously; and along with them the art of flying blind has become almost an exact science, especially since radar has given the airplane pilot eyes in the dark and fog. Planes are comfortably heated now; and regulations (though they are occasionally ignored) require every possible provision for the safety of passengers in emergency landings.

Yet even today the hazards are not gone. Lulled into a sense of security by a dependence on instruments, the modern flyer is perhaps losing the resourcefulness demanded under earlier flying conditions. The short gliding angle of the big modern plane gives very little leeway in picking out good spots for forced landings. And on the ground, the city-bred pilot of today often lacks the knowledge of the bush that was part and parcel of the boyhood of almost every Canadian a generation or two ago.

Two specific examples might be given here of technical developments that have had their effect on the character of air transport into isolated parts of Canada.

TECHNICAL CHANGES

In the 1930's few really large airplanes, by present standards, were used in Canada. It was still unusual to see a multi-engined airplane in flight. The rarity of such craft in Canada can best be judged by the fact that in 1928 only three were registered, apart from one in service with the R.C.A.F. All three were tri-motored, two being Fords and the other a Fokker.

One Ford, registered G-CARC, was owned and operated by Sky View Lines, Limited, Chippawa, Ont., the other, G-CATX, belonged to B.C. Airways, Limited, Victoria, B.C. Western Canada Airways owned the Fokker, G-CASC. The B.C. Airways Ford went into regular passenger service over a scheduled route covering Vancouver–Victoria–Seattle. She had a very brief life. Shortly after being commissioned, the machine vanished into the sea. She disappeared in misty weather between Victoria and Seattle on August 24, 1928, taking

"PAT" REID AND THE BEECHCRAFT
The Imperial Oil seaplane that covered many thousands of miles between 1939 and 1948

341

THE FLYING BOX-CAR

A doped horse is loaded aboard the big Junkers for transportation into northern Quebec

to their death five passengers and the pilot. Many years afterwards, part of the metal fuselage was brought up in a fisherman's net.

In 1931, when "big stuff" was seldom seen aloft, Canadian Airways purchased their largest airplane. Although it was not multi-engined, it had a wing span of 96 feet 8 inches. This all-metal Junkers monoplane was fitted to use wheels, floats, or skis. The Junkers was shipped from the German factory at Dessau, and on arrival at Montreal, the huge cases it had been packed in were hauled out to the Fairchild Aircraft Company plant at Longueuil. There the JU52 was assembled. It was allotted Canadian registration CF-ARM on October 26, 1931. RM weighed four and a quarter tons empty. During full-load tests carried out at St. Hubert airport, Montreal, on December 1, 1931, it lifted an additional 7,590 pounds into the air, in a take-off time clocked at 17½ seconds. The machine was later flown to Winnipeg.

For many years the "flying box-car," as RM was dubbed, was the largest single-engined airplane in Canada. Carrying a two-ton pay load, and with extra fuel aboard, she could fly 1,000 miles without making a landing. A large door at the side, and another in the roof of the fuselage, allowed bulky items to be loaded aboard, the floor being reinforced to take a load of approximately 750 pounds per square foot. Metal rings were placed conveniently on the floor and walls to aid in lashing goods firmly inside the cabin. The entire plane was well designed for the multiple jobs it performed after it went into service.

In 1936, the original German B.M.W. 600 h.p. engine was replaced by a Rolls Royce Buzzard of considerably greater horsepower. One job performed in that year was the transportation from Gold Pines to Casummit Lake, Ont., a total of 500 tons of miscellaneous freight. The operation consisted of a seventy-mile flight each way across otherwise inaccessible country. This freighting was done for the Argosy Gold Mines, Limited, and proceeded without a hitch.

An extraordinary assortment of items was flown in for the building of the Lake Manouan dam. Between August 11 and November 1, 1940, a number of planes carried over 626 tons of freight and 494 passengers on the 110-mile flight to the construction area. All the bulky, heavy stuff was transported by RM. It included the major parts of 11 tractors; a 20-foot inboard motorboat; 4 oxen; 8 horses; and a huge Diesel-engined shovel, the base and bucket of which alone weighed 4,000 pounds!

Some 4,000 tons of freight were flown by various aircraft to the Shipshaw project of the Aluminum Company of Canada, in Quebec. This major transport job was most vital to the war effort, for no railroad had been completed to Shipshaw at that time.

The era of aerial freighting in Canada has an immense future, but the splendid work done by RM and her pilots will stand for a long time.

CONVERTED TO SEAPLANE
The Junkers JU52 at Winnipeg

Introduction of the Oil Dilution System

In 1938, experiments in winter engine operations in Canada were made which revolutionized the methods then in use.

Up to that time, aircraft operating in the north in sub-zero temperatures had to have all the oil drained from their engines the instant the motors stopped. If this was not done, engines would quickly cool off and freeze solid. To start them again required two to three hours' heating with a blow-torch—a risky procedure. All through the 1930's, airplane motors were babied along in this manner during northern flying. In 1936, it was brought to the attention of T. W. Siers, engineer and superintendent of maintenance for Canadian Airways, that the United States Army Air Corps was using an oil dilution system, by which lubricant was thinned to a low viscosity. This allowed engines to be started in zero weather without difficulty.

Siers was at first quite sceptical of the idea because the dilution was produced with the aid of gasoline, and he feared a fire hazard. Later, at a meeting at Wright Field, Dayton, Ohio, he learned that cold oil when diluted to proper viscosity by the addition of gasoline would provide satisfactory lubrication. The thinning of the oil occurs prior to stopping the engine, and later allows the motor to be turned over quite freely, even after standing idle in sub-zero temperatures. In addition, the diluted oil provides proper lubrication immediately after starting. Best of all, the method does away with the usual long warming-up periods.

Siers placed the matter before G. A. Thompson, general manager of Canadian Airways, who made arrangements with Canadian Vickers Limited, of Montreal, to procure the Canadian rights. In November 1938 the first installation of an oil dilution system in a Canadian aircraft was made in a Norseman, CF-BDC. This plane unfortunately went through the ice when making a landing and was not salvaged for a couple of months. In the meantime, Canadian Vickers Limited were instructed to fabricate a Worth Oil Dilution system for use in a Junkers, CF-AQW. The work was completed by February 1939, and was thoroughly tested in the aircraft by W. E. Catton, pilot, and Rex Terpening and E. W. Chapman, engineers. The lowest temperature recorded during the tests was 44° below zero, and the experiments were fully successful. Dilution had no harmful effect on engines; in fact, the motors were in better condition after hundreds of hours of operation than they would have been under the old blowtorch and hot oil treatment. The system became one of the greatest aids to cold weather flying operations in Canada.

For his enterprise in investigating and applying the idea, Tommy Siers was awarded the McKee Trophy for the year 1940.

THE WORTH OIL DILUTION
SYSTEM

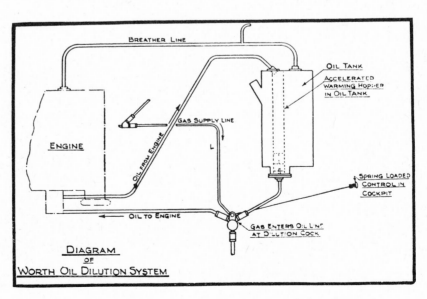

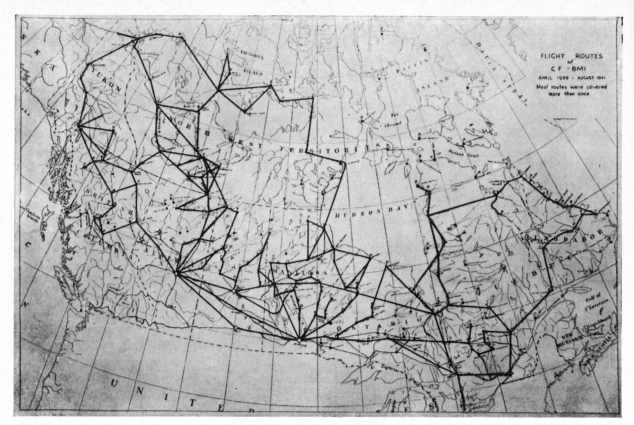

SEVEN-LEAGUE BOOTS

This map shows the immense distances covered by the Hudson's Bay Company Beechcraft during 28 months' service in Canada

344

35. Retrospect and prospect

LOOKING back over these pages I have written I don't know whether to be glad or sorry. Certain misgivings still disturb me. Have I, by using the details of my own story from time to time, thrown out of perspective the parts played by the chief actors on the stage of Canadian aviation? Surely, though, it was the only way I could tell the story of all of those minor actors whose bit parts, played only once, have been all but forgotten. For, after all, I was fortunate enough to be there: and I owe it to those who were not to convey, not only the facts, but the feel of those days.

Since 1921 I, too, have been a member of the audience, while the play proceeded; for thirty years I have been a bus-driver in the service of West Vancouver's Municipal Transportation System, carrying on as a hobby correspondence with fellow Early Birds. This correspondence grew into articles and finally into a book. To look back, with those of my generation—flyers and non-flyers alike—is a fascinating, perhaps even an awe-inspiring, experience. How much has been accomplished in so short a time!

Of the first two Canadians to make a flight in a heavier-than-air machine, both are now deceased. Casey Baldwin died at Baddeck, N.S., in 1948, and McCurdy passed away at Montreal, P.Q., in 1961. Bell, an old gentleman even at the time when flight by airplanes began, is long dead, and so is Turnbull, the great Canadian theorist, whose invention of the controllable pitch propeller marked a great step forward in aviation.

Of the rest who have appeared in this story, but whose own stories are incomplete, I have included what I could in the appended material following this chapter. Many are living and to them I extend my warmest gratitude for information and pictures so freely given. For those who are dead these pages must stand as tribute.

Of the pioneer airmen in the Canadian West my early flying pal Blakely also is dead. Most of the American barnstormers are dead—all too many of them at the height of their career. Curtiss, whose part—though he was not a Canadian—is woven deep into the fabric of Canadian aviation, from the days of the Aerial Experiment Association to the Curtiss flying school and factory in Toronto, and on again through his organization's sturdy aircraft to the present day, is gone, too.

Only a few of the great figures of the early hey-day of bush flying remain, and of them fewer still have any active connection with flying. Stevenson, the earliest "great" of them all, has passed on. Of the others—but, perhaps there is a better way of looking back over Canada's flying heritage of the late twenties and the thirties, at the same time telling the story of a unique award—the McKee trophy.

Originated by American pilot Dalzell McKee, in appreciation of the assistance and courtesy shown to him when he made the long-distance flight in Canada in 1926 already recorded, the McKee award was first named the Trans-Canada Trophy. Only after the death

345

DALZELL McKEE

for the gallant midwinter flight that he and a fellow airman, Vic Horner, made to isolated Fort Vermilion, with desperately needed antitoxin for the diphtheria epidemic at Little Red River.

In 1930 the emphasis turned to air mail, going to J. H. Tudhope for his pioneer air-mail flights to and from Rimouski to make connections with outgoing and incoming liners bound to and from Europe.

Another type of bush flying was recognized in the 1931 award. George H. R. Phillips, superintendent of eastern flying services for the Ontario Provincial Air Service whose many hazardous forest fire patrols that year had made aviation history, received the trophy.

Maurice "Moss" Burbidge was singled out in the following year for outstanding work in training pilots with the Edmonton and Northern Alberta Aero Club, where, after years as an instructor, he had chalked up the im-

of its donor in a plane crash in eastern Canada did the massive silver trophy come to be designated by his name.

The stated purpose of the award is to stimulate aviation in Canada, and the honour is conferred upon those whose efforts are judged to have contributed most substantially to the progress of aviation in Canada in any given year. At first only qualified pilots were considered eligible, but latterly it has been awarded to other persons, or even groups, who were adjudged to have contributed substantially in any way to the advancement of Canadian aviation. The judges, who represent the R.C.A.F., the Department of Transport, and the Royal Canadian Flying Clubs,[1] are appointed each year.

To survey the list of those who have received the award since its inception is to view the essential character of Canada's progress in the air. Three pioneer bush pilots head the list of recipients: H. A. "Doc" Oaks, 1927, who not only made a name for himself as a bush pilot but was the founder of a flying company whose name became a byword for efficient air transport in the thirties—Western Canada Airways; C. H. "Punch" Dickins, 1928, in recognition of his pioneer flight over the Barren Lands; and Edmonton's W. R. "Wop" May, 1929, chiefly

[1]Because of its contribution to the war effort, the Association was given in 1944 the right to use the word "Royal."

THE McKEE TROPHY

NON-STOP ACROSS CANADA

Five key members of the team. *Left to right*: Flying Officer Jolicœur, chief pilot; Sergeant Edwards, radio operator; Corporal Dewan, flight engineer; Flight Lieutenant Brown, co-pilot; Flight Lieutenant Lafferty, navigator

pressive total of 15,000 flying hours, without injury to a single pupil who had passed through his hands.

The exceedingly popular award of 1933 went to Walter E. Gilbert, bush pilot *par excellence,* whose many mercy and exploration flights had made his name known from end to end of Canada.

The roll of famous bush pilots and the record of the work they did continues with Elmer Fullerton, trophy winner for 1934. As pilot of the Imperial Oil Limited Junkers machine which in 1921 was the first aircraft ever to fly to Fort Norman in the Northwest Territories, he had thoroughly earned the honour. Back in the permanent R.C.A.F. at the time of the award, Fullerton was the first Canadian officer to be so honoured.

When W. M. Archibald, manager of the Consolidated Mining and Smelting Company in British Columbia, inaugurated the use of a fleet of aircraft for company business in the mountainous country of that province, he had no thought of the larger contribution he was making to progress across Canada. For his foresight, and for proving the worth of airplanes in rough country, he became the McKee Trophy winner for 1935.

So the story goes: A. M. (Matt) Berry became the outstanding airman in Canada for 1936, for his magnificent aerial rescue flights in the high north under the severest of mid-winter conditions. To Romeo Vachon went the 1937 award as a veteran pilot and mechanic who distinguished himself particularly by his efficient organization of radio and weather reporting for the safety of airline operations. The following year the trophy was made a general tribute to all members of the maintenance department of Trans-Canada Air Lines for their excellent record during the year; a recognition of non-flying personnel that met with strong favour throughout all flying circles as a gesture long overdue to the men on the ground on whom the pilot's success ultimately depended.

An entirely different phase of aviation featured the choice for 1939. Murton A. Seymour was honoured, not as a pilot—though he had learned to fly in Vancouver in 1915—but for fostering the progress of the Canadian Flying Clubs through organizing and other activities connected with them over a period of eight years.

Here we reach the end of the pre-war McKee awards: and here I am nearing the end of my self-appointed task.

What, we might ask at this point, is the greatest contribution Canadian aviation has

THE NORTH STAR

From the R.C.A.F. Experimental and Proving Establishment on the first non-stop flight across Canada, January 14-15, 1949

made to the world in the years preceding World War II?

The history of Canadian flying up to the end of World War I is inextricably interwoven with that of the United States and during the war years with that of Great Britain. The Bell organization was fruitful in results because it was international, and the western experiments remained unproductive largely because they were so isolated. But in the period between the two world wars Canadian airmen came into their own. And, if this book has laid more than a little emphasis on the bush pilot, it is because out of our own rugged hinterland have come the men whose experience and resourcefulness have given to Canadian aviation its unique character.

Perhaps too much has been made of the "firsts," though these the reader wants to know. I keep thinking of two French flyers, Coli and Nungesser, who flew westward from Paris on May 8, 1927, along the Channel, past England; the ears of the world keyed for news of the first east-west trans-Atlantic flight. When these men set out no one had made the Atlantic crossing since Alcock and Brown in 1919. Less than three weeks later Lindbergh startled the world with his famous solo flight. Yet some day the last weathered shreds of Coli and

Nungesser's wrecked *Whitebird* may be discovered in the wilds of Labrador or Quebec and identified as theirs, changing the records of every Atlantic flyer but the first. Of such flimsy stuff is fame made, when measured in terms of competition.

The whole story of Canadian flying during the years of World War II has yet to be written; and much material has accumulated since then. What of the future? Is a new trend discernible in the McKee awards of 1940 to 1946?

Tommy Siers's award in 1940 for his contribution to the oil dilution system has been mentioned in the previous chapter.

When A. D. McLean was awarded the 1941 trophy he had been working for two years on the stupendous job of preparing air bases for the British Commonwealth Air Training Plan. His background as acting superintendent of airways and airports during 1934 when he paved the way for the establishment of the Northwest Staging Route ideally qualified him for this task.

In 1942 and 1943 the winner was the late T. M. "Pat" Reid, another of the hardy band of early bush pilots, who had been appointed aviation fuels sales manager for Imperial Oil Limited, and had earlier shown his organizing ability as tour leader, in 1931, of the Trans-Canada Air Pageant.

There is not a pilot nor an engineer in Canada who does not know the name of John A. Wilson, 1944 winner. Though he had never piloted a plane, he had grown up with Canadian aviation, and from 1919 until 1941 had been the well-liked controller of civil aviation.

Grant W. G. McConachie, another veteran pilot of northern flying, who had organized his own flying company, United Air Transport, Limited, in the early days, was general manager of Canadian Pacific Airlines in 1945, when the award went to him, and has since become president. Mr. McConachie won distinguished rank as a McKee Trophy winner by his far-sighted achievements in extending air service from Edmonton through northern British Columbia to the Yukon. In July 1937 he pioneered the first air-mail service between Edmonton and White Horse, Y.T. As he piloted the big tri-motored Ford on its trail-blazing flight, he was already envisioning this route as the first stage of a short Great Circle air route to the Orient. Today, this "over-the-top" route from Canada to Japan and China is plied by the four-engined "Empress" airliners bearing the insignia of Canadian Pacific Airlines.

It was appropriate that at the end of World War II the 1946 trophy should honour the R.C.A.F. in general and Group Captain Z. L. Leigh in particular. Leigh, as an early Canadian Airways bush pilot was later prominently

THE FIRST TO FLY

John McCurdy stands beside the first of the 1134 Mosquito fighter-bombers produced during World War II by De Havilland at Toronto

connected with the formation of the Transport Command which aided so tremendously in ferrying aircraft across the Atlantic to Europe and Africa during the war.*

In these six awards a new trend is observable. The days of the great individual pilots are passing: the new times, as thousands of trained young men take expertly to the air, and huge air transport systems such as T.C.A. and C.P.A. continue to grow, call for administrative and organizational ability of the highest order. There will always be scope for the individual in flying; but new conditions call for collective and co-operative activity of a new kind.

In 1947 an organization was re-activated in western Canada which had its origin as far back as 1927. Now known as the British Columbia Aviation Council, it is a body of individuals whose primary object is the help and sponsorship of civil aviation as applied to private flying and its varied branches. This is a unique group, completely dedicated to voluntary aid in the cause of safer flying, and already through its

*See Appendix C for list of McKee Trophy awards.

FLIGHT LIEUTENANT MEL LEE

In the De Havilland Vampire in which he flew the 488 air miles from Vancouver to Calgary in 54 minutes

THE SHAPE OF THINGS TO COME

Michael Cooper-Slipper, Avro test pilot, climbs into a Canadian CF-100 all-weather fighter wearing the latest type of pressure suit

efforts many high tension wires across rivers and wide valleys in British Columbia have had suitable black markers attached to them, to draw the attention of low flying pilots to the hazard. Credit for the idea goes to Mr. A. H. Wilson, of the Department of Transport, and a B.C.A.C. member. Other contributions in aid of operators of small aircraft were efforts to abolish a road tax which previously was charged towards aviation gasoline, great improvement in Customs aviation clearances and regulations, the establishment of new airstrips, and the marking of all unlicensed air strips throughout the province.

The B.C.A.C. has a representative membership drawn from almost every city, town, and village in British Columbia, as well as individuals from government branches and private and commercial air industries and operations. The Council's example is one which should be followed in every province in Canada to champion the cause of all private air operators—from the owners of small aircraft and gliders to the

devotees of the sport of parachute and sky diving.

The original research for this book came to an end, with a few brief exceptions, when publication began early in 1954, and since that time spectacular air history has been made in Canada. A large volume could be written to cover such happenings, but as that cannot be indulged in here a brief résumé of some of the most outstanding will help to bring this history up to date.

In 1948, a Trans-Canada Air Lines four-engined North Star made the flight between Vancouver and Montreal in a flying time of 6 hours and 52 minutes, and on August 31, 1948, a Royal Canadian Air Force De Havilland Vampire Jet aircraft, piloted by Flight Lieutenant Mel Lee, flew from Vancouver to Calgary in 52 minutes. With the advancement of aviation since World War II, it is worthy of note that the first truly non-stop flight from coast to coast took place on January 14–15, 1949. Captained by Flying Officer J. A. F. Jolicœur, with a 14-man crew, a North Star of the R.C.A.F. made the flight over the Great Circle route from Vancouver to Halifax. The distance flown, 2,785 miles, was covered in 8 hours and 32 minutes, a speed of 329 m.p.h. In July of the same year, Canadian Pacific Air Lines inaugurated their passenger air route from Canada to Japan, and by September it was fully established.

From the huge factory of A. V. Roe of Canada there emerged on January 19, 1950, the first of many CF-100 twin-jet intercepter fighter aircraft constructed for the R.C.A.F. and also for sale to other countries. Powered by two Rolls-Royce Avon jet engines, it was put through its paces by company test pilot Squadron Leader W. A. Waterton.

With the Korean war in full swing, the R.C.A.F. joined the airlift on July 27 with No. 426 Squadron taking over the long air haul to Japan and back twice weekly. Their job was to ferry top personnel, priority items, wounded men, and men returning from the Orient on leave.

The vast plant of Canadair Limited, at Montreal, makers of the four-engined North Stars, entered the jet age in 1950, when they unveiled the first of many hundreds of Sabre Jet, single-seater, fighter aircraft they were to make for the R.C.A.F. and for sale abroad. The job of testing the craft was conducted by pilot A. J. "Al" Lilly. With their production, Canada's R.C.A.F. No. 410 Squadron in 1951 became the first unit to be fully equipped with jets. This was followed in October of 1951 by the first delivery to the R.C.A.F. of a CF-100 all-purpose twin-jet intercepter by A. V. Roe of Canada. A highlight for the latter company followed in 1953, with the production of their first large Orenda jet engine, the first of its type to be designed and fully produced in Canada. It was one more large milestone in the country's air history. The true jet era was fast approaching at this point in our air progress, and rumour had become confirmed fact that it would not be long before the large airlines would be ordering huge, multi-jet passenger airliners to supersede the existing prop-driven types then in operation.

In 1954, with cold war scares in both the press and the public mind, Canada and the United States released the joint statement that they were to co-operate in building a vast network of mid-Canada early warning radar sites, to detect the possible approach of unwelcome aircraft from Russia. Although three of these systems are now spread across our northland, we hope they will never be needed in war. At this time also, jet aircraft were beginning to demonstrate what they were really capable of accomplishing and on April 11, 1954, Squadron Leader R. G. Christie flashed almost across Canada, from Vancouver to Ottawa, in an R.C.A.F. Sabre Jet in 3 hours and 46 minutes. Interest in other spheres of aviation was also shown by the establishment, in July 1954, of the Canadian Aeronautical Institute, an organization which is having far-reaching effect on Canadian air industry.

It was during 1955 that turbo-prop airliners —which use propellers driven by jet-type engines—made their first appearance commercially in Canadian skies. On April 1 Trans-Canada Air Lines introduced their newest acquirement, the Vickers Viscount, and as these modern craft entered their service they became the first of their type to ply the air lanes anywhere on the North and South American continents.

The echoing whine of high-flying jet aircraft was becoming an accustomed sound in the heavens by 1956, and cross-country records by them have been established and re-established ever since. An outstanding one took place on January 16, 1956, when Squadron Leader Lou Hill and Flight Lieutenant Alex Bowman made the trip from Vancouver to Dartmouth in an R.C.A.F. T-33 trainer jet in the spectacular time of 6 hours and 52 minutes. Records are made to be broken, and on August 30 the same year the T-33's time was shattered when an R.C.A.F. Sabre Jet, manned by Flight Lieutenant R. H. Annis and Flying Officer R. J. Childerhose, flew the 2,740 miles in 5 hours and 30 minutes. The Royal Canadian Navy, long a partner with the R.C.A.F. in air activities, launched into a bold project of its own on January 17, 1957, with the commissioning of the large aircraft carrier, H.M.C.S. *Bonaventure*, obtained from Great Britain. Not to be outdone by T.C.A. in the use of turbo-prop airliners, Canadian Pacific Air Lines, in February 1958, took delivery from England of their first Bristol Britannia, a fleet of which they now operate on domestic and foreign routes.

It was also in the spring of 1958 that A. V. Roe of Canada rolled of their assembly the twin-engined jet, all-purpose fighter intercepter, the CF-105, with which it was then expected Canada's Air Force would be equipped. It immediately underwent exhaustive tests conducted by company pilot Janusz Zurakowski. The results of the huge delta-wing craft were fabulous, and speeds of well over 1,000 m.p.h. were announced by the company. It then seemed certain that the CF-105 was assured of a bright future, but that was not to be. The government suddenly announced that all further construction of additional CF-105's had been cancelled. The machine was probably the finest aircraft ever built in Canada, and to have it relegated to scrap heap almost overnight was the greatest blow ever experienced in Canada's air industry. A nation-wide controversy developed between opposing factions as to why the work on it was stopped, but whatever the correct cause, the decree remained final.

The versatile aircraft known as the Beaver and the Otter, which have been produced in large numbers by the De Havilland Company at Toronto over the years, are still rolling off their assembly lines. Those two types of aircraft are now used on a multitude of jobs all over the world. When the Korean war was in progress, the United States government purchased over 600 Beavers from the De Havilland Company against the stiff opposition of

DE HAVILLAND BEAVER
With insignia of the United States
Army

their own aircraft manufacturers. Once the
Beaver becomes the property of the United
States government, it loses its Canadian identity, and thereafter when in use with their
armed forces becomes simply the L-20.

Since 1959, numerous types of airplanes have
been designed and built by various Canadian
companies, and all are making their mark in
the progress of air history both in Canada and
abroad. De Havilland have developed a large

air carrier, the Caribou, which features rear-
end loading facilities. Having proved itself
commercially, it is being purchased widely by
many countries, as well as Canada. Among
other fine aircraft that Canadair Limited of
Montreal have produced, one of their latest is a
"swing-tail" freight-carrying model completed
in 1960. It is something quite new in aviation,
with its unique ability to have the hinged tail
section swung completely to one side when on

FIRST "SWING-TAIL" CARGO PLANE
Unveiled at the Montreal plant of Canadair Limited, August, 1960

the ground to speed up unloading and loading of the thousands of pounds of freight which such craft are capable of handling. At the present time air freighting is becoming a fully established means of quick and reliable transport for goods over world-wide air routes, and Canadian aircraft manufacturers are playing their part in supplying the planes to accomplish it.

Now, in 1961, it may seem that the zenith of passenger air transport is near at hand, with the announcements by both of Canada's largest air companies, Trans-Canada Air Lines and Canadian Pacific Air Lines, that this year will see their newly acquired 127-passenger DC-8 jet airliners taking over their major routes. It may be many years before greater strides are made.

The DC-8's are pure jet-driven craft, having no propellers, and are fitted with four powerful engines, which are in keeping with the size of the airliners themselves: length, 150 feet, 6 inches; wing span, 142 feet; tail height from ground when at rest, 42 feet; gross take-off weight 300,000 pounds—150 tons! When a DC-8 takes off on a long non-stop flight with its full load of 127 passengers and a crew of from eight to nine, depending on the route being flown, it carries a full load of 18,300 imperial gallons of fuel for a maximum non-stop flight of 4,160 statute miles. On such journeys, it flies at a height of approximately 40,000 feet, at speeds of well over 500 m.p.h. A huge below-decks cargo compartment will carry 5 tons of freight, express, and mail. They are truly monster machines in every sense, including their cost of $6,000,000 each. They require extra long runways as their take-off distance is in the close vicinity of 8,500 feet, and landing 6,600 feet minimum. So once again we have topped a new peak in aerial transport, and marvellous as it may now appear to be, who can say what greater progress the future holds in store.

The advent of the big jetliners has had great impact on the main airports throughout the nation. What were considered to be runways of adequate length in the early 1950's have now shrunk to small proportions, compared with the vast stretches of hard surface now being required. At Montreal's Dorval Airport, Toronto's Malton Airport, Winnipeg's International Airport (once Stevenson Field), at Regina, Calgary, Edmonton, and Vancouver, the runways are reaching out into distances which seem fantastic. At all of these airports, hordes of passengers come and go yearly, and freight and mail stack up in great piles on equal terms with most large railway depots. Air terminal buildings, too, have been undergoing a face lifting in the past few years, and from the small and dingy buildings of less than fifteen years ago are emerging beautifully designed, spacious modern structures in steel and glass.

TRANS-CANADA AIR LINES DC-8 JETLINER

Passenger flying has come to stay, and it is fast becoming the accepted mode of travel by people in every walk of life, for business and for pleasure. Look what has happened to Atlantic travel. It is only a little over forty years ago since Alcock and Brown flew across non-stop for the first time. Compare that episode with the statistics for 1960 in the months of July, August, and September, for example. During those three months, the number of paying passengers who flew east or west over scheduled airline routes totalled 670,124, as compared with 523,031 in the same months of 1959. Over the same period in 1960, 25,722,941 pounds of cargo was flown over the Atlantic, and 8,853,686 pounds of mail. Of course, these figures include all airlines flying the ocean, Canadian and otherwise.

Although aircraft have their terrible aspects during wartime, in peace they have proved to be one of the biggest boons to mankind since the world began. Think back a little for the moment to the year 1959 to an event which reviewed the past in a most dramatic style. February 23, 1959, was a day of comparisons in aviation which was brought impressively to the notice of hundreds of thousands of Canadians by the miracle of television. On that date, at Baddeck, Cape Breton, Nova Scotia, a re-enactment of the first airplane flight in Canada by J. A. D. McCurdy took place. It was a completely realistic affair, as the airplane used was a full-sized replica of the famous *Silver Dart*, the original of which was flown at that spot

exactly fifty years before. On this Golden Anniversary, the machine used, the *Silver Dart II*, had been constructed by Leading Aircraftsman Lionel McCaffrey with the help of Flight Lieutenant William Bell and Leading Aircraftsman J. Trimm, all three members of the R.C.A.F. The work was accomplished in the Trenton, Ontario, workshops of the R.C.A.F. mostly on the men's own time, but the cost of the celebrations and the staging of the enactment of the flight at Baddeck in 1959 was borne chiefly by the Canadian government.

As did the original *Silver Dart* on February 23, 1909, the replica took off from the ice-covered surface of Lake Bras d'Or on the afternoon of February 23, 1959, under the able control of Wing Commander Paul Hartman. A strong wind was blowing, making it very difficult to fly with crude controls, but although the craft suffered some damage on landing, the whole historical affair was a huge success. Mr. J. A. D. McCurdy was there in person, along with hundreds of officials and other interested spectators who had gathered there from many parts of Canada and the United States.

Now as I complete these pages, and this book, what do the historical events contained signify? They prove but one thing: inevitably mankind will progress. The doubting Thomases and the dour crêpe hangers who look to the future with a jaundiced eye have only to glance back at the past to learn what the future may hold in store.

My memory goes back to the Jules Verne and H. G. Wells stories I read as a youngster. A

AERIAL PASSENGER BUS OF TOMORROW

Okanagan Helicopters Limited, Vancouver, B.C., plan to have this British-built Fairey "Rotodyne" in operation by 1962

trip to the moon, a visit to Mars, and a marvellous mythical airship, *The Flying Fish*, whose crew explored this world and outer space. At that date, the first frail, man-made airplanes were only just faltering in the air; but there lived men with foresight who knew that flight was possible, while the writers dreamed of soaring to the moon, and far beyond to other planets in our galaxy, and even to the stars.

Considering the fair measure of success we have gained in the past fifty years, we have only to look at the youngsters today, as they eagerly read their comic and science-fiction books, to realize that for them space travel seems inevitable. More significantly, many highly intelligent adults are accepting interplanetary travel as a serious possibility.

The picture of the past is also a preview of the future. No matter how careful the planning, or what pains are taken, there will be surprises, frustrations, miscalculations, and failures. Men will die, and others will survive to fly again. There will be other Turnbulls and Gibsons to design the early human space rockets, and other Baldwins and McCurdys to pilot them. Some will work out and some will not. Some, like the Underwood creation, will never get a fair trial. But in the end there will come success.

Will some of the pioneers be Canadians? Let us hope so, and that they will not only embark on such thrilling journeys, but that they will come back to tell us about them. Thus will be added still greater laurels which, in their turn, will do their part in enriching Canada's already magnificent flying heritage.

CANADA'S "MANNED MISSILE"

The first of two hundred CF-104 aircraft being built by Canadair Limited, Montreal, for the R.C.A.F. No. 1 Air Division operating with NATO forces in Europe. The airplane is a single-seat, strike-reconnaissance craft capable of supersonic speeds at low or high altitudes, day or night, in all weather

APPENDIXES

OUTSTANDING EVENTS IN CANADIAN AVIATION

1879	First balloon flight in Canada. Montreal, Que., July 31, 1879	Richard Cowan, Charles Grimley, Charles Page
1902	First wind tunnel in Canada. Rothesay, N.B., 1902	Wallace R. Turnbull
1906	First internal combustion gasoline aircraft engine imported into Canada. Rothesay, N.B., 1906	Wallace R. Turnbull
	First airship flight in Canada. Montreal, Que., July 12, 1906	Lincoln Beachey, pilot
1907	First towed glider flight in Canada. Montreal, Que., August 1907	Larry J. Lesh, pilot
	First man-lifting kite in Canada. Krugerville, Alta., summer of 1907	John Underwood
	Man-carrying flight with Bell tetrahedral kite, *Cygnet I.* Baddeck, N.S., December 6, 1907	Lieutenant Thomas Selfridge
1908	First British subject to fly a heavier-than-air machine. Hammondsport, N.Y., March 12, 1908	F. W. (Casey) Baldwin, pilot. Aerial Experiment Ass'n machine, the *Red Wing*
	First development of the aileron (for lateral stability in flight). Hammondsport, N.Y., May 18, 1908	F. W. (Casey) Baldwin, pilot. Aerial Experiment Ass'n machine, the *White Wing*
	First official flight in the world of one kilometre. Hammondsport, N.Y., July 4, 1908	Glenn Curtiss, pilot. Aerial Experiment Ass'n machine, the *June Bug*
1908–1909	First water-cooled aero engine used anywhere in the world. Hammondsport, N.Y., December 6, 1908, and Baddeck, N.S., February 23, 1909	Fitted to the Aerial Experiment Ass'n machine, the *Silver Dart*, and flown exclusively by John A. D. McCurdy
1909	The first successful controlled flight by a British subject at any point in the British Commonwealth, in a heavier-than-air machine. Baddeck, N.S., February 23, 1909	John A. D. McCurdy, pilot. Aerial Experiment Ass'n machine, the *Silver Dart*
	First passenger flights in an airplane in Canada. Petawawa, Ont., August 2, 1909	John A. D. McCurdy, pilot; F. W. (Casey) Baldwin, passenger
	First demonstration flights of an airplane for military use in Canada. Petawawa, Ont., August 2, 12, and 13, 1909	John A. D. McCurdy, pilot; F. W. (Casey) Baldwin, passenger. The *Silver Dart* and *Baddeck I*
	First advertised exhibition flight in Canada, and at any point in the North American continent. Toronto, Ont., September 2, 1909	Charles F. Willard, pilot. First Curtiss airplane made, the *Golden Flyer*
1910	First successful airplane engine fabricated in Canada. Victoria, B.C., completed March 10, 1910	William W. Gibson. Fitted in the *Gibson Twin-plane*
	First flight of a heavier than-air machine in western Canada. Vancouver, B.C., March 25, 1910	Charles K. Hamilton, pilot. Curtiss-type machine
	First official Canadian altitude record established — 1,650 feet. Montreal, Que., June 27, 1910	Walter Brookins, pilot
	First airplane flight in the province of Quebec. Montreal, June 27, 1910	Walter Brookins, pilot . Wright seaplane
	First airplane flight directly over any Canadian city. Montreal, Que., July 2, 1910	Count Jacques de Lesseps, pilot. Blériot monoplane, *La Scarabée*
	First model airplane contest held in Canada. Montreal, Que., July 4, 1910	First prize winner, J. H. Parkin
	First flight of an airplane in the province of Manitoba. Winnipeg, July 15, 1910	Eugene B. Ely, pilot. Curtiss biplane
	First Canadian subject to obtain a pilot's licence. Aero Club of America, No. 18, dated October 5, 1910	John A. D. McCurdy
	First Canadian woman to become a passenger in an airplane. Belmont Park, New York, October 25, 1910	Grace Mackenzie of Toronto; Count Jacques de Lesseps, pilot. Blériot monoplane
1911	First airplane flight in the province of Alberta. Edmonton, April 29, 1911	Hugh A. Robinson, pilot. Curtiss biplane
	First airplane flight in the province of Saskatchewan. Saskatoon, May 19, 1911	"Lucky Bob" St. Henry, pilot. Curtiss biplane
	First passenger flight in western Canada. Winnipeg, Man., July 22, 1911	Frank Coffyn, pilot; W. C. Power, passenger. Wright biplane
	First cross-country air race in Canada. Hamilton to Toronto, Ont., August 2, 1911	Contestants, John A. D. McCurdy and Charles Willard. Curtiss and Farman machines
	First recorded "letter dropping" from the air. Quebec City, September 1, 1911	George Mestach, pilot. Blériot monoplane

1912	First woman passenger to be flown in an airplane in Canada. Vancouver, B.C., April 24, 1912	Mrs. Olive Stark of Vancouver, passenger; William M. Stark, pilot. Curtiss biplane	
	First parachute jump from an airplane flying in Canada. Vancouver, B. C., May 24, 1912	Charles Saunders, parachutist; Phil Parmalee, pilot	
	First flight by a hydroplane (seaplane) in Canada. Port Stanley, Ont., July 17, 1912	Walter Brookins, pilot. Burgess-Wright seaplane	
	First passenger flight in a seaplane in Canada. Port Stanley, Ont., July 17, 1912	Lorne Bradley, passenger; Walter Brookins, pilot. Burgess-Wright seaplane	
	First mile-high altitude flight recorded in Canada (6,000 feet). Winnipeg, Man., July 19, 1912	Jimmie Ward, pilot. Curtiss biplane	
	First woman passenger in an airplane in eastern Canada, and the first woman to be flown in a seaplane in Canada. Port Stanley, Ont., July 20, 1912	Dora Labatt, passenger; Walter Brookins, pilot. Burgess-Wright seaplane	
	First airplane flight in the province of New Brunswick. Saint John, September 2, 1912	Cecil Peoli, pilot. Pusher biplane, the *Red Devil*	
1913	First flight to be made by a woman pilot in Canada. Vancouver, B.C., July 31, 1913	Mrs. Alys McKey Bryant. Curtiss-type biplane	
	First death of a pilot caused by aerial misadventure in Canada	Johnny Bryant. Curtiss-type biplane	
	Canada's first long cross-country flight. Montreal, Que., to Ottawa, Ont., October 8, 1913	W. C. Robinson, pilot. Lillie biplane, engine of pilot's design and make	
	First newspapers delivered by air in Canada. From Montreal, Que., to Choisy, Que., Caledonia Springs, Ont., Leonard, Ont., and Ottawa, Ont., October 8, 1913	W. C. Robinson, pilot. Same machine	
1914	First flying-boat owned or flown in Canada. Toronto, Ont., early in 1914	W. A. Dean, owner; Theodore Macaulay, pilot. Curtiss flying-boat, *Sunfish*	
	First passenger to be flown from one Canadian city to another. Toronto, Ont., to Hamilton, Ont., and return, May 15, 1914	Lou E. Marsh, passenger; Theodore Macaulay, pilot. Curtiss flying-boat, *Sunfish*	
	Earliest known photography from an airplane flying in Canada. Toronto, Ont., May 1914	Norman Pearce, photographer; Theodore Macaulay, pilot. Curtiss flying-boat, *Sunfish*	
	First pilot to loop the loop and accomplish inverted flying in Canada. Winnipeg, Man., July 11, 1914	Lincoln Beachey. Curtiss biplane	
1915	Canada's first flying field (unofficial), Long Branch, Ont.	The land school of the Curtiss Aviation School	
	Canada's first seaplane base (unofficial), Hanlan's Point, Toronto, Ont.	The seaplane training branch of the Curtiss Aviation School	
	The first two flying pupils to graduate as pilots in Canada. Toronto, Ont., July 11, 1915	H. Strachan Ince and F. Homer Smith. Curtiss Aviation School	
	Aero Club of Canada officially formed, Toronto, Ont., July, 1915		
	First aircraft manufacturing plant in Canada. Toronto, Ont., 1915	Curtiss Aeroplanes & Motors, Limited; John A. D. McCurdy, manager	
	First twin-engined flying-boat to be flown in Canada. Toronto, Ont., November 10, 1915	Theodore Macaulay, pilot. Curtiss flying-boat	
1917	The Royal Flying Corps established in Canada, January 1917		
	First Canadian-built airplanes for export to the United States. Toronto, Ont.	Built by Canadian Aeroplanes, Limited	
1918	First twin-engined aircraft (flying-boats) to be built in Canada. Toronto, Ont.	Built for the United States government by Canadian Aeroplanes, Limited	
	First official air-mail flight in Canada. Montreal, Que., to Toronto, Ont., June 24, 1918	Captain Brian A. Peck, R.A.F., pilot; Corporal Mathers, R.A.F., passenger. Curtiss JN4	
	Royal Canadian Naval Air Service established, with two bases in Nova Scotia, September 5, 1918. Disbanded December 5, 1918		
1919	First civil airplane flight in Newfoundland. St. John's, January 14, 1919	Major Cotton, pilot. Martinsyde biplane	
	First official air mail flown between Canadian and United States points. Vancouver, B.C., to Seattle, Wash., March 3, 1919	Eddie Hubbard, pilot; W. E. Boeing, passenger. C3 Boeing seaplane	
	Earliest air-mail attempt across the Atlantic. Started from St. John's, Nfld., May 18, 1919 (forced down in mid-Atlantic)	Harry Hawker, pilot; Lieutenant Commander K. Mackenzie-Grieve, navigator. Sopwith biplane, the *Atlantic*	
	First non-stop Atlantic flight. St. John's, Nfld., to Clifden, Ireland, June 14–15, 1919	Captain John Alcock, pilot; Lieutenant Arthur Whitten Brown, co-pilot and navigator. Vickers Vimy biplane	
	First air mail to be flown across the Atlantic. St. John's, Nfld., to Clifden, Ireland, June 14–15, 1919	Captain John Alcock, pilot; Lieutenant Arthur Whitten Brown, co-pilot and navigator. Vickers Vimy biplane	
	First parachute jump by a Canadian from an airplane flying in Canada. Crystal Beach, Ont., July 5, 1919	Frank H. Ellis, parachutist; Don Russell, pilot. Curtiss JN4 biplane	

	First flight across the Canadian Rockies. Vancouver, B.C., to Calgary, Alta., *via* Lethbridge, Alta., August 7, 1919	Captain Ernest C. Hoy, pilot. Curtiss JN4 biplane
	First use of aircraft for forest fire patrol. Over the St. Maurice Valley, Que., summer of 1919	Stuart Graham, pilot; Bill Kahre, mechanic. HS2L flying-boat
	First International Air Races, Toronto–New York, and return, August 25–26, 1919	Winner of the Canadian event, Major R. W. Schroeder
	First airplane to fly in the province of Prince Edward Island. Truro, N.S., to Charlottetown, P.E.I., September 24, 1919	Lieutenants Laurie Stevens and L. L. Barnhill, pilots. Curtiss JN4 biplane (wheel-equipped)
	First multi-passenger flight in Canadian history (twelve persons). Parrsboro, N.S., October 9, 1919	Majors Brackley and Gran, pilots. Handley-Page bomber
1920	First private pilot's certificate issued. No. 1. January 24, 1920	Issued to James Stanley Scott, Ottawa, Ont.
	First licensed air engineer certificate issued. No. 1. April 20, 1920	Issued to Robert McCombie, Regina, Sask.
	First commercial aircraft to be registered in Canada. Lettering G–CAAA. April 20, 1920	A Curtiss JN4 biplane, owned by Aerial Service Company, Ltd., Regina, Sask.
	Order-in-Council constitutes the new Canadian Air Board, with full powers to act. April 19, 1920	Air regulations set up in 1919 come into force
	Provisional establishment authorized of the Canadian Air Force with a personnel to total 1,340 officers and 3,905 men. June 30, 1920	
	First commercial air pilot's certificate issued. No. 1. July 31, 1920	Issued to Roland J. Groome, Regina, Sask.
	First aircraft to penetrate the Yukon Territory. Wrangell, Alaska, to White Horse, Y.T., August 16, 1920	DH aircraft and men of the United States Army Air Service
	First trans-Canada flight. Halifax, N.S., to Vancouver, B.C., October 7 to 17, 1920	Aircraft and personnel of the Canadian Air Force
	First use of an airplane by a Canadian newspaper for fast news coverage. Winnipeg, Man., to Winkler, Man., and return, October 13, 1920	Hector Dougall, pilot; Frank H. Ellis mechanic; Winnipeg *Free Press* reporter, Cecil Lamont, passenger. Avro biplane
	First international official air-mail route established on the North American continent. Seattle, Wash., to Victoria, B.C. Commenced service October 15, 1920	Eddie Hubbard, pilot. C3 Boeing seaplane
	First commercial passenger flight into the Canadian north country. Winnipeg, Man., to The Pas, Man., October 15–17, 1920	Hector Dougall, pilot; Frank H. Ellis, mechanic; Frank J. Stanley, passenger. Avro biplane
1921	First aircraft to penetrate into the Northwest Territories. Peace River Crossing to Upper Hay River post, March 22, 1921	Two Imperial Oil Ltd., Junkers aircraft, crew, E. Fullerton, G. Gorman, P. Derbyshire, W. Hill; passengers, Sgt. Thorne, W. Waddell
1922	First commercial amphibious aircraft to be used in Canada	A Vickers Viking owned by the Laurentide Air Services
1923	The Department of National Defence created. January 1, 1923	
	His Majesty the King approved of the Canadian Air Force being designated "Royal" Canadian Air Force. February 15, 1923	
1924	First successful use of aircraft in Newfoundland for spotting seal herds. Over the Atlantic off Newfoundland, March 1924	Roy S. Grandy, pilot; H. C. Wallis, mechanic. Baby Avro with skis for take-offs and landings on the ice-fields
	First flight between Greenland and Canada. August 31, 1924	Four airmen of the U.S. Army Air Service in two Douglas Air Cruisers, during their round-the-world flight
	First use of aircraft for carrying treaty money to Indian reserves in Northern Ontario. Remi Lake to points on James Bay. August 1924	Roy S. Grandy, pilot; B. McClatchey, mechanic; Mr. Aurie, official of the Department of Indian Affairs, passenger. Vickers Viking (G–CAEB)
	First regular air-mail route established in Canada. Rouyn gold fields area. September 11, 1924	Aircraft owned by Laurentide Air Services
1925	First aerial northern mineral exploration flights in Canada —northern British Columbia and the Yukon. Commenced from Prince Rupert, B.C., June 1, 1925	J. Scott-Williams, pilot; C. S. ("Jack") Caldwell, co-pilot and engineer. Vickers Viking (G–CAEB)
1926	First mineral exploration by air in the Northwest Territories, from Lac la Biche, Alta., to points in the N.W.T. Commenced June 23, 1926	C. S. ("Jack") Caldwell, pilot; Irenée Vachon, engineer. Vickers Viking (G–CAEB)
	First "ground landing" of an amphibious aircraft in Canada. High River R.C.A.F. Station, September 4, 1926	C. S. ("Jack") Caldwell, pilot; Irenée Vachon, engineer. Vickers Viking (G–CAEB)
	First seaplane flight across Canada. Montreal, Que., to Vancouver, B.C., September 11, 1926 to September 19, 1926	J. Dalzell McKee, pilot and owner; Squadron Leader Earl Godfrey, R.C.A.F., co-pilot. Douglas seaplane

1927   First award of the McKee Trophy — H. A. ("Doc") Oaks, first recipient

Formation of the light airplane clubs in Canada

First major use of aircraft for insecticide dusting from the air. Cape Breton Island, June 1927 — Aircraft and personnel of the R.C.A.F.

First successful test in the air of a controllable pitch propeller. Camp Borden, Ont., June 6, 1927 — Wallace R. Turnbull, inventor; Flight Lieutenant G. G. Brookes, R.C.A.F., pilot. Avro biplane

First Canadian Atlantic flight attempt. From Harbour Grace, Nfld., July 7, 1927 (lost at sea) — Terrence Bernard Tully and James V. Medcalf, pilots. Stinson-Detroiter, the *Sir John Carling*

First flight of a Canadian-owned aircraft in the high north. Port Burwell, Hudson Strait, July 27, 1927 — Squadron Leader T. A. Lawrence and Flight Lieutenant A. A. Leitch, R.C.A.F., pilots. De Havilland Moth seaplane

First air mail to be dropped by parachute in Canada. Quebec City, December 21, 1927 — Romeo Vachon, pilot

1928   First Canadian recipient of the Harmon Trophy (the trophy for 1927) — Awarded posthumously to Captain Frederick J. ("Steve") Stevenson, Winnipeg, Man

First Canadian woman to receive a private pilot's certificate. No. 77. Dated March 13, 1928 — Eileen Vollick, Hamilton, Ont.

First east-to-west non-stop flight across the North Atlantic. Dublin, Ireland, to Greenly Island, Strait of Belle Isle, April 12–13, 1928 — Baron Huenefeld, Captain Koehl, and Major Fitzmaurice, pilots. Junkers monoplane, the *Bremen*

First aerial circumnavigation of Hudson Bay. August 1928 — T. M. ("Pat") Reid, pilot. Loening amphibian

First aerial search and rescue in the high north. Northwest Territories, August 1929 — T. M. ("Pat") Reid, search pilot

First flight from Canada to Greenland. Cochrane, Ont., to a landing on the Greenland ice-cap, August 18, 1928 — B. R. J. ("Fish") Hassell and Parker D. Cramer, pilots, Stinson monoplane

First flight across the northern Barrens. Baker Lake to Stony Rapids, September 3, 1928 — C. H. ("Punch") Dickins, pilot; William Nadin, engineer; Lieutenant Colonel C. D. H. MacAlpine and Richard Pearce, passengers

1929   First commercial northern winter flights. To Richmond Gulf, on Hudson Bay. January 1929 — Two aircraft, piloted by H. A. ("Doc") Oaks, and T. M. ("Pat") Reid

First commercial aircraft to be equipped with radio transmission and receiving set in Canada. March 1929 — Charles Sutton, pilot. Fairchild aircraft (CF–AAO)

First aerial invasion of the Barren Lands under winter conditions. March-April 1929 — Two aircraft, piloted by Stan McMillan and Charles Sutton (with a number of passengers). Ski-equipped Fairchilds

First Canadian woman to fly an aircraft across the international border. Winnipeg, Man., to Minneapolis, Minn., and return. April 29, 1929 — Eileen S. Magill

First emergency parachute jump to take place in Canada. Montreal, Que., May 17, 1929 — Airman whose life was saved, C. S. ("Jack") Caldwell. Irvin parachute

First airman to reach any point by air along the western Canadian Arctic coastline. Aklavik, N.W.T., July 1, 1929 — C. H. ("Punch") Dickins, pilot

First airmen to reach Coppermine on the Arctic coast. September 1, 1929 — Two aircraft, piloted by T. M. ("Pat") Reid and Jimmy Vance

Canadian Flying Clubs Association formed. November 1, 1929

1930   First woman pilot to establish an altitude record of over 15,000 feet in Canada. Saskatoon, Sask., June 8, 1930. Altitude reached, 16,000 feet. — Nellie Carson. De Havilland Moth

First visit of a major airship to Canada. Cardington, England, to Montreal, Que., July, 1930 — Commander R. S. Booth. R–100

First successful Canadian attempt to span the North Atlantic. St. John's, Nfld., to the Scilly Isles, October 9–10, 1930 — J. Erroll Boyd, pilot; Harry P. Connor, navigator. Bellanca monoplane

1931   First Trans-Canada Air Pageant — Leader, T. M. ("Pat") Reid

First Canadian-built aircraft to be flown from Canada to England *via* the South Atlantic. October 20, 1931 to December 7, 1931 — Bert Hinkler (Australia), pilot. Canadian-built De Havilland Puss Moth monoplane

Canada's first "flying boxcar." Montreal, Que., October 26, 1931 — Junkers aircraft, owned by Canadian Airways (CF–ARM)

1932   First Webster Trophy award — Edward G. Cox, first recipient

First east-to-west non-stop solo flight from Great Britain to a landing on the Canadian mainland. Partmarnock, Ireland, to Pennfield Ridge, N.B., August 18–19, 1932 — Jimmy Mollison, pilot. De Havilland Puss Moth

1933   First regular use of an ambulance plane in Canada. Toronto, Ont., 1934 — Capreol & Austin Airways. Waco biplane (CF–AVN)

First non-stop flight across Canada. Vancouver, B.C., to Kingston, Ont., August 25–26, 1933 — Captain Frank Hawks, pilot. Northrop Delta all-metal monoplane

1934	First west-to-east flight from Canadian soil to a landing in England. Wasaga Beach, Ont., to Heston, Middlesex, England, August 8–9, 1934	James R. Ayling and Leonard G. Reid, pilots. Twin-engined De Havilland Dragon biplane
1937	Aviation Section of the Royal Canadian Mounted Police formed	
	First dawn-to-dusk flight accomplished in Canada. Montreal, Que., to Vancouver, B.C., July 30, 1937	J. H. Tudhope, pilot; J. D. Hunter, co-pilot; Lew Parmenter, engineer; Hon. C. D. Howe and other government officials, passengers
1939	First aircraft to fly official mail across the North Atlantic. Between Foynes, Ireland, and Botwood, Nfld., July 5–6, 1939	The four-engined Short flying-boat, *Caledonia*—pilot in charge, Captain S. Wilcockson. The four-engined Sikorsky Clipper flying-boat, *Clipper III*—pilot in charge, Captain H E. Gray
1940– 1946	War years witnessed tremendous increase in aircraft manufacture and military activity	
1947	British Columbia Aviation Council re-activated	
1948	Four-engined T.C.A. North Star flew non-stop from Vancouver to Montreal, 6 hours, 52 minutes	
	R.C.A.F. Vampire Jet flown from Vancouver to Calgary in 52 minutes, August 31, 1948	Flight Lieutenant Mel Lee, pilot
1949	First non-stop flight completely across Canada. R.C.A.F. four-engined North Star, from Vancouver to Halifax, January 14–15, 1949	Flying Officer J. A. F. Jolicœur, captain, with crew of fourteen
	July, Canadian Pacific Air Lines inaugurated air route between Canada and Japan	
1950	First CF-100 twin-jet intercepter military aircraft turned out at A. V. Roe plant at Malton, Ontario	Test pilot, Squadron Leader W. A. Waterton
	Squadron 426, R.C.A.F., commenced airlift between Vancouver and Tokyo, in connection with Korean war operations	
1953	First fully jet engine built in Canada	Orenda, designed and constructed by A. V. Roe, at Malton, Ontario
1954	The Canadian Aeronautical Institute incorporated	
1955	First turbo-prop airliner goes into service over T.C.A. routes, April 1, 1955. First use of this type of aircraft on the North American continent	Vickers Viscount
1956	Outstanding flight by R.C.A.F. T-33, two-man jet trainer, from Vancouver to Dartmouth, in 6 hours and 52 minutes, January 16, 1956	Squadron Leader Lou Hill and Flight Lieutenant Alex Bowman
	The Alexander Graham Bell Museum opened at Baddeck, Cape Breton Island, Nova Scotia, August 18, 1956	Curator, C. M. Bowman
1957	Royal Canadian Navy commissioned their first aircraft carrier, H.M.C.S. *Bonaventure*, January 17, 1957	
1958	A. V. Roe, Malton, Ontario, unveils the CF-105 all-weather fighter-intercepter capable of speeds of over 1,000 m.p.h.	Test pilot, Janusz Zurakowski
	Canadair Limited, Montreal, establishes an air museum	
	First Canadian parachutist takes part in World Parachute and Sky Diving Championships. Prague, Czechoslovakia, August, 1958	Parachutist—sky diver Floyd Martineau, St. Catharines Parachute Club
1959	Golden Anniversary of Flight year in Canada, and re-enactment of first flight with replica aircraft, the *Silver Dart II*, at Baddeck, Cape Breton Island, Nova Scotia, February 23, 1959	Pilot of the *Silver Dart II*, Wing Commander Paul Hartman, R.C.A.F.
1960	Canada's National Air Museum officially opened at Ottawa Airport, October 25, 1960	Curator, K. M. Molson
	World's first "swing tail," multi-engined, all-purpose aircraft designed to speed up loading and unloading when craft is used as an air freighter	Canadair Limited, Montreal
1961	Modern, four-engined, pure jet, passenger airliners, the DC-8's, now in full operation over all major routes of Trans-Canada Air Lines and Canadian Pacific Air Lines, at home and abroad	
	First rotary wing passenger air bus, the Fairey "Rotodyne," on order by Okanagan Helicopters, Ltd., Vancouver, B.C., the world's largest operators of rotary wing aircraft	To be used in air transit across the Gulf of Georgia, between Vancouver and Victoria

APPENDIX B

NOTEWORTHY AIR-MAIL EVENTS IN CANADA TO 1939

The list which follows is compiled to include air-mail flights and incidents connected with Canada and Newfoundland which are outstanding from the historical standpoint. While the list may be of some help to those who wish to know a little about first-flown covers in Canada, philatelists who desire full knowledge of all air-mail routes which have been established in Canada must refer to regular stamp catalogues. A great number of such routes have been inaugurated since official air mail has operated under government regulations.

Date	Description	Pilot or Company involved	Where flown
Sept. 1, 1911	Earliest known message dropped from a plane in Canada	George Mestach	Quebec City
Sept. 12, 1911 (dated postal cancellation)	One of the earliest pieces of flown mail received by post by a resident in Canada	Addressed to Mrs. H. E. Leveson-Gower, Traynor, Sask.	Flown from London to Windsor, England, then by ordinary mail to Canada
June 24, 1918	First official air-mail flight in Canada	Captain Brian A. Peck, R.A.F., pilot; Corporal Mathers, R.A.F., passenger	Montreal, Que., to Toronto, Ont.
July 9, 1918	Earliest official air mail in western Canada	Katherine Stinson	Calgary, Alta., to Edmonton, Alta.
March 3, 1919	First official international air mail	Eddie Hubbard, pilot; W. E. Boeing, passenger	Vancouver, B.C., to Seattle, Wash.
May 18–19, 1919	First letters flown across Puget Sound and return	Lieutenant Robert Rideout, pilot; Lieutenant H. Brown, co-pilot	Victoria, B.C., to Seattle, Wash., and return
June 14–15, 1919	First mail flown across the Atlantic, non-stop	Captain John Alcock, pilot; Lieutenant Arthur Whitten Brown, co-pilot and navigator	St. John's, Nfld., to Clifden, Ireland
Aug. 7, 1919	First mail flown over the Rocky Mountains in Canada	Captain Ernest C. Hoy	Vancouver, B.C., to Calgary, Alta., via Lethbridge, Alta.
Aug. 16, 1919	Earliest mail flown on Vancouver Island	James Gray and Gordon Cameron	Victoria, B.C., to Nanaimo, B.C.
Aug. 25, 1919	Earliest mail flown between Toronto and New York	Aircraft connected with the International Air Race	Toronto to New York
Sept. 24, 1919	Earliest mail flown between Nova Scotia and Prince Edward Island	Lieutenants Laurie Stevens and L. L. Barnhill	Truro, N.S., to Charlottetown, P.E.I.
Oct. 7–17, 1920	First trans-Canada air-mail flight	Canadian Air Force	Halifax, N.S., to Vancouver, B.C.
Oct. 15, 1920	First international official air-mail route established on the North American continent	Eddie Hubbard	Seattle, Wash., to Victoria, B.C.
Feb. 26, 1921	First mail flown between St. John's and St. Anthony, Nfld.	F. S. Cotton	
March 10, 1921	First air-mail flight between St. John's and Fogo Island, Nfld.	F. S. Cotton	
Feb. 8 to May 16, 1923	First air-mail flight from Newfoundland to Labrador (delayed)	F. S. Cotton	St. John's, Nfld., to Cartwright, Labrador
Aug. 23, 1923	First unofficial flight with mail to the Magdalen Islands	R.C.A.F.	Dartmouth, N.S., to Grindstone Harbour
Sept. 11, 1924	First regular air-mail route established in Canada. Rouyn gold fields area	Laurentide Air Services	
March 6, 1926	First mail flown into the Red Lake gold fields area	J. V. Elliot Air Service	
Sept. 12, 1927	First successful flight with mail transferred to outgoing liner at Rimouski, Que.	Canadian Airways, Ltd. H. S. Quigley, pilot	Montreal, Que., to Rimouski, Que.
Sept. 16, 1927	First successful flight with mail transferred from an incoming liner at Rimouski, Que.	Government aircraft. H. S. Quigley and Stuart Graham, pilots	Rimouski, Que., to Montreal, Que.

Date	Description	Pilot or Company involved	Where flown
Nov. 11, 1927	First mail to be flown in the Yukon Territory	Yukon Airways and Exploration Co. Ltd. Andy Cruickshank, pilot; Mrs. Esmée Cruickshank, passenger	White Horse to Mayo, *via* Dawson, non-stop
Nov. 30, 1927	First mail flown to Pelee Island in Lake Erie	London Air Transport, Ltd. F. I. Banghart and Wm. Drury, pilots	Leamington, Ont., to Pelee Island, Ont.
Dec. 21, 1927	First air mail dropped by parachute in Canada	Canadian Transcontinental Airways. Romeo Vachon, pilot	Dropped at air field as plane passed over Quebec City
Dec. 25, 1927	Inaugural flight with air mail along the north shore of the St. Lawrence River	Canadian Transcontinental Airways. Romeo Vachon, pilot	La Malbaie to Seven Islands, and points *en route*
Sept. 5–8, 1928	First officially franked air mail flown across Canada	R.C.A.F. Squadron Leader Earl Godfrey, pilot; Sergeant Major M. Graham, engineer	Ottawa, Ont., to Vancouver, B.C.
Sept. 21, 1928	First air-mail stamps issued by authority of Canadian Post Office Department (placed on sale Oct. 1, 1928)		
Jan. 23, 1929	First mail flown north in western Canada	Western Canada Airways Ltd. C. H. ("Punch") Dickins, pilot; Mickey Sutherland, engineer	Fort McMurray to Fort Simpson
April 2, 4, 1929	First mail dropped from the air in the north country (free drop)	Dominion Explorers	Dropped at Tavani and at Chesterfield Inlet, *en route* from Fort Churchill to Baker Lake, N.W.T.
Dec 30, 1929	First official air mail flown into the Arctic	Commercial Airways Ltd. W. R. ("Wop") May and I. Glyn-Roberts, pilots	Fort McMurray to Aklavik, and points between
March 3, 1930	Actual inauguration of the Prairie Air Mail service	Western Canada Airways	Between Winnipeg, Regina, Saskatoon, Calgary, Edmonton
Aug. 25–26, 1933	First mail flown across Canada from west to east	Frank Hawks	Vancouver, B.C., to Quebec City
July 7, 1937	First official air-mail route established to the Yukon	United Air Transport, Ltd. Grant W. G. McConachie, pilot	Edmonton, Alta , to White Horse, Y T.
Aug. 4, 1938	First direct official air-mail route established from Vancouver to the Yukon	United Air Transport, Ltd. Grant W. G. McConachie, Ginger Coote, Sheldon Luck, pilots	Vancouver, B.C., to White Horse and Dawson, *via* Fort St. John
March 1, 1939	Full-scale official air-mail schedules inaugurated east and west across Canada	Trans-Canada Air Lines	
July 5–6, 1939	Regular official Atlantic air mail started	Imperial Overseas Airways and Pan American Airways	Between Foynes, Ireland, and Botwood, Nfld.

Appendix C

McKEE TROPHY AWARDS

	Name of recipient	*Basis of award*
1927	H. A. Oaks. Pioneer bush pilot, and founder of Western Canada Airways (inaugural award)	For early flying endeavours which helped greatly in opening up the north country by air
1928	C. H. Dickins. Pioneer bush pilot. Awarded the O.B.E. in 1935	For early bush flying, and for his flight across the Barren Lands in 1928
1929	W. R. May. Pioneer bush pilot. Awarded the O.B.E. in 1935	Chiefly for his mid-winter flight carrying diphtheria antitoxin from Edmonton to Fort Vermilion in 1929
1930	J. H. Tudhope. Early member of the Canadian Air Force, and experimental air-mail route pilot	For experimental air-mail flights conducted from Rimouski in 1927, and for supervision of prairie air-mail routes in 1930
1931	George H. R. Phillips. Early bush pilot with Ontario Provincial Air Service	Received chiefly for outstanding forest fire air patrol work, and for flights made in connection with Ontario Provincial Police and Health authorities
1932	Maurice Burbidge. Pilot and flying instructor with the Edmonton and Northern Alberta Aero Club	For his exceedingly fine record as a flying instructor
1933	Walter E. Gilbert. Early bush pilot. Awarded F.R.G.S. in 1932	Awarded for impressive flying in the north country, and for photographic exploration of King William Island in 1930
1934	Flight Lieutenant Elmer G. Fullerton. Pioneer bush pilot. Later continued service with the R.C.A.F.	Earned by reason of outstanding work with the Royal Canadian Air Force. First permanent Air Force officer to be honoured thus
1935	W. M. Archibald. Pilot, and manager of Consolidated Mining and Smelting Company	For his establishment of the use of aircraft for communication and transport in connection with mining operations
1936	A. M. Berry. Early bush pilot	For his practical flying in the north, and for two hazardous aerial rescues accomplished in 1936
1937	J. Romeo Vachon. Pioneer air engineer and bush pilot	Awarded for his work in organizing radio and weather reporting for the safety of air transport when manager of Quebec Airways in 1934
1938	Trans-Canada Air Lines	Awarded to the maintenance department of T.C.A. as a whole, for their splendid performance during the year
1939	Murton A. Seymour. Connected with Canadian aviation since 1915, as pilot and in many other capacities	The reward was made to him for 1939 chiefly for his splendid work in connection with the formation and operation of the Canadian Flying Clubs
1940	T. W. Siers. Early air engineer	Chosen for 1940 for his outstanding work as an engineer, and for his part in establishing the Worth Oil Dilution system, for cold weather flying in Canada
1941	A. D. McLean. Early pilot, and later with Civil Aviation Branch of the Department of Transport	For outstanding efforts in connection with the establishment of early air-mail routes and other aviation developments in Canada, and particularly for preparing bases for the Commonwealth Air Training Plan
1942-3	T. M. Reid. Early bush pilot, and later pilot and aviation fuel sales manager for Imperial Oil Limited	For his many flights and efforts in the advancement of flying in Canada
1944	J. A. Wilson. Controller of Civil Aviation in Canada from 1919 to 1941	For exceptional service and devotion to the progress and safety of aviation in Canada during his years of office
1945	Grant W. G. McConachie. Early bush pilot, and continued connections with Canadian aviation	For establishment and development of air routes in the north, particularly in the Yukon
1946	Group Captain Z. L. Leigh. Connected with flying in Canada since 1929, and for many years a member of the R.C.A.F.	First Chief Test Pilot of T.C.A. and for outstanding work as an administrative officer in the R.C.A.F.
1947	Barney A. Rawson. Pilot since 1928, and connected with T.C.A. since its inception	For outstanding efforts in connection with his work with T.C.A.
1948	Flying Officer R. B. West. Active member of the R.C.A.F.	For his exceptionally fine flying in connection with air search and rescue operations, and as a gesture to the entire staff of the R.C.A.F. engaged in such undertakings
1949	Dennis K. Yorath. Pilot since 1928, and since actively associated with the Royal Canadian Flying Clubs in various executive capacities	For his unstinted help in and efforts towards the progress of private flying in Canada

H. A. OAKS

(McKee Trophy for 1927)

C. H. DICKINS

(McKee Trophy for 1928)

W. R. MAY

(McKee Trophy for 1929)

J. H. TUDHOPE

(McKee Trophy for 1930)

G. H. R. PHILLIPS

(McKee Trophy for 1931)

M. BURBIDGE

(McKee Trophy for 1932)

W. E. GILBERT

(McKee Trophy for 1933)

E. G. FULLERTON

(McKee Trophy for 1934)

W. M. ARCHIBALD

(McKee Trophy for 1935)

MATT BERRY

(McKee Trophy for 1936)

J. ROMEO VACHON

(McKee Trophy for 1937)

M. A. SEYMOUR

(McKee Trophy for 1939)

T. W. SIERS

(McKee Trophy for 1940)

A. D. McLEAN

(McKee Trophy for 1941)

T. M. REID

(McKee Trophy for 1942–3)

J. A. WILSON

(McKee Trophy for 1944)

G. W. G. McCONACHIE

(McKee Trophy for 1945)

Z. L. LEIGH

(McKee Trophy for 1946)

B. A. RAWSON
(McKee Trophy for 1947)

R. B. WEST
(McKee Trophy for 1948)

DENNIS K. YORATH
(McKee Trophy for 1949)

CARLYLE C. AGAR
(McKee Trophy for 1950)
Astray Photographs, Vancouver

PHILIP C. GARRATT
(McKee Trophy for 1951)
Karsh, Ottawa

KEITH R. GREENWAY
(McKee Trophy for 1952)

FRANKLIN I. YOUNG
(McKee Trophy for 1953)

JERAULD G. WRIGHT
(McKee Trophy for 1954)

GERALD L. MacINNIS
(McKee Trophy for 1955)

ROBERT T. HEASLIP
(McKee Trophy for 1956)

J. G. SHOWLER
(McKee Trophy for 1957)

JANUSZ ZURAKOWSKI
(McKee Trophy for 1958)

J. A. D. McCURDY
(McKee Trophy for 1959)

W. G. LEACH
(McKee Trophy for 1960)

	Name of recipient	*Basis of award*
1950	Carlyle C. Agar. Pilot since 1929. Top helicopter pilot in Canada	For his outstanding work in the development of helicopter operations in Canada, with which, in later years, all his activities have been connected
1951	Philip C. Garratt. Original student of the Curtiss Flying School in Toronto in 1915, and connected with aviation since that date	For his lengthy connection with aviation in Canada with the De Havilland Aircraft Company, and the progress which he has helped to foster as their manager
1952	Squadron Leader Keith R. Greenaway. Active member of the R.C.A.F.	For his work in aiding air navigation in the Arctic, and for perfecting the twilight computor, now used as standard equipment in the R.A.F. and the R.C.A.F.
1953	Franklin I. Young. Bush and commercial pilot since 1927, and later connected with Trans-Canada Air Lines	Earned for his outstanding work in development of the National Air Show at Toronto in 1953

1954	Wing Commander Jerauld G. Wright. Member of the R.C.A.F. since 1940, and an outstanding navigational officer	For his invention of the R-Theta Computer, a unique navigational instrument adopted for use by the R.C.A.F. in long-range, high-speed aircraft
1955	Captain Gerald Lester MacInnis. Maritime Central Airways, Mont Joli, P.Q.	For his meritorious work as pilot and captain of aircraft with Maritime Central Airways in establishing the Distant Early Warning Line in the Canadian Eastern Arctic
1956	Wing Commander R. T. Heaslip. Active member of the R.C.A.F., and expert helicopter pilot	For his exceptionally fine contributions in connection with helicopter operations during the construction of the mid-Canada radar line
1957	Wing Commander J. G. Showler. Active member of the R.C.A.F., Officer Commanding, No. 408 Photographic Squadron	For his outstanding work as commanding officer in connection with the vast photographic survey of 3½ million square miles of the Canadian northland, 1954 to 1957
1958	Janusz Zurakowski. Chief experimental and test pilot, Avro Aircraft Limited, Malton, Ontario	For his contributions as a test pilot on various types of aircraft, and particularly for his ability in 1958 as test pilot on the CF-105 intercepter.
1959	J. A. D. McCurdy. Pioneer Canadian pilot and aviation executive	For his many fine contributions to the advancement of flying and in connection with the country's air industry, since he became the first Canadian to fly in Canada, in February, 1909
1960	Wing Commander W. G. Leach. Medical Officer, R.C.A.F., Institute of Aviation Medicine, Toronto	For his personal courage, displayed in pursuit of his research, beyond the call of duty which has resulted in greater safety for people the world over who fly high-altitude aircraft

WEBSTER TROPHY AWARDS

M. RENE SIMARD
(Webster Trophy for 1939)
Douglas Paisley, Sarnia

C. WILSON
(Webster Trophy for 1947)

J. H. BLACKBURN
(Webster Trophy for 1948)

D. R. JACQUES
(Webster Trophy for 1949)

EDWIN MONA
(Webster Trophy for 1950)

DONALD G. FISHER
(Webster Trophy for 1951)

HAPPY WELLS
(Webster Trophy for 1952)

GLEN E. McLARTY
(Webster Trophy for 1953)

ROBERT P. PURVES
(Webster Trophy for 1954)
Paramount Studio, Winnipeg

Recipient	Club		Recipient	Club
1932–1934	Edward C. Cox (inaugural award)	Montreal Light Aeroplane Club	1949	Donald R. Jacques	Aero Club of British Columbia
1935–1936	Gordon R. McGregor	Kingston Flying Club	1950	Edwin Mona	Aero Club of British Columbia
1937	B. J. Bourchier	Toronto Flying Club	1951	Donald G. Fisher	Kingston Flying Club
1938	Gordon R. McGregor	Kingston Flying Club	1952	Happy T. A. Wells	Chilliwack Flying Club, B.C.
1939	René Simard	Montreal Light Aeroplane Club	1953	Glen McLarty	Central Airways Company, Toronto
			1954	Robert Purves	Winnipeg Flying Club

1940–1946 The Webster Trophy contests were discontinued during World War II, and were not reorganized or held again until 1947

1947 Charles Wilson — Aero Club of British Columbia

1948 J. H. Blackburn — Edmonton and Northern Alberta Aero Club (Edmonton Flying Club)

The Webster Trophy Competitions were discontinued after the contest and award in 1954, and the opinion in aviation circles is that there is little likelihood of the competitions ever being renewed.

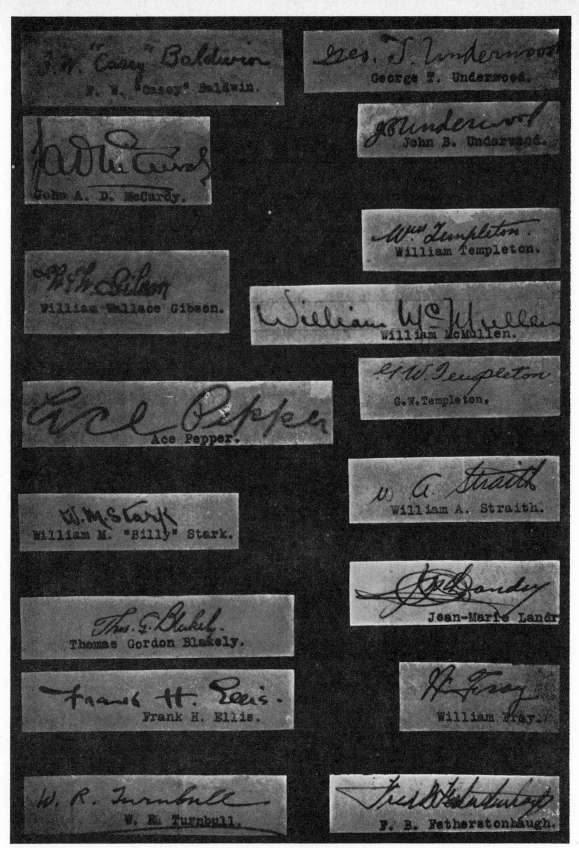

AUTOGRAPHS, NOW IN THE POSSESSION OF THE AUTHOR, OF ALL THE CANADIANS WHO PIONEERED AIRCRAFT CONSTRUCTION, OR WERE PILOTS, PRIOR TO THE OUTBREAK OF WORLD WAR ONE

APPENDIX E

FULL LIST OF THE GRADUATE PUPILS OF THE CURTISS AVIATION SCHOOL, TORONTO, DURING THE ENTIRE PERIOD OF ITS OPERATION, 1915–16

Name	Graduated	Served with	Name	Graduated	Served with
A. S. Ince	July 11, 1915	R.N.A.S.*	G. N. Hughes	October 31, 1915	R.N.A.S.
F. H. Smith	July 11, 1915	R.N.A.S.	A. M. Shook	November 5, 1915	R.N.A.S.
C. I. Van Nostrand	July 12, 1915	R.N.A.S.	G. G. MacLennan†	November 8, 1915	R.N.A.S.
G. A. Gooderham	July 12, 1915	R.N.A.S.	E. B. Waller	November 8, 1915	R.N.A.S.
D. A. Hay†	July 12, 1915	R.N.A.S.	G. S. Abbott	November 8, 1915	R.N.A.S.
C. N. Geale	July 20, 1915	R.N.A.S.	J. K. W. Aird	November 8, 1915	R.F.C.
E. H. McLachlin	July 20, 1915	R.F.C.‡	E. G. Ryckman†	November 11, 1915	R.F.C.
D. G. Joy	July 20, 1915	R.F.C.	A. G. Knight†	November 11, 1915	R.F.C.
C. MacLaurin	July 20, 1915	R.N.A.S.	M. M. Mowat†	November 11, 1915	R.F.C.
W. H. Peberdy†	July 20, 1915	R.N.A.S.	A. J. Boddy†	November 11, 1915	R.N.A.S.
R. D. Delamere	July 30, 1915	R.N.A.S.	J. A. Page†	November 11, 1915	R.N.A.S.
A. J. Nightingale	July 30, 1915	R.N.A.S.	R. T. Griffin	November 14, 1915	R.F.C.
A. T. N. Cowley	July 30, 1915	R.N.A.S.	G. M. Murray	November 23, 1915	R.F.C.
R. A. Logan	August 29, 1915	R.F.C.	D. Cushing	November 23, 1915	R.F.C.
B. S. Wemp	August 31, 1915	R.N.A.S.	J. H. N. Drope	November 23, 1915	R.F.C.
T. G. M. Stephens	September 1, 1915	R.N.A.S.	A. W. C. Kidner	November 23, 1915	R.N.A.S.
C. H. Darley	September 1, 1915	R.N.A.S.	N. H. McDiarmid	November 23, 1915	R.N.A.S.
G. R. S. Fleming†	September 1, 1915	R.N.A.S.			
R. A. Courtnage	September 1, 1915	R.N.A.S.	J. J. Malone†	July 15, 1916	R.N.A.S.
F. C. Henderson	September 1, 1915	R.N.A.S.	F. R. Johnson	July 19, 1916	R.N.A.S.
H. J. Page†	September 3, 1915	R.N.A.S.	R. M. Keirstead	July 19, 1916	R.N.A.S.
G. K. Williams†	September 3, 1915	R.N.A.S.	J. R. Allan†	July 20, 1916	R.N.A.S.
E. Potter	September 3, 1915	R.N.A.S.	R. A. Blyth†	July 22, 1916	R.N.A.S.
G. L. E. Stevens	September 4, 1915	R.N.A.S.	J. A. Morell	July 27, 1916	R.N.A.S.
A. S. Todd†	September 4, 1915	R.N.A.S.	S. E. Ellis†	July 29, 1916	R.N.A.S.
E. R. Grange	September 20, 1915	R.N.A.S.	A. C. Reid	August 1, 1916	R.N.A.S.
G. C. V. Hewson	September 23, 1915	R.N.A.S.	H. H. Booth	August 7, 1916	R.N.A.S.
C. Day	September 23, 1915	R.N.A.S.	A. H. Lofft	August 9, 1916	R.N.A.S.
G. E. Hervey	September 30, 1915	R.N.A.S.	W. A. Curtis	August 11, 1916	R.N.A.S.
J. Robinson	September 30, 1915	R.N.A.S.	O. J. Gagnier	August 15, 1916	R.N.A.S.
T. W. Webber	September 30, 1915	R.N.A.S.	R. McDonald†	August 15, 1916	R.N.A.S.
K. M. Van Allen†	September 30, 1915	R.N.A.S.	D. H. Wollatt	August 15, 1916	R.N.A.S.
L. E. Smith†	September 30, 1915	R.N.A.S.	M. C. Crerar†	August 17, 1916	R.F.C.
R. J. McDougall	October 3, 1915	R.N.A.S.	S. H. McCrudden	August 17, 1916	R.N.A.S.
A. O. Brissendon	October 3, 1915	R.N.A.S.	A. M. Walton	August 18, 1916	R.N.A.S.
R. K. Shives†	October 3, 1915	R.F.C.	D. M. Shields	August 22, 1916	R.N.A.S.
G. Thom	October 3, 1915	R.N.A.S.	G. H. G. Smyth	August 25, 1916	R.N.A.S.
A. W. Kilgour†	October 4, 1915	R.F.C.	R. C. Steele	August 25, 1916	R.F.C.
A. S. Shearer	October 6, 1915	R.N.A.S.	N. D. Hall	August 26, 1916	R.N.A.S.
D. Whittier†	October 6, 1915	R.N.A.S.	A. R. Knight	August 26, 1916	R.N.A.S.
A. Goulding	October 6, 1915	R.F.C.	A. E. Cuzner†	September 3, 1916	R.N.A.S.
L. W. Nurse	October 6, 1915	R.N.A.S.	C. J. Clayton	September 6, 1916	R.N.A.S.
J. J. Lynch	October 13, 1915	R.F.C.	F. McP. Bryans†	September 26, 1916	R.N.A.S.
J. W. Hobbs	October 13, 1915	R.N.A.S.	A. C. Burt	September 26, 1916	R.N.A.S.
D. A. H. Nelles	October 15, 1915	R.N.A.S.	C. E. S. Lusk	September 26, 1916	R.N.A.S.
V. P. Cronyn	October 15, 1915	R.F.C.	G. R. Marshall	October 2, 1916	R.N.A.S.
J. K. Waugh	October 17, 1915	R.N.A.S.	J. P. Hales†	October 12, 1916	R.N.A.S.
S. B. Lee	October 17, 1915	R.F.C.	H. C. M. Fitton	October 16, 1916	R.N.A.S.
H. G. Leslie	October 19, 1915	R.N.A.S.	G. A. Trorey	October 23, 1916	R.F.C.
			G. D. Eckardt	October 24, 1916	R.F.C.

*Royal Naval Air Service.
†Killed in action or through air accident.
‡Royal Flying Corps.

			C. G. Brock	October 28, 1916	R.N.A.S.
			G. B. G. Scott†	October 28, 1916	R.N.A.S.
			A. G. A. Spence	October 28, 1916	R.N.A.S.

GRADUATES OF CURTISS AVIATION SCHOOL

Name	Graduated	Served with	Name	Graduated	Served with
E. C. R. Stoneman	October 28, 1916	R.N.A.S.	L. J. Dunham†	December 4, 1916	R.N.A.S.
R. T. Eyre	November 5, 1916	R.N.A.S.	H. H. S. Fowler	December 7, 1916	R.N.A.S.
W. N. Smith	November 5, 1916	R.N.A.S.	T. L. Glasgow†	December 7, 1916	R.N.A.S.
F. S. Strathy†	November 5, 1916	R.N.A.S.	H. L. Webster	December 15, 1916	R.N.A.S.
R. F. P. Abbott	November 7, 1916	R.N.A.S.	N. G. Fraser	December 18, 1916	R.N.A.S.
W. C. Johnston	November 7, 1916	R.N.A.S.	J. E. Greene†	December 19, 1916	R.N.A.S.
N. I. Larter†	November 7, 1916	R.N.A.S.	L. L. Lindsay	December 19, 1916	R.N.A.S.
R. E. Carroll	November 10, 1916	R.N.A.S.	R. C. Tyler	December 19, 1916	R.N.A.S.
P. W. Jenckes	November 10, 1916	R.N.A.S.	C. W. L. Calvert	December 20, 1916	R.N.A.S.
F. A. Major	November 10, 1916	R.N.A.S.	C. G. Davis	December 20, 1916	R.N.A.S.
T. C. May†	November 21, 1916	R.N.A.S.	T. B. Holmes	December 20, 1916	R.N.A.S.
P. Wickens	November 21, 1916	R.N.A.S.	O. F. McGregor	December 20, 1916	R.N.A.S.
E. S. Campbell	November 22, 1916	R.N.A.S.	K. V. Turney†	December 20, 1916	R.N.A.S.
C. H. Weir†	November 26, 1916	R.N.A.S.	R. C. Whitfield	December 20, 1916	R.N.A.S.

Appendix F

KNOWN PUPILS CONNECTED WITH THE AERO CLUB OF BRITISH COLUMBIA, TO THE END OF 1915

	Home town	Date of graduation as pilot
Murton A. Seymour	Vancouver, B.C.	November
Philip H. Smith	Vancouver, B.C.	November

Pupils who trained in November but did not graduate

Arthur H. Allardyce	Vancouver, B.C.	Charles ("Cy") Perkins	Port Kells, B.C.
"Chad" Chadwick	United States	J. V. W. ("Vick")	
J. B. Crawford	United States	Phillips	New Westminster, B.C.
W. E. Damer	Vancouver, B.C.	Charles Raynor	Vancouver, B.C.
Gerry Hodgson	Vancouver, B.C.	Cyril Scott	Vancouver, B.C.
Robert Main	Vancouver, B.C.	John F. Stephen	Vancouver, B.C.
William G. McRae	Agassiz, B.C.		
Sidney Mowat	Vancouver, B.C.		

APPENDIX G

THE CATERPILLAR CLUB

The Caterpillar Club, originated by the Irvin Air Chute Company of Buffalo, N.Y., began in 1920. Persons of any nationality who have made an emergency jump wearing an Irvin parachute are eligible for membership. The word, caterpillar, and the gold badge depicting a silkworm are trade-mark features of the Irvin Air Chute Company. The earliest parachutes were fabricated from pure silk, but nylon has now replaced silk in the manufacture of most modern parachutes.

Between the date of the first emergency parachute jump and the end of 1939, eleven lives were saved in Canada by Irvin parachutes. No other type was then in use in Canada.

The first airman to be saved was C. S. ("Jack") Caldwell, who figured prominently in the history of Canadian aviation during the twenties. On May 17, 1929, while in the employ of Canadian Vickers Ltd. at Montreal, Caldwell was testing a

C. S. ("JACK") CALDWELL

new Vickers Vedette flying-boat above the St. Lawrence River when the machine went into a spin. Unable to regain control, Caldwell bailed out, 500 feet from the ground, to drift down to a safe landing in the river. The aircraft was a total loss. The wind kept the 'chute inflated after the landing, and towed the airman along the surface until he was able to reach a small island, whence he was later rescued. This episode earned for Caldwell the first Canadian membership in the Caterpillar Club.

A list of the eleven who were admitted to membership in the club by the end of 1939 is given below:

Date	Name
May 17, 1929	Jack Caldwell, chief test pilot, Canadian Vickers, Ltd.
August 27, 1929	Sergeant J. M. Ready, R.C.A.F.
June 2, 1930	Lieutenant M. F. Slaght, U.S.A.A.C.
September 28, 1931	Sergeant F. B. Brisco, R.C.A.F.
December 19, 1931	E. C. Burton
July 26, 1932	Flying Officer F. M. Gobiel, R.C.A.F.
May 27, 1936	G. C. Upson
December 15, 1938	Flying Officer R. M. McKay, R.C.A.F.
December 15, 1938	Corporal P. Wilkinson, R.C.A.F.
August 30, 1939	Air Gunner T. Perley-Martin, R.C.A.F.
November 18, 1939	Flying Officer J. D. Morrison, R.C.A.F.

After the outbreak of World War II, a great many airmen, both in Canada and overseas, became eligible for membership. None of the firms who make parachutes could possibly have compiled complete lists of the lives saved under such circumstances—many who went down to safe landings in enemy territory, by parachute, will never have their names recorded. However, gold caterpillar lapel pins were presented to all the airmen eligible for membership that could be traced.

As an early parachute jumper by choice, I have the utmost respect for a true "Caterpillar." It is one thing to make a planned jump from a plane flying under perfect control and in ideal weather, but quite another to be forced to bail out of a disabled aircraft under stress of combat, perhaps in the dead of night, with the plane spinning groundward out of control, and sometimes on fire to boot. With only split seconds in which to act, the only hope is a good 'chute in working order.

The hazardous experience of Sergeant D. A. Clarke of West Vancouver, B.C., was typical of many whose lives were saved during World War II by Irvin parachutes. As a wireless operator serving with the R.C.A.F. in a Halifax bomber squadron, Sergeant Clarke was on a mission over Germany on the night of June 21, 1943, when his Halifax was hit and caught fire. With a burning crash inevitable, the seven-man crew endeavoured to bail out. Five survived. The gold caterpillar pin presented to Sergeant Clarke by the Irvin Air Chute Company is a tangible reminder to him of his moments of intense peril in the air, by night.

APPENDIX H

LICENSING OF AIRCRAFT AND PERSONNEL

AIRCRAFT REGISTRATION

Before January 1, 1929, all aircraft registered in Canada bore the prefix letter G, followed by a dash, the letter C, and three additional letters.

When the original international allocation of national markings for aircraft began—it became effective in Canada in 1920—the British Commonwealth was given the nationality letter G. The second letter specified the country to which the aircraft belonged. E was for England, C for Canada, and so on, with different letters following.

In 1927, at the Radio Telegraph Convention held in Washington, D.C., groups of call-signs, each group consisting of five letters, were allotted to different countries for their civil aircraft. These letters, painted on the sides and wings of an aircraft, also constitute its letters of registration. Thirteen of these national indicators were allotted to Canada, and of them, the group beginning with CF was chosen as the first series to be employed. Only one series can be selected for use in civil aircraft at one time by any one nation. At some remote date, when the last three-letter combinations which complete the group beginning with CF become exhausted, two new prefix letters will replace the CF now in use. Then the three following letters will begin all over again, starting with AAA.

The change from G to CF became effective in January 1929, the G gradually disappearing as aircraft bearing it were withdrawn from use. If you should see an aircraft with a G at the head of its registration lettering today, you would be looking at a real old-timer.

To meet growing requirements, and to prevent confusion between military and civil aircraft, all military machines in Canada came under the lettering system from May 22, 1948, but were allotted a different prefix series. The following blocks were chosen: VG–AAA to VG–ZZZ for the use of the Royal Canadian Navy, and VC–AAA to VC–ZZZ for the Royal Canadian Air Force.

The first commercial airplane registered under the old system was a Curtiss JN4 biplane owned by Aerial Service Company, Ltd., of Regina, Sask. It received its registration G–CAAA on April 20, 1920.

The first civil airplane to be registered in Canada under the new system was a De Havilland Gypsy Moth. It received the lettering CF–CAK on January 28, 1929. The machine that actually bore the first lettering, CF–AAA, was also a Moth owned by Sir James MacBrien. To stimulate flying interest throughout the country, Sir James undertook a very energetic programme of coast-to-coast flying trips, and he and his dainty AAA became a familiar sight in many parts of Canada. Up to June 30, 1931, he had flown over 50,000 miles which included two trans-Canada journeys by air.

PILOT AND AIR ENGINEER CERTIFICATES

Throughout the entire pioneer period of flying in Canada, until the year 1920 to be exact, no law existed which forced a pilot to obtain a certificate as proof of his flying ability. Neither were aircraft or airships registered.

However, membership in the Aero Club of America certainly carried high prestige. The tests were not at all severe, but a pilot was required to take off and land within specified points, and was not considered eligible for a licence unless certain basic evolutions could be carried out in flight. The two major ones were a circular flight and what was then termed a figure-eight. A pilot was also expected to be reasonably conversant with engine and airplane maintenance. A great many of the pioneer pilots did not trouble to obtain a licence, the chief reasons being either the fees involved, or the inconvenience of travelling to a point where examinations by Aero Club officials could be conducted.

Canada had only four licensed pilots living within her domain when World War I broke out in August 1914. They were John A. D. McCurdy of Baddeck, N.S., and Toronto, Ont., who possessed No. 18, dated October 5, 1910; William M. Stark of Vancouver, B.C., with No. 110, issued March 22, 1912; William Fray, of Mirror, Alta., with licence No. 306, dated July 12, 1914; and Jean Marie Landry, who obtained his as a pupil at the Blériot School in France, receiving Fédération Aéronautique Internationale licence No. 1659, dated June 20, 1914, under authority of the French Aero Club. The first three obtained their licence through the Aero Club of America.

When a pilot's licence became a must for airmen in 1920, the various types as they originated began with the number one. Private pilot's certificate No. 1 was issued to James Stanley Scott, of Ottawa, Ont., and was dated January 24, 1920.

The No. 1 commercial air pilot's certificate went to Roland J. Groome of Regina, Sask., being dated July 31, 1920.

Air engineers also came under government supervision beginning in 1920, and the No. 1 licence issued in this classification was received by Robert McCombie of Regina, Sask., bearing the early date of April 20, 1920.

In 1935 a new type of commercial licence was issued known as the limited commercial pilot's certificate. This was to allow pilots of smaller aircraft to operate without being obliged to sit for examinations for the more complicated fully commercial pilot's licence. The initial one of this class, C–1301, was issued to R. E. Hadfield on September 20, 1935. By the middle thirties, the vast increase in passenger travel by air, and the use of large multi-engined aircraft, brought stricter examination of those who flew on this type of work, as the safety of passengers required a multitude of additional flight rules and regulations. To cover this phase of flying the transport pilot's certificate came into existence in 1936, and the No. 1 endorsed for this class was procured by Douglas Joy of Ottawa, dated April 1, 1936.

AIR NAVIGATION CERTIFICATE

The first air navigation certificate obtained in Canada, marked No. 1, was issued to Major Robert A. Logan, in 1921. It was signed by Mr. J. A. Wilson, Controller of Aviation, and was good for only one year. Up to that date there had been no standard of rules governing examinations for obtaining such a licence. Major Logan at the time was officer in charge of Ground Instructional School of the Canadian Air Force at Camp Borden.

MUSEUMS

In England, the United States, and a number of other countries, where interest in early aviation achievements has been kept alive, gratifying success has been realized in gathering together relics of the past.

The Science Museum at South Kensington, London, has collected some splendid exhibits and records, numerous pioneer aircraft being on display. One of the prized items is the Vickers Vimy biplane in which Alcock and Brown did the first non-stop flight across the Atlantic.

In the National Air Museum, a part of the Smithsonian Institution at Washington, D.C., a very extensive collection has been formed, and they have a mass of material still in storage, awaiting enlarged building space. In one of the great halls can be seen the Wright brothers' biplane which made world flying history at Kitty Hawk in 1903. A few yards behind, also suspended from the high ceiling, is the *Spirit of St. Louis*, the craft that Lindbergh flew from New York to Paris in 1927. Still another famous exhibit is the *Winnie Mae*, in which Wiley Post twice circled the world by air. A second well-established collection in the United States is at Greenfield Village in Dearborn, Mich., where the Henry Ford

Aeronautical Museum is located. One of their most valued items, strange to relate, is the German Junkers monoplane, the *Bremen*, first airplane to span the Atlantic from east to west.

That such priceless aircraft have been preserved for posterity reflects great credit on the countries involved.

Now let us turn the spotlight on Canada. What has been done here in this respect? The answer, disappointing, but true, is "very little."

True, the National Research Council has an Aeronautical Museum. The collection was started some years ago, when numerous items were donated by F. W. (Casey) Baldwin, John A. D. McCurdy, Wallace R. Turnbull, and by members of Alexander Graham Bell's family, and it was set up in 1937 in one of the exhibition halls of the National Research Council building on Sussex Street, Ottawa, under the sponsorship of the Associate Committee of Aeronautical Research. It is now under the direction of J. H. Parkin. Material has been added since, though it is all chiefly connected with more recent times. A few relics from the pioneer days have since been found in widely scattered parts of Canada and placed

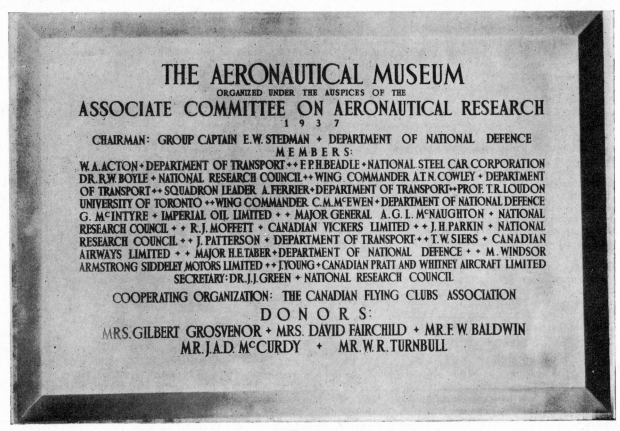

DEDICATION PLAQUE AT THE NATIONAL MUSEUM, OTTAWA

on view. However, too much time has now elapsed since the early days for many additional items to have survived, and it is very apparent that Canada's exhibit of her own aviation development will always remain woefully incomplete.

In Winnipeg the Hudson's Bay Company own a splendid museum which is open to the public, but its value as a collection is chiefly for the history of Canada in general. Nevertheless, a few interesting flying relics have come into the possession of the Company and are now on view.

Still more recent is the establishment of the Western Development Museum, which has exhibits at Saskatoon, North Battleford, and Yorkton. Its interest is confined to the aviation history of the province of Saskatchewan, but it is a very worthy effort and deserves every help.

The members of Canada's Air Industries and Transport Association have avowed an interest in the possibility of establishing a national air museum. If they carry through their ideas, they will obtain great credit not only for themselves, but for Canada.

The pioneer machines have vanished, of course, but it is quite practicable for exact scale models to be made to take their place, similar to those shown in the illustrations contained in chapters 2 and 4. Such small replicas show all the details quite well, and the public at large would at least be able, by examining them, to envisage the appearance of the original machines. Models would certainly serve a purpose in reminding the public of the almost forgotten past of our flying history. Models of later machines, now unobtainable, could well be included.

Another efficient method of bringing the history of a country to the attention of the public is by means of impressively designed murals. Canada has not branched out into this sort of thing very seriously as yet. Of recent years it has made some headway, but not as much as it should, given such an adventurous history.

The use of murals for this purpose is very noticeable throughout the United States, where many magnificent works of art adorn the walls of government and public buildings. Some of them impressively depict various phases in the progress of American aviation.

There are thousands of huge blank walls in Canada's government and public buildings which could well be adorned with murals depicting events in Canadian aviation. We have artists of high ability, and if they were to work in close conjunction with aviation historians, the murals which they created would be small additional museums.

THE ALEXANDER GRAHAM BELL MUSEUM
BADDECK, CAPE BRETON ISLAND, NOVA SCOTIA

Since the original text for this book was compiled and published in 1954, events closely connected with our air history have taken place in Canada which prove that interest has finally developed to the point of influencing governmental action in the setting aside of funds for museum purposes. The first to be completed and opened to the public was the Alexander Graham Bell Museum at Baddeck, Nova Scotia, which houses a large collection of plans, drawings, models, and parts of original items connected with the career of this illustrious Canadian. The museum is not entirely confined to aviation relics and records; nevertheless, they do play a large part in the exhibit. Moreover, although he is world famous for the invention of the telephone, his work in connection with developing mechanical and touch systems for the teaching of the deaf and the blind and other contributions towards public health are not too well known by the general public.

The museum was completed early in 1956 under the able administration of the Honourable Jean Lesage, Minister of Northern Affairs and National Resources. Under the curatorship of Mr. C. M. Bowman since its official opening in August of 1956, by Dr. Bell's daughters, Mrs. Gilbert Grosvenor and Mrs. Daniel Fairchild of Washington, D.C., the museum has welcomed thousands of people through its doors. More and more are making the pilgrimage to Baddeck each year to indulge in the opportunity of seeing first hand the fine display it contains and to stand in wonder before the many exhibits developed by the great man for the benefit of mankind.

THE CANADAIR MUSEUM
MONTREAL, P.Q.

In the spacious plant of Canadair Limited, Montreal, designers and skilled workers have turned out many types of commercial and military aircraft which have contributed greatly to the advancement of aviation in Canada and throughout the world. With such activity taking place, it is encouraging to know that the company's executives have not been too busy, or indifferent, to commemorate the part that Canada as a whole has played in air history.

Since 1958 a splendid and permanent air museum has been established at their Montreal plant under the able direction of Mr. G. Haynes Davies, Manager of the Information Service. Through his efforts and with company sanction, the museum has become one of the outstanding displays of its kind in Canada, with a great many models ranging from the earliest pioneer airplanes which were built or flown in Canada to replicas of modern aircraft connected with present-day flying activities. The author was commissioned by Canadair to construct numerous models of the pioneer airplanes flown in Canada and they are included in the representative display.

The public is more than welcome to inspect the museum during plant hours and a trip to see its exhibits is a must for all air-minded visitors to Montreal who have an hour or so to spare.

THE NATIONAL AVIATION MUSEUM
OTTAWA

On October 25, 1960, a ceremony took place at Ottawa which marked the culmination of the hopes and efforts of numerous individuals and organizations who for years had been advocating the establishment of a permanent and public air museum for Canada. It was considered that the time was long overdue for centralizing a pictorial record relating Canada's splendid aviation progress.

The first portion of this Appendix was written in 1951, when a National Air Museum was still a project of wishful thinking, and it is very gratifying to be afforded the opportunity here of including the facts of its establishment.

The museum is housed in a section of the new buildings erected at Ottawa's International Airport, under the able direction of Mr. K. M. Molson, Canada's first National Aviation Museum Curator. Special credit goes to Mr. J. H. Parkin, retired Director of the National Research Council, General A. G. L. McNaughton, Professor T. R. Loudon, the late Air Vice Marshal E. W. Stedman, and many others, together with numerous organizations, who originally sparked the idea, when the first small museum was opened in 1937.

And now it has become an established fact. It should be borne in mind by all Canadians that it is their museum, and no effort should be spared to contribute items to it relating to our air history. The story of our pioneer years is fairly well represented by photographs and written descriptions, but unfortunately none of the Canadian-built pioneer airplanes survived the years to find an honoured resting place in the museum.

It is a young museum at this moment, but as the years pass its popularity will grow at a fast pace. No one residing in Ottawa should miss the pleasure of viewing the many exhibits. Visitors too, should not forego the chance of browsing around the museum, to get first-hand knowledge of Canada's adventurous past from the time the country's air pioneers placed their trust in such frail wings up to the events which have made Canada the great air nation it has become today.

APPENDIX J

CAIRNS AND MONUMENTS

The cairns and monuments erected in Canada to commemorate famous events in aviation are few to this date, and all of them are in eastern Canada. Nevertheless those that have been built mark well the achievements they are dedicated to, and are worthy records of a permanent nature.

Strangely enough, the first memorial to Canadian airmen on this continent was not raised in Canada.

During the winter of 1917–1918, when the 42nd and 43rd wings of the Royal Flying Corps were transferred from Ontario to Texas, where it was believed that their training could continue unabated, twenty-nine officers and men lost their lives while on duty, mainly through air accidents. Before the Corps returned to Canada in the spring of 1918, it set up an impressive memorial at Benbrook Field. The memorial remained standing there for a number of years until the area was no longer used as an air field. Then it was removed to Greenwood Cemetery where eleven of the men of the R.F.C. lie buried. The plot upon which it centres has been dedicated to Canada, and is kept in permanent neatness by Texan citizens, who are members of the British-American Association of Fort Worth.

It is to the credit of the Clarke Steamship Company, Limited, Montreal, that they raised a suitable cairn on Greenly Island, in the Strait of Belle Isle, to commemorate the arrival there of the German aircraft, the *Bremen*, the first airplane to fly non-stop across the North Atlantic from east to west. The cairn was erected entirely at the expense of the Company, and was the first of its kind to mark an outstanding Atlantic flight.

The dedication ceremony was performed on July 25, 1928, and the unveiling was done jointly by Monseigneur Laventoux, Roman Catholic Bishop of the North Shore, Reverend Father Gallix, and the Anglican representative, Canon S. F. G. Scott. Captain J. A. Brie, Master, and the officers and crew of the Clarke Line vessel S.S. *North Shore*, attended, together with Louis T. Blais, personal representative for Desmond A. Clarke, president of the steamship company which bears his name. Cruise passengers from Canada and the United States were also on hand, together with many representative fishermen and their families from various localities near by.

The words on the plaque affixed to the cairn are as follows:

On this island landed the Bremen *on Friday April 13, 1928 after the first non-stop flight of the Atlantic from east to west having left Ireland at dawn Thursday April 12. Members of its crew were, Baron Gunther von Huenefeld of Germany, Captain Hermann Koehl of Germany, Major James C. Fitzmaurice of Ireland. Erected by Clarke Steamship Co. Ltd, Montreal, Quebec in recognition of a very gallant feat.*

Overlooking the village of Gaspé from the slopes above, an imposing ten-foot monument designed by the eminent Montreal sculptor Henri Hébert was erected on August 16, 1932.

This, the first memorial to be placed on Canadian soil as a mark of remembrance to airmen flying in Canada was to record the passing of a famous French airman and his engineer, who were lost at sea during 1927 while engaged in photographic work over the Gaspé area.

Should you ever visit Gaspé to enjoy its rural atmosphere and the glory of its summer beauty, do not fail to climb the green-clad hill that rises behind the village, to the place where the marble tribute stands. Thereon you may read inscribed in the French language the brief history of a pilot of international fame.

Translated, the inscription reads:

To the memory of Jacques de Lesseps, Chevalier of the Legion of Honour, Croix de Guerre, Distinguished Service Cross (U.S.A.). Born in Paris in 1883. The second person to fly across the English Channel by airplane, in 1910. The first person to fly over Montreal and Toronto. Cited four times in the Orders of the Day, 1914–1918. Lost at sea on October 18, 1927, during a flight taking photographs for the map of the Gaspé. Cited in the Order of the French Nation. His body is buried in the Gaspé graveyard. Theodor Chickenko. Born in 1894, at Gourilouka, Russia. Enlisted in the French Army during the Great War. Lost at sea with Jacques de Lesseps.

On the sunlit afternoon of August 16, 1934, a double guard of honour comprised of Cape Breton Highlanders and a company of sailors from H.M.C.S. *Saguenay* took up position in front of the court house at Baddeck, N.S., to assist in the unveiling ceremonies of a cairn and tablet to commemorate the work accomplished by the Aerial Experiment Association, and the first airplane flight in Canada.

The ceremony was under the auspices of the Canadian Flying Clubs Association, working in conjunction with the Baddeck Board of Trade and the Cape Breton Flying Club. The memorial itself was supplied by the Association. Four thousand persons were on hand to witness the unveiling, at which Walter H. Cavert, K.C., Lieutenant-Governor of Nova Scotia, officiated. Brief addresses were delivered by J. A. Wilson, Controller of Civil Aviation, and Lieutenant Colonel George A. Drew, to which both Mr. Baldwin and Mr. McCurdy replied modestly, receiving prolonged cheers and applause.

The bronze tablet on the cairn bears the following inscription:

First Aeroplane Flight in the British Empire
Commemorating the work of the Aerial Experiment Association, which resulted in the first aeroplane flight within the British Empire, made by J. A. D. McCurdy at Baddeck, Nova Scotia, on February 23rd, 1909. The Aerial Experiment Association was formed at Baddeck, by Dr. Alexander Graham Bell. The other members were, F. W. Baldwin, whose first successful flight was made at Hammondsport, N.Y. on March 12th, 1908; Glenn H. Curtiss; J. A. D. McCurdy and Thos. E. Selfridge. Their contributions to the advancement of aeronautical science are gratefully acknowledged.

What can be considered a fitting climax to the history of the Aerial Experiment Association, particularly as it applies to Casey Baldwin and John McCurdy, were the events which took place in August 1952 at Petawawa military camp forty-three years after the flights made there by the two Canadian airmen.

Plans made under the auspices of the Historic Sites and Monuments Board of Canada culminated in the erection

THE UNVEILING OF THE *BREMEN* MEMORIAL CAIRN

THE REID-AYLING MEMORIAL AT WASAGA BEACH   DE LESSEPS MONUMENT AT GASPÉ

ROYAL FLYING CORPS PLOT, GREENWOOD CEMETERY, FORT WORTH, TEXAS

THE ALCOCK-BROWN MEMORIAL   THE PETAWAWA MEMORIAL

Courtesy The Evening Telegram, St. John's, Newfoundland

of a suitable cairn and plaque; the dedication ceremonies took place on August 12, 1952, the unveiling being conducted by Hon. Brooke Claxton, Minister of National Defence.

Among the large number of guests and spectators gathered to witness the affair was John A. D. McCurdy, Lieutenant-Governor of Nova Scotia, together with some of Canada's senior Navy, Army, and Air Force officers.

The smart guard of honour for the occasion was composed of fifty officers and men drawn from the ranks of the Royal Canadian Regiment. Heading the additional troops lined up to the rear of the area was a detachment of the Royal Canadian Engineers, who were given the priority in this instance because it was men of that regiment who made arrangements and helped in supervising the ground activities during the Petawawa flights in 1909.

The wording on the bronze plaque attached to the cairn is as follows:

First Military Test Flights

The first military demonstration of aircraft flight in Canada was given at Petawawa Camp in August, 1909, by J. A. D. McCurdy and F. W. Baldwin, with the assistance of the Royal Canadian Engineers.

On the morning of the 2nd August they made four successful flights in the "Silver Dart." This aircraft was destroyed in landing after the fifth flight.

Further flights were made in the "Baddeck No. 1" on the 12th and 13th of August. The tests were terminated on the 13th when this aircraft was damaged.

Beneath the large plaque is a smaller bronze tablet, on which is depicted the *Silver Dart* in full flight.

On the afternoon of November 26, 1952, an event took place at St. John's Nfld., which had been planned and discussed for many years, when a stone memorial and bronze plaque were unveiled in the small park off Le Marchant Road in tribute to the memory of Alcock and Brown, first men to fly across the Atlantic Ocean non-stop.

Thirty-three years after the daring flight, the monument was finally erected and dedicated through the efforts of the Newfoundland Historical Society, in conjunction with the Historic Sites and Monuments Board of Canada. The monument stands close by the spot from which the historic flight began.

The ceremony was attended by His Honour the Lieutenant-Governor of Newfoundland, Sir Leonard Outerbridge; C. E. A. Jeffery, the Newfoundland representative of the Historic Sites and Monuments Board; Acting Premier L. R. Curtis; His Worship Mayor H. G. R. Mews of St. John's; and members of the government and the Newfoundland Historical Society. A host of spectators were also on hand, many of whom had seen the actual start of the Alcock and Brown flight.

A full guard of honour was formed by members of the Newfoundland branch of the Air Cadet League of Canada. The memorial was unveiled by Captain Douglas Fraser, a well-known Newfoundland airman.

The inscription on the bronze plaque reads:

Historic Sites and Monuments Board of Canada
ALCOCK-BROWN TRANSATLANTIC FLIGHT

On 14th. June, 1919, Captain John Alcock and Lieutenant Arthur Whitten Brown of the R.A.F. took off nearby on the first non-stop Transatlantic flight in a Vickers Vimy aeroplane at 12:58 P.M. Newfoundland time. Sixteen hours and twelve minutes later they landed at Clifden, Ireland, a distance of 1800 miles.

In addition to the massive stone memorial, a small bronze tablet prepared by the Newfoundland Historical Society has been placed at the point off Blackmarsh Road, to mark the exact spot from which the Vickers Vimy became airborne

on the adventurous journey which remains one of the greatest feats in the history of world aviation.

On August 8, 1934, James R. Ayling and Leonard G. Reid, two British airmen, took off from Wasaga Beach in a twin-engined De Havilland biplane, their objective, Baghdad. Trouble *en route* forced them to bring their flight to an ending at Heston airport, near London, England, after they had accomplished the first west-to-east, non-stop flight across the Atlantic, from a point on the Canadian mainland to one in England.

To mark their splendid achievement, a cut-stone monument, ten feet high, was erected at Wasaga Beach under the combined sponsorship of the Historic Sites and Monuments Board of Canada and the Huronia Historic Sites and Tourist Association.

For eight years plans for the monument slowly progressed, chiefly through the untiring efforts of R. W. Thom and Professor F. Landon, both of Collingwood.

The unveiling ceremony took place on August 24, 1954, and was attended by a large crowd of local residents and visitors. Addresses were delivered by Dr. J. R. K. Main, Assistant Controller of Civil Aviation for Canada, by Dr. Fred Landon, Chairman of the Historic Sites and Monuments Board of Canada, and by R. W. Thom. A fly-past by members of the R.C.A.F. lent added interest to the proceedings.

The bronze plaque on the face of the monument bears the following brief tribute to a brilliant achievement:

HISTORIC SITES AND MONUMENTS BOARD OF CANADA

From Wasaga Beach on the morning of 8th August, 1934, James R. Ayling and Leonard G. Reid took off in their plane, "The Trail of the Caribou", headed for Bagdad. Due to adverse weather conditions and exhaustion of fuel supplies, they were forced to land at Heston Airfield, London, England, on the afternoon of 9th August, having flown 3700 miles in 30 hours, 55 minutes.

After Norway was attacked and overrun by force of German arms during World War II, the Royal Norwegian Air Force was re-established and re-organized with its headquarters and training centres in Canada. Many young Norwegians escaped from their homeland, and arrangements were set up to funnel those who wished to become pilots to Canada to undergo training. By the end of the war, 3,000 Norwegian men had been taught to fly, and most of them had returned to the battlefronts to help in the final release of their homeland. Three centres were established in Ontario: one near Toronto, a second at Gravenhurst, and a third near Huntsville—called Vesle Skaugum after the residence of the Crown Prince of Norway situated near Oslo.

When the war was over all three camps were closed, but at Vesle Skaugum a plain cairn of stone was erected as a gift from the Force's commander, Unger Vtelese, to keep fresh the memory of Norwegian airmen who had died for their country.

It is not a forgotten cairn, as many others seem almost to be. Once a year the children of Norwegian parents or ancestry living in Toronto and the surrounding areas make a pilgrimage to Huntsville and to Camp Kiwanis, the present name of the location of the original Camp Vesle Skaugum. There, in the peace of rural Ontario, home-made wreaths are laid at the base of the memorial, and during additional ceremonies the stirring Norwegian war-poem, "All for Norway," is read in English.

The inscription on the cairn's bronze plaque, beneath a raised insignia of the Royal Norwegian Air Force, reads as follows:
In memory of the members of the Royal Norwegian Air Force who gave their lives for freedom in the war, 1940–45. "As screams of eagles linger in the air so shall their names ring across the seas a message of liberty to the hills of home."

INDEX

INDEX

INDEX

Trans-Canada Air Lines, 322–325, 347, 349, 350, 351, 353, 363, 365, 366
Trascolini, Tosca, 294
Tredwell Yukon Company, 235, 318, 331
Tregenza, W. E., 131
Trenholme, R. G., 240
Treris, Bert, 335–336
Trimm, Leading Aircraftsman J., 354
Tripp, Leonard J., 294
Trorey, G. A., 375
Tudhope, J. H., 313, 323, 324, 346, 363, 366
Tully, T. B., 237, 268–270, 362
Tupper, Sir Charles, 115
Turnbull, Wallace Rupert, 3, 12–15, 345, 359, 362, 380
Turner, Roscoe, 286
Turney, K. V., 376
Turpin, Clifford, 89–91, 97
Twinplane, Gibson, 25–28, 359
Tylee, Lieutenant Colonel Arthur, 183–185
Tyler, R. C., 376

UNDERWOOD, JOHN, 359
Underwood brothers, 16–21
United Air Transport Ltd., 192, 322, 365
University of Toronto, 121, 238, 302
Upson, G. C., 378

VACHON, IRENÉE, 162, 212, 213, 361
Vachon, Romeo, 237, 251, 315, 317, 348, 362, 365, 366
Vail, Mrs., 207
Valiquette, G. H., 218, 225
Van Allen, K. M., 375
Van der Byl, R. I., 214, 331
Van Dyk, Evert, 266
Van Nostrand, C. Innes, 110, 111, 375
Vance, James, 248, 260, 362
Vancouver Aerial League, 1919-1920, 155
Vanoni, M., 69, 71
Vergne, Gertrude de la, 294, 296
Vernon, Victor, 112
Vezines, 75
Vic, 203–211
Vickers Vanessa, 313, 314
Vickers Vedette, 238
Vickers Viking Mark IV, 211–214, 238, 361
Vickers Vimy, 175, 176, 360, 380, 385
Vollick, Eileen, 294, 295, 362
Von Gronau, Commander Wolfgang, 275–276, 277, 278–279
Von Hindenburg, 274
Vtelese, Commander Unger, 385

WACKER, HENRY, 148
Waddell, W., 204, 205, 208, 361
Wade, Lieutenant Leigh, 193
Wakefield, Sir Charles, 250
Wakefield of Hythe, Lord, 307
Wakefield Trophy, 307
Wakeman, G., 323
Wallace, O. C., 317
Waller, E. B., 375
Wallis, H. E., 216, 355, 361
Walsh, Charles F., 78, 97; at Halifax 1912, 95–96
Walsh, Joe, 329–333
Walton, A. M., 375
Wann, Clyde G., 241
Ward, Captain, 115
Ward, Jimmie, at Winnipeg, 91–92, 360; at Regina, 93
Wasson, Everett, 329–333
Waterton, Squadron Leader W. A., 363
Watson, T., 86

Watts, J., 46
Waugh, Keith, 375
Waukessa, Chief, 133
Weber brothers, 300
Weber, H. J., 118
Webber, T., 375
Webster, John, 291
Webster, H. L., 376
Webster Trophy, 291–294, 362, 372, 373
Wedell, Jimmy, 286–288
Weir, C. H., 376
Wells, Happy T. A., 373
Wemp, Bert, 111, 375
West, Flying Officer R. B., 366
West Wind, 1, 39–46
Western Canada Airways Ltd., 232, 235, 245, 248, 251; air mail contracts, 316, 318, 365; incorporation, 239; merger with Canadian Airways, 319
Wheeler, 248
White, Fred, 300
White, Pauline, 242–243
White Wing, 6, 7, 9, 359
Whitebird, 348
Whitfield, R. C., 376
Whyte, Lieutenant Kenneth, 285, 286
Whittier, D., 375
Wickens, P., 376
Wigle (Postmaster), 132
Wilcockson, Captain A. S., 320, 363
Wilcox (Inspector), 218
Wilcox, E. M., 68, 73, 79
Wilkins, Sir Hubert, 338
Wilkinson, Corporal P., 378
Willard, Charles F., 61–63, 64, 67, 79, 359
Williams, Claude, 160
Williams, G. K., 111, 375
Williams, J. Scott, 212, 361
Wilson, Cadet, 129
Wilson, A. H., 350
Wilson, Charles, 373
Wilson, Ellwood, 161, 162
Wilson, Captain H. A., 180, 181
Wilson, J. A., 182, 218, 323, 348, 366, 379, 382
Wilson, Robert F., 282
Windsor, Duke of, *see* Edward, Prince of Wales
Wings Ltd., 325
Winnie Mae, 266, 268, 380
Winnipeg Flying Club, 290
Winny, Harold, 339, 340
Wollatt, D. H., 375
Wood, Phil, 269, 271
Wood, Robert, 272
Worth Oil Dilution Systems, 343
Wright, Jack, 115
Wright, Wing Commander Jerauld G., 371
Wright Aeronautical Corporation, Canadian branch, 249
Wright brothers, 3, 6, 12, 16, 21, 34, 61, 69; Wilbur, 28; Orville, 10, 380
Wyatt, Chief Mechanic, 177

YORATH, DENNIS K., 366
Young, Sergeant, 182, 183
Young, Franklin I., 370
Young, Lieutenant W., 149
Yukon Airways & Exploration Company, 235, 240, 365
Yukon Southern Air Transport Ltd., 322, 325

ZIMMER, EDUARD, 276
Zimmerman, Mrs. L., 102
Zurakowski, Janusz, 351, 363, 371

398